Green Energy: Technologies and Systems

Green Energy: Technologies and Systems

Editor: David McCartney

R CALLISTO
REFERENCE

www.callistoreference.com

Callisto Reference,
118-35 Queens Blvd., Suite 400,
Forest Hills, NY 11375, USA

Visit us on the World Wide Web at:
www.callistoreference.com

ISBN: 978-1-63239-993-9 (Hardback)

Cataloging-in-Publication Data

Green energy : technologies and systems / edited by David McCartney.
    p. cm.
Includes bibliographical references and index.
ISBN 978-1-63239-993-9
1. Clean energy. 2. Power resources. I. McCartney, David.
HD9502.5.C542 G74 2018
333.79--dc23

Table of Contents

Permissions

List of Contributors

Index

Preface

The purpose of the book is to provide a glimpse into the dynamics and to present opinions and studies of some of the scientists engaged in the development of new ideas in the field from very different standpoints. This book will prove useful to students and researchers owing to its high content quality.

Green energy refers to the use of renewable sources for energy production. Some major technologies related to green energy are wind power, bio energy, hydropower, photovoltaics, etc. The topics included in this book on green energy are of utmost significance and bound to provide incredible insights. It will also provide innovative topics for research which interested readers can take up. As this field is emerging at a rapid pace, the contents of this book will help the readers understand the modern concepts and applications of the subject.

At the end, I would like to appreciate all the efforts made by the authors in completing their chapters professionally. I express my deepest gratitude to all of them for contributing to this book by sharing their valuable works. A special thanks to my family and friends for their constant support in this journey.

Editor

Flexible Mode Control of Grid Connected Wind Energy Conversion System Using Wavelet

Bhavna Jain, Sameer Singh, Shailendra Jain, and R. K. Nema

MANIT, Bhopal 462003, India

Correspondence should be addressed to Bhavna Jain; jain_bhavna69@yahoo.com

Academic Editor: Ahmet Z. Sahin

Small wind turbine systems offer services to critical loads during grid faults and also connected back to grid in normal condition. The connection of a wind energy conversion system to the grid requires a robust phase locked loop (PLL) and continuous monitoring of the grid conditions such as overvoltage, undervoltage, overfrequency, underfrequency, and grid outages. This paper describes a flexible control operation to operate a small wind turbine in both stand-alone mode via planned islanding and grid connected mode as well. In particular, a proper monitoring and control algorithm is required for transition between the modes. A wavelet based energy function is used for detection of grid disturbances as well as recovery of grid so that transition between the modes is made. To obtain good power quality LCL filter is used to reduce ripples. PLL is used for synchronization whenever mode changes from stand-alone to grid connected. Simulation results from a 10 kW wind energy conversion system are included to show the usefulness of the proposed methods. The control method is tested by generated gate pulses for single phase bridge inverter using field programmable gate array (FPGA).

1. Introduction

Due to awareness for pollution free environment, depletion of conventional energy sources, and growing demand of energy worldwide, the government of different countries has been promoting electricity generation from renewable sources. Renewable energy sources are becoming an important option to meet the growing demand of energy and simultaneously help in controlling of greenhouse gas emission caused by conventional energy sources. Another major benefit of renewable energy source is that it can work in stand-alone mode to fulfil the customer demand and can work as grid connected system to supply extra power generated to grid, hence, increasing the reliability of power supply.

Out of all renewable sources, wind power generation technology has been grown up, in the beginning with few KW to multi-MW capacity wind turbine manufactured and installed. An attractive idea of universal mode of small wind turbine (SWT) has been implemented in which they can either operate as stand-alone mode or can work in grid connected mode [1, 2]. Small wind turbines can be used in rural areas in developed countries where grid is not available [3].

The amplitude and frequency of generated voltage of wind plants vary according to speed of wind [4]. Hence, power electronics interfaces are used to convert generated voltage to a fixed dc voltage, which can later be either stored or converted into required ac voltage and frequency. Grid connected wind energy conversion systems are more in trend. In case of grid faults, WECS system connected to grid can be safely islanded to serve critical loads connected to it.

When a small wind turbine unit is connected to the grid, the voltage and frequency at the point of common coupling are controlled by the grid. However, in case of weak grids, voltage sags and disturbances may occur when WECS is interfaced to grid. In such situation, the wind unit must support to the grid voltage. A wind energy conversion system can work in two different modes; they are grid connected and stand-alone modes. The stand-alone/islanding mode is a situation in which the WECS is isolated from the utility grid when grid disturbances due to network fault are cropped up. A monitoring unit is used to achieve it. This unit will help in islanded operation of WECS in a planned manner when voltage magnitude crosses the threshold value.

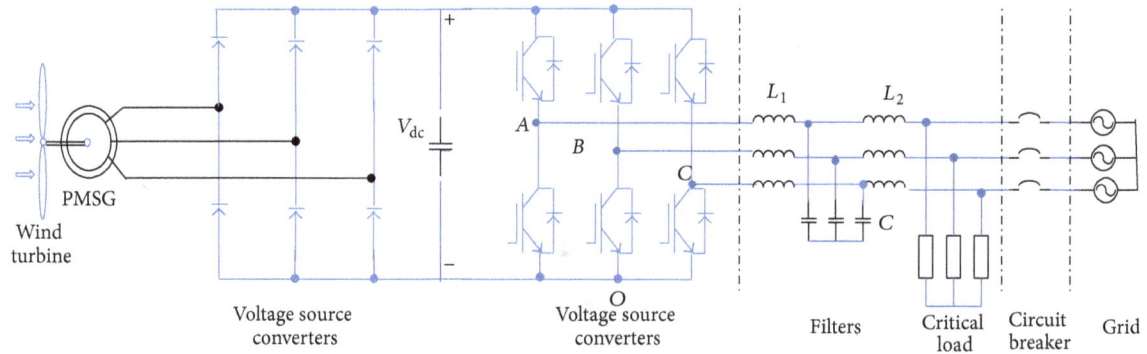

FIGURE 1: Schematic diagram of WECS interfacing to grid.

Authors in [5] verified experimentally the control algorithm applied in SWT working in both stand-alone and grid connected modes. Performance of synchronization algorithm is also checked which is used to connect DC/AC converter back to grid after recovery from disturbance. A novel method novel based on PLL is proposed by Teodorescu and Blaabjerg for grid failure detection and flexible mode switching automatically in which the phase difference between the grid and the inverter is used to determine grid failure and recovered from fault [3]. Jang and Kim presented three papers starting in an algorithm improved successively which utilizes four system parameters voltage magnitude, frequency, phase, and total harmonic distortion (THD) of current for islanding detection. The method monitors changes in four parameters and detects islanding by logical rules [6–8]. During disturbances many system parameters change significantly. Hence selection of most vulnerable system parameter and selection of threshold value is very challenging task for effective detection of disturbances. Many authors have implemented different methods based on wavelet transform (WT) for detection of power quality disturbance [9–14]. Wavelet energy entropy, variance, standard deviation, mean, and wavelet energy are various statistical features suitable for detection of power quality disturbances and an islanding event.

In this paper a WECS is developed using SIMULINK with a flexible mode control strategy and synchronization algorithm to allow dual mode operation of it as and when required. A seamless transfer between the modes is realized by opening and closing of the circuit breaker as shown in Figure 1 which disconnects/connects the WECS from/to the main grid [15]. Once the WECS is isolated from the main grid, WECS will be responsible for maintaining the voltage and frequency while supplying to load. During autonomous operation it is essential that inverters should not be overloaded. Simultaneously system must ensure that the changes in load are handled by inverters properly in a control manner.

The DC/AC converter of WECS is connected to grid to inject active and reactive power. Mainly current-controlled voltage source converter (VSC) is used in grid connected mode [16, 17]. Conversely, voltage controlled VSC is used when the WECS works in the stand-alone mode. To regulate voltage and maintain it constant is the major responsibility of control method used in isolated mode of WECS. During grid disturbances, the detection method based on wavelet energy function is used to change of mode of operation for grid connected to isolated mode of WECS and vice versa. A circuit breaker is used for this and will switch between the modes on the basis of signal received from control method used.

The main objective of this paper is to control the flexible mode operation of control grid connected wind energy conversion system using wavelet energy based function. Since the wind energy conversion system has been competent to operate in both grid connected mode and stand-alone mode according to the grid conditions, the control design is a bit tricky. This paper is presented as follows. Section 2 describes the inverter control in stand-alone mode of operation. Then, control methods for grid connected operating mode including the PLL design and current regulation of the inverter are explained in Section 3. Use of wavelet energy function for detecting the status of grid is described in Section 4. In Section 5 the verification of the control methods is done through simulation results. The control scheme implemented in WECS is tested using FPGA in Section 6. Finally, conclusions are drawn in Section 7.

2. Inverter Control in Stand-Alone Mode of WECS

Electrical power available at the electrical generator output of the wind energy conversion system is not sinusoidal in nature. To get the sinusoidal voltage at supply frequency and to keep the output power optimally constant, power electronic interfacing is done between generator and grid/load as shown in Figure 2. In general, a power electronic interface device is a combination of a rectifier, an energy storage device to regulate the DC-link voltage and an inverter.

Voltage source inverter of load side is responsible for providing controlled output voltage in terms of frequency and amplitude [18, 19]. At load side inverter, appropriate control method is applied for generating switching pulses of inverter to produce output of required magnitude and frequency. To achieve it the control method has an output voltage controller using any modulation technique. Here space vector modulation method is implemented. The schematic diagram of the control method is shown in Figure 3.

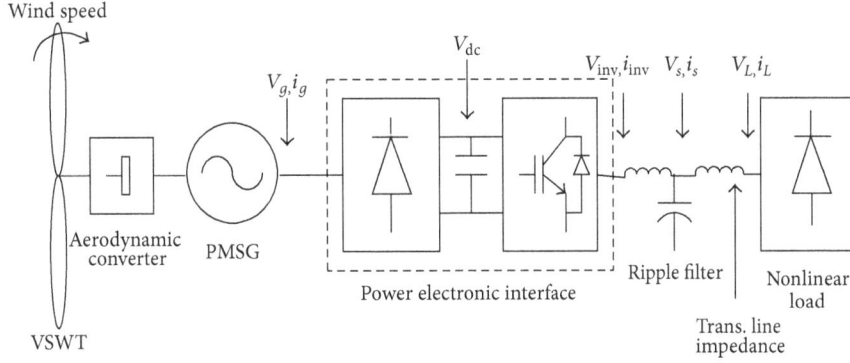

FIGURE 2: Power transfer stages in isolated WECS.

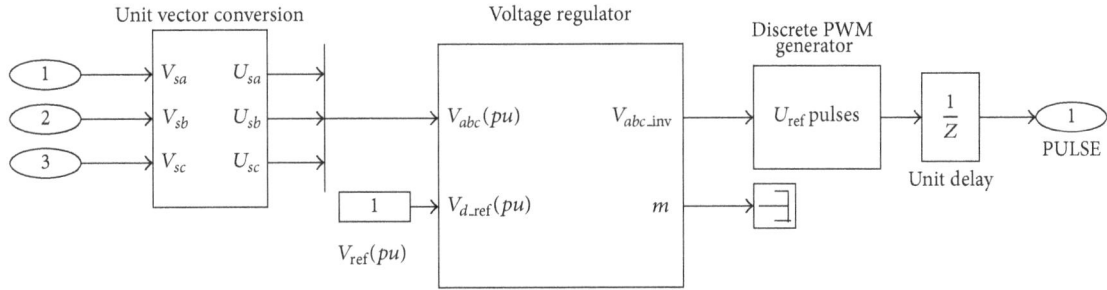

FIGURE 3: Simulation block diagram for SVPWM to generate gate pulses.

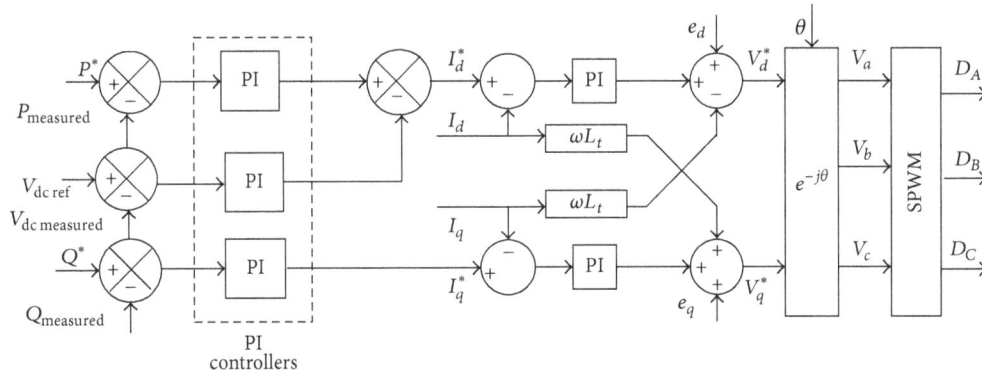

FIGURE 4: Grid side inverter control.

In space vector PWM, 3-phase sinusoidal modulating signal gets transformed into a revolving voltage vector with a constant magnitude and angular frequency. Here, the constant voltage magnitude is magnitude of desire voltage to be produced and angular frequency is the sampling frequency. In space vector based PWM, instead of three modulating signals for 3-phase, a revolving voltage vector is used as a voltage reference. This voltage reference vector is sampled once in every subcycle T_s and sampled voltage vector gives the voltage command for the given subcycle.

3. Inverter Control of WECS in Grid Connected Mode

Voltage oriented control (VOC) is mostly used for grid side voltage source inverter as shown in Figure 4. A phase locked

loop (PLL) is used to find out grid angle θ which is used for transformation of inverter output currents and output voltages in synchronous reference frame. To obtain better response of inverter, it has been selected to decouple active and reactive power. The active power depends on the d-axis current component. Similarly, reactive power and q-axis current component are directly related. Therefore, the d-axis PI controller controls active power, and q-axis PI controller controls reactive power.

Grid currents are converted in synchronous reference frame currents i_d, i_q to provide separate control for active and reactive power. High power factor and sinusoidal grid currents can be obtained by doing so [20].

In order to operate under synchronization with grid, the system uses three PI controllers. The DC-link voltage controller is used for calculating d-axis reference current to

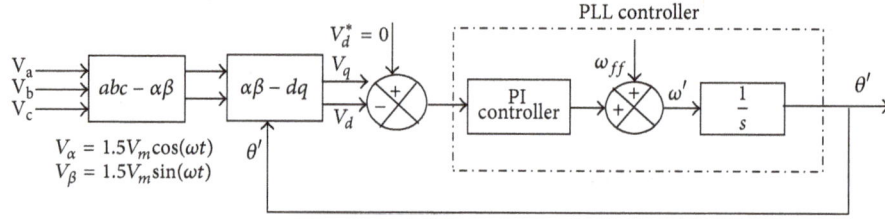

FIGURE 5: Block diagram of phase locked loop based on synchronous reference frame.

control active power. The q-axis reference can be set to zero to get unity power factor. By using PI controllers for controlling the errors in d-axis and q-axis currents, the control voltages are generated for the 3-Ø voltage source inverter in PQ control method as shown in Figure 4 and are given by

$$V_d^* = V_d + \omega L_t i_q - e_d$$
$$V_q^* = V_q - \omega L_t i_d - e_q, \tag{1}$$

where L_t is the total inductance on the grid side inverter and e_d and e_q are d-q components of grid side voltage vector.

Tuning of PI controllers must be done accurately to obtain better control of DC-link voltage, active power, and reactive power. The feedforward and cross coupling terms used in generation of reference voltage vector in synchronous reference frame help out in system linearization and make controller design easier. The reference voltages V_d^* and V_q^* are further transformed into and used to generate inverter gate pulses through a SVPWM algorithm. LCL filter is used to improve the power quality at the inverter output.

4. Grid Status Monitoring and Transition of Modes

4.1. Phase Locked Loop. The utility grid status monitoring must be done continuously in real time, to ensure good quality power supply to loads. The grid status includes sensing fault, overvoltage, and undervoltage conditions. Outage detection is carried out in every sampling cycle by comparing the instantaneous grid voltage. The block diagram of the three-phase PLL used for synchronization in the grid connected mode is as shown in Figure 5. A resonant filter can be added to make standard PLL more robust in case of unbalance and voltage harmonics.

In case of grid failure, islanded cannot be avoided. Hence suitable method must be used to detect grid failure. For connected to grid again when grid returns to its normal condition, use of a synchronization algorithm is necessary prior to transition of mode from stand-alone to grid connected. The flowchart shown in Figure 6 is proposed for islanding detection and further for transition from grid connected mode of WECS to stand-alone mode. In case of grid disturbances,

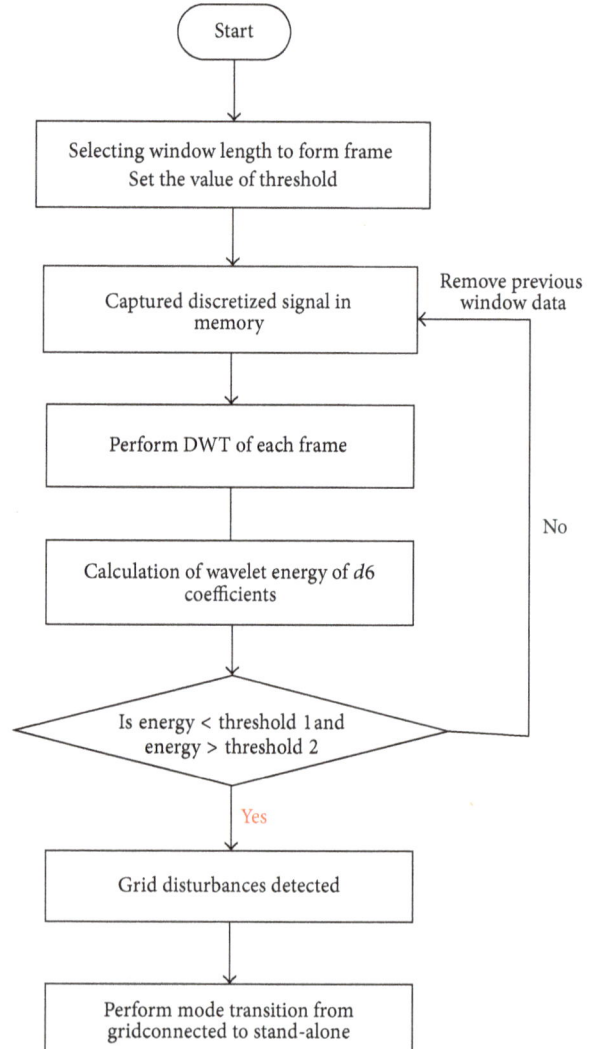

FIGURE 6: Flowchart for detection of grid faults/islanding.

mode change of WECS from grid connected to stand-alone is performed in the following steps.

(1) Identify grid condition using power quality monitoring.

(2) Generate a signal to turn off circuit breaker in case of grid fault.

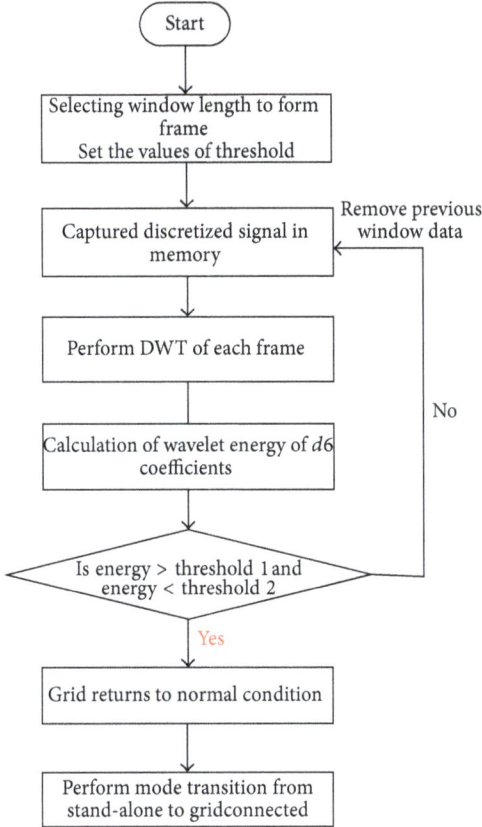

FIGURE 7: Flowchart for returning to grid normal operation.

(3) Mode transition can be done as soon as the circuit breaker turns off. Change the grid connected mode of WECS to stand-alone mode.

(4) In case of stand-alone mode, the control strategy will be voltage controlled. In voltage controlled mode of load side inverter, the reference value used for voltage will be last value of grid voltage when mode transition takes place.

The flowchart shown in Figure 7 is proposed for detection of grid recovery and transition from stand-alone mode to grid connected mode of WECS.

4.2. Discrete Wavelet Transform. Discrete wavelet transform (DWT) converts a time domain discretized signal into its corresponding wavelet domain. Principally, the discrete wavelet transformation has two-phase determination of wavelet coefficients and calculation of detailed and approximated version of the original signal, in different scales of resolutions in the time domain. In filtering process the original signal is passed through two complementary filters and produces approximate and detail coefficients. To extend the frequency resolution, decomposition of signal is done repeatedly and signal can be realized into two lower frequency ranges. This process is known as multiresolution analysis (MRA) and goal of MRA is to represent a complex signal by several simple signals to study them separately.

4.3. Frame Length. Coefficients of wavelet transform represent the energy of the signal. These coefficients will be used to measure the magnitude of the disturbance in distorted signal. In real time application wavelet transform can be used as a monitoring tool when it becomes essential to detect disturbances in minimum time. For such cases distorted signal is processed through time window of fixed length frame. Length of the frame means the number of sample points of discrete data signal for which wavelet energy has to be calculated. The time window move forward along the signal and wavelet energy is calculated for each frame. Frame length decides the response time of the method. If length of the frame is long it will take more time in calculation and response time will get delayed. Sampling frequency, size of buffer, and level of decomposition are three main factors which must be wisely selected according to application. A fixed frame length of sample points 128 is used in this paper to obtain fast response time. The sampling frequency selection has been done according to Parseval's theorem and decomposition has been done into 6th level.

4.4. Wavelet Energy. The discrete wavelet divides a signal into approximated and detailed version of the original signal, in different scales of resolutions in the time domain using low-pass and high-pass filters. Decomposition of approximate version can be repeated to obtain signal in required frequency subbands with number of approximate and detail coefficients. Sum of coefficients square at a particular level represent the energy of the signal at that level. These coefficients will be used to compute the level of the disturbance in distorted signal. Wavelet energy measure based on wavelet analysis is able to observe the unsteady signal and complexity of the system at time-frequency plane. The mother wavelet function selected is db and scale factor 2 that is according to literature reviews. The signal is decomposed into 6th level. Hence, cD6 coefficients will represent the fundamental frequency component of the signal and coefficients energy will be calculated by using

$$E_j = \sum_{k=1}^{N} \left| D_{jk} \right|^2, \quad j = 1, 2, \ldots, l, \tag{2}$$

where D_{jk} is the value of wavelet detail coefficients obtained in decomposition from level 1 to level J. N is the total number of the coefficients at each decomposition level and E_j is the energy of the detail coefficients at decomposition level j.

4.5. Deciding Threshold. The most important part of monitoring algorithm is deciding the setting for threshold level. The value should be selected to change mode of WECS whenever voltage of any phase crosses the standard limits such as voltage dip of less than 0.8 pu or voltage swell of magnitude more than 1.2 pu Simultaneously, it should not cause unnecessary false tripping of circuit breaker in case of small voltage dip or swell. It is the value of wavelet energy calculated for output voltage signal (grid voltage) under normal grid condition plus a variation allowed as per standards. For calculating the threshold a reference signal of same frame

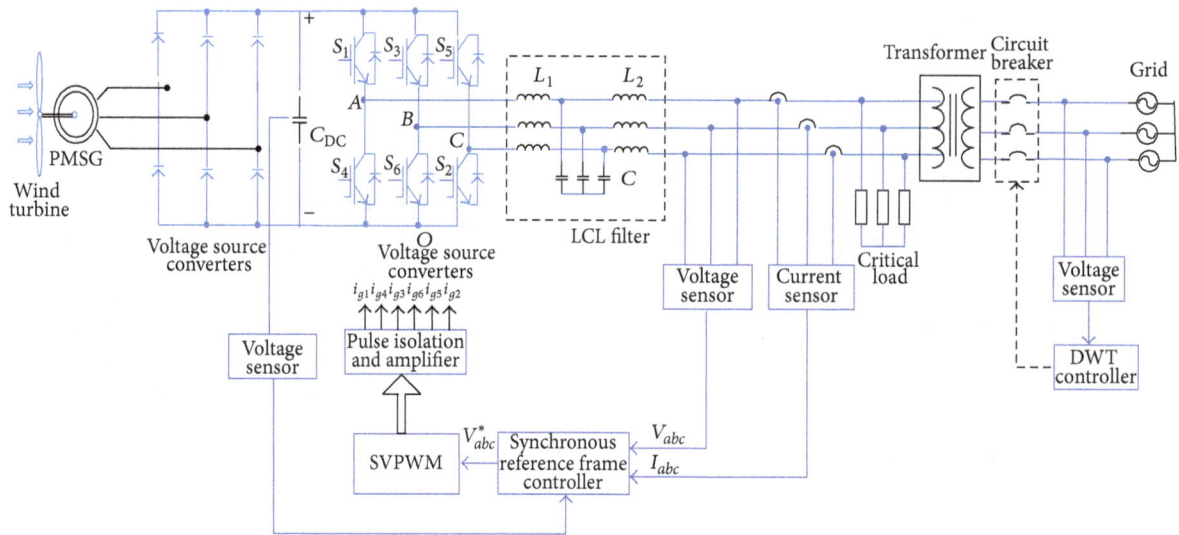

FIGURE 8: Schematic diagram of WECS connected to grid.

TABLE 1: Parameters selection for proposed method.

Parameter	Peak voltage level (V)	Wavelet energy (V^2)	Normalized wavelet energy
Peak voltage of grid	325	$2.3 * e6/10^5$	23
Permissible limit of voltage swell	390	$4.0 * e6/10^5$	40
Permissible limit of voltage sag	260	$11 * e6/10^5$	11

length decomposed in 6th level using same mother wavelet function db2 and coefficients energy of cD6 is calculated. Permissible variation in reference signal is considered and the entire procedure is repeated to calculate lower and upper threshold settings. Table 1 is listing the wavelet energy for different cases which helps in deciding lower and upper threshold settings.

5. Schematic Diagram of System, Results and Discussions

A schematic diagram of grid connected WECS consists of 3-Ø PMSG, full-bridge rectifier, DC-link capacitor, a 3-Ø IGBT based full-bridge inverter, critical load, LCL filter, transformer, and circuit breaker and 400 volt, 50 Hz ac source is shown Figure 8. The system parameters used in simulation are given in Table 2. Simulation model of the system is developed in MATLAB/simulink environment. All the values given in Table 2 have been calculated during mathematical modelling of WECS and grid connected WECS.

5.1. Before Fault. Figure 9 shows the output voltages under normal grid condition. It shows that the implemented control method of voltage source inverter is maintaining the output in desired form. Figure 10 shows load voltage and load current under normal grid condition.

5.2. During Fault. It can be seen from Figure 11 that, in case of fault in phase B, grid voltages of phase A and phase C have

TABLE 2: Design parameters for simulation.

Description	Symbolic representation	Value
Capacitor	C_{dc}	2200 μF
DC-link voltage	V_{DC}	650 volts
AC output voltage	V_s	230 volts rms
AC output frequency	f	50 Hz
Inverter side filter inductor	L_1	0.2 mH
Grid side filter inductor	L_2	0.1 mH
Filter capacitor	C_1	10 μF
Inverter switching frequency	f_s	3 kHz

been inceased. Grid fault occurs at $t = 0.45$ sec and continues till $t = 0.7$ sec. Monitoring algorithm is constantly monitoring the grid condition. Such cases must be detected and reported to utility intactive inverter at the earliest so that supply to critical load will be continued by intentional islanding of WECS. The load is supplied from WECS and grid current is zero in case of grid fault. Source current and load currents (local and shared load both) which are now supplied from WECS in case of grid not present can be seen from Figure 12.

Grid fault causes voltage variations in all the three phases. Voltage waveform of all the three phases and corresponding coefficients energy plots are displayed in Figures 13, 14, and 15, respectively, in red, green (istead of yellow for improved visibility), and blue color for phase A, phase B, and phase C. Wavelet energy is normalized as calculated value has very

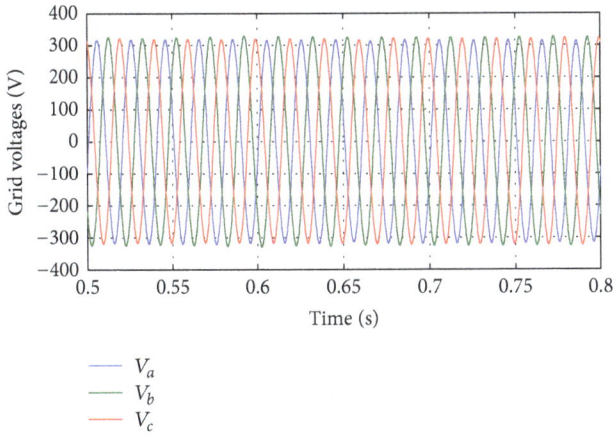

FIGURE 9: Three-phase grid voltages before fault.

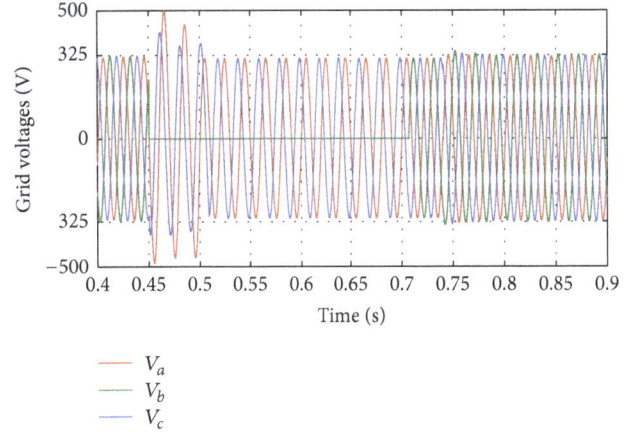

FIGURE 11: Three-phase grid voltages during disturbance.

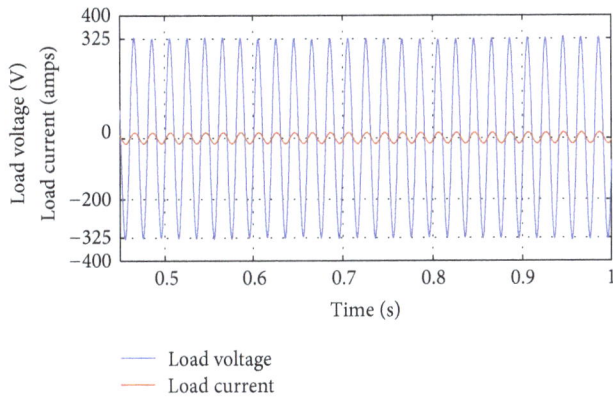

FIGURE 10: Output waveform of load voltage and load current before fault.

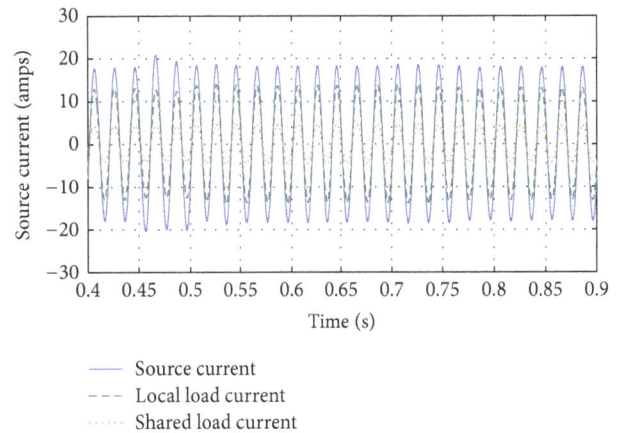

FIGURE 12: Source current, local load current, and shared current.

large value and even a large change in coeficients energy cannot be clearly observed from actual plots of wavelet energy. Grid voltages and corresponding coefficients energy plot are graphical representation of change in wavelet coefficients energy in case of grid disturbances. Threshold level with calculated threshold further helps in transition of modes from grid connected to stand-alone and vice versa.

At $t = 0.45$ sec grid fault occurs; wavelet based monitoring algorithm detects it at 0.48 sec and changes the mode of WECS from grid connected to stand-alone. Many distribution systems use autoreclosing to clear temporary faults and hence it is essential to detect grid faults before the autorecloser operates to avoid out phase reclosing. Grid fault causes voltage variations in all the three phases which can be observed clearly in Figure 16.

Transition of mode from grid connected to stand-alone occured at 0.48 sec which is the total time taken by the proposed method in detection of grid disturbance and further isolate the WECS by operting circuit breaker which disconnects it from grid. When mode trasition occurs, controller of VSI changes its mode and continues to feed power to the load connected to it. The grid monitoring algorithm detects fault and changes mode of operation from grid connected

mode to stand-alone mode at 0.48 sec when circuit breakers opens. The timings of mode change and circuit breaker status (1 means WECS grid connected, 0 means disconnected from grid) can be seen from Figure 17. Figure 18 displays the lower threshold value due to which transition occurs.

Grid voltage of all the phases and corresponding coefficients energy plots are shown in Figures 19, 20, and 21, respectively, of phase A, phase B, and phase C. Grid voltages and corresponding coefficients energy plot are graphical representation of wavelet coefficients energy. A small change in voltages at load terminal (PCC) can be observed in the signal and associated wavelet energy plot in Figures 19, 20, and 21 that is due to change of load because of mode transition. WECS is now fed power to critical load and local loads connected to it. Monitoring is still continued at PCC to check for grid condition. On the basis of threshold value of wavelet energy transition occurs from stand-alone mode to grid connected mode, which can be observed from Figures 19(a) and 19(b) at time 0.7 sec when fault is cleared.

The power conditioning module for both the modes is based space vector pulse width modulation. Instead of conventional SVM, triangular carrier based SVM is used to

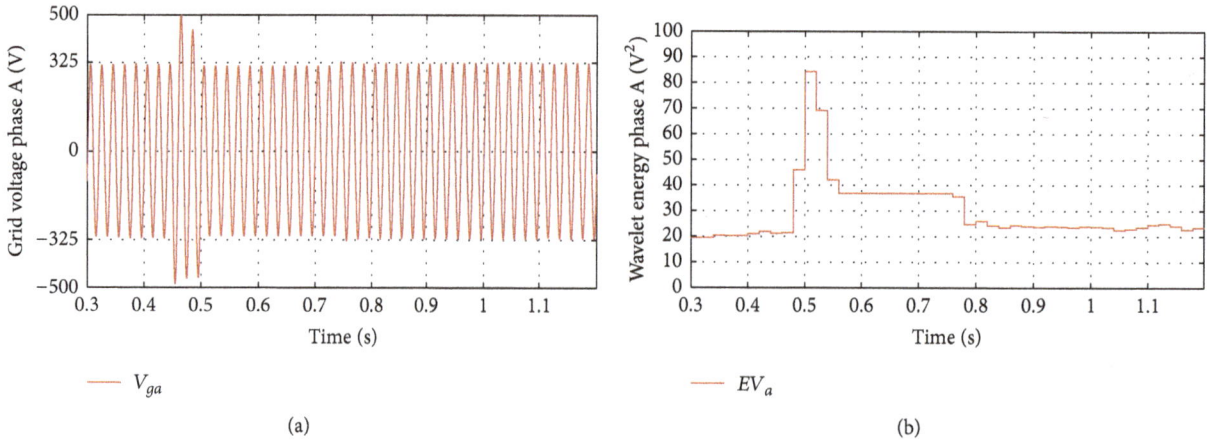

FIGURE 13: (a) Voltage signal and (b) corresponding wavelet energy plot of phase A.

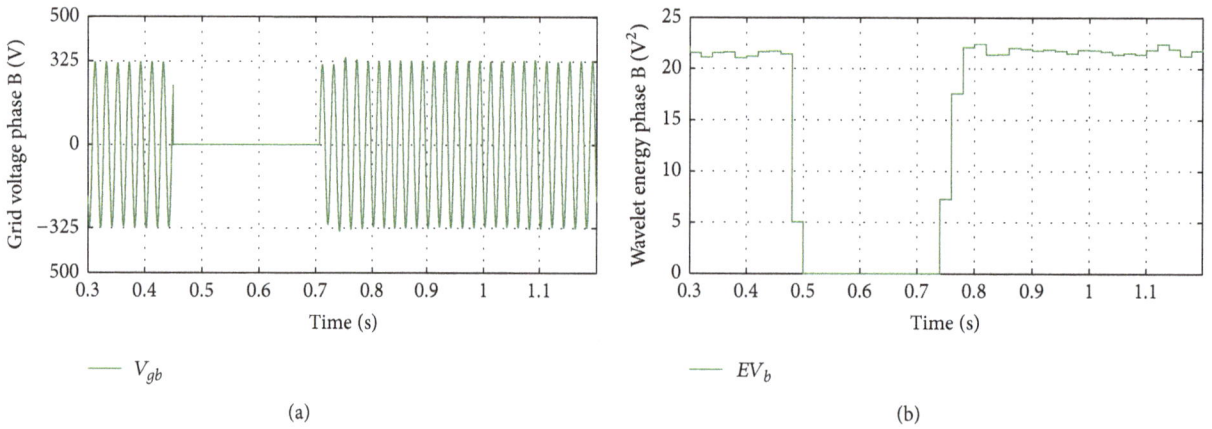

FIGURE 14: (a) Voltage signal and (b) corresponding wavelet energy plot for fault in phase B.

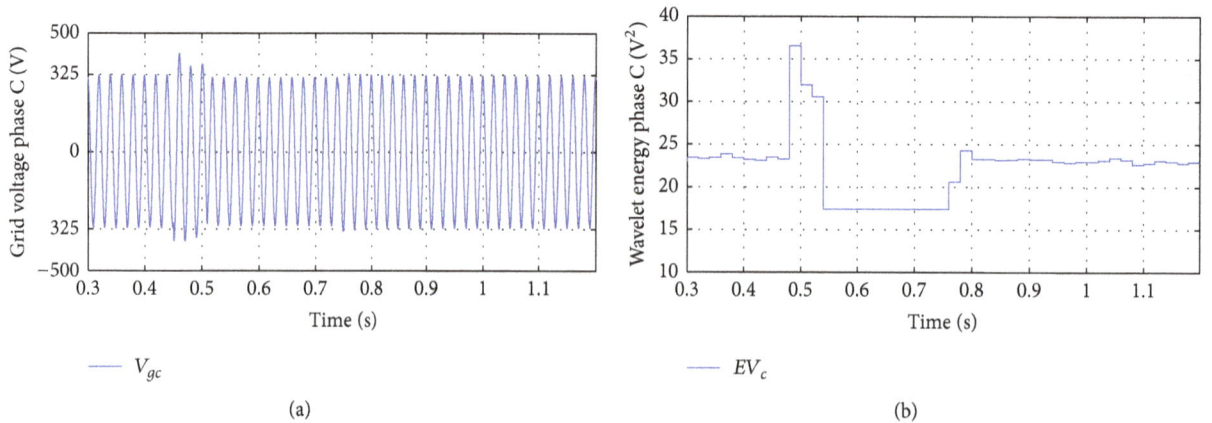

FIGURE 15: (a) Voltage signal and (b) corresponding wavelet energy plot of phase C.

reduce computation burden as sector determination is not required in implementation. The benefits of using DWT and wavelet energy based algorithm are accurate detection of start and end time of occurrence of any event or varaitions. Accurate and quick detection can also be helpful for protection of equipments as well as for the safety and stability of the system.

6. Generation of Gate Pulses for Single Phase Bridge Inverter Using FPGA

In grid connected mode as well as in stand-alone mode of WECS, space vector pulse width modulation scheme has been used for inverter control. Its performance is checked

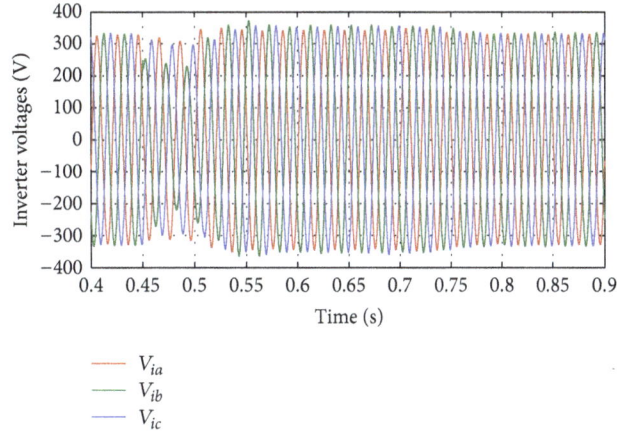

FIGURE 16: Three-phase voltages at PCC.

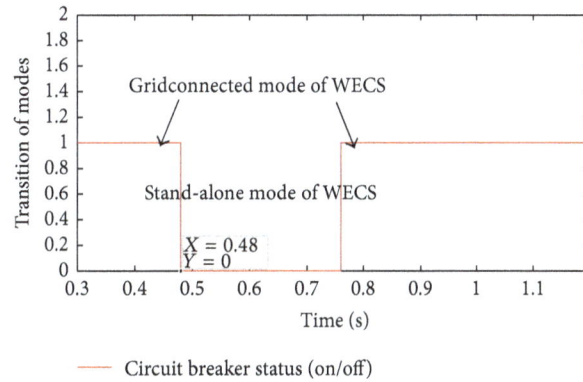

FIGURE 17: Transition of modes.

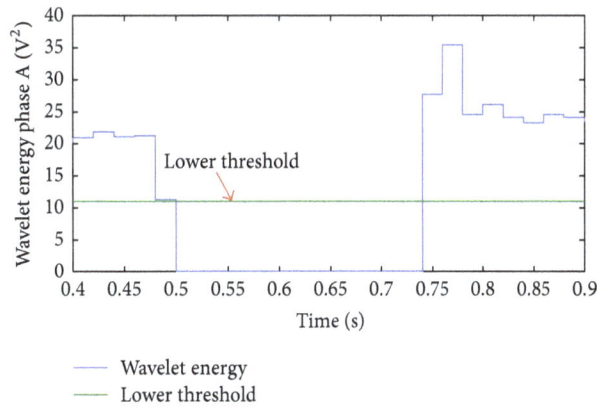

FIGURE 18: Threshold settings.

using FPGA with Altium NB 3000, Xilinx Spartan 3AN processor. Step-by-step procedure for generating inverter gate pulse is shown in Figure 22. XILINX ISE design suite 14.5 is used for model based design for PWM pulse generation for single phase bridge inverter and Altium designer software is used for FPGA project design. Number of steps has been performed for bit file generation using Altium design software. It generates the programming file that is required for

downloading the design to the physical device. A detailed procedure for project design using FPGA is given in [21].

Schematic diagram to test SVPWM control method for generating gate pulses using FPGA is shown in Figure 23.

Sine waveform and triangular carrier waveform of frequency 500 Hz are given as input by ADC-SPI port and inverter gate pulses are obtained by user I/O port which is shown in Figures 24 and 25, respectively.

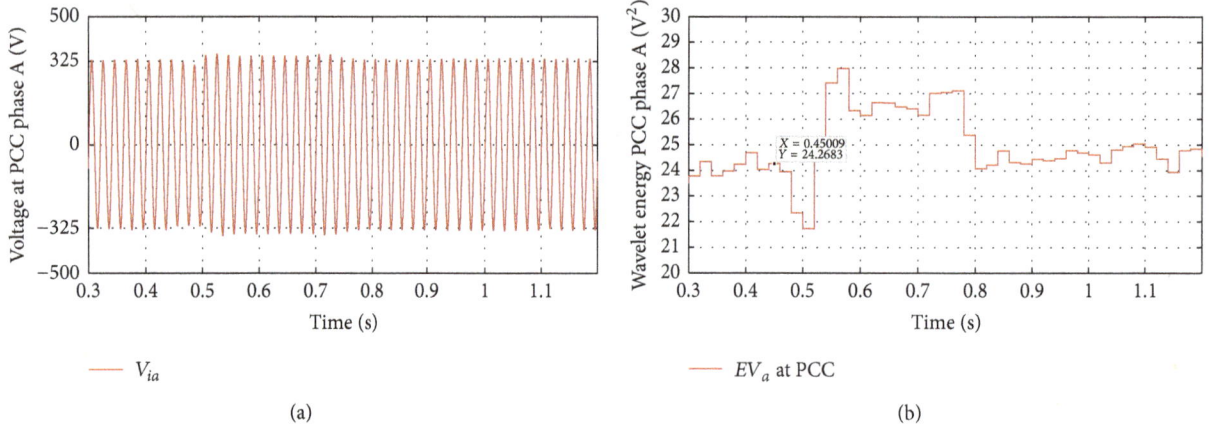

(a)   (b)

FIGURE 19: (a) Voltage signal and (b) wavelet energy plot at PCC for phase A.

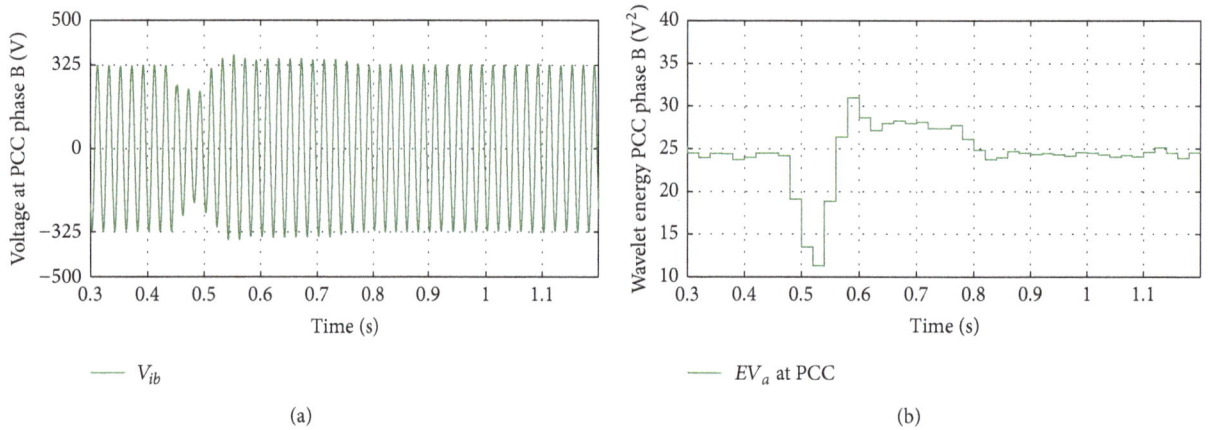

(a)   (b)

FIGURE 20: (a) Voltage signal and (b) wavelet energy plot at PCC for phase B.

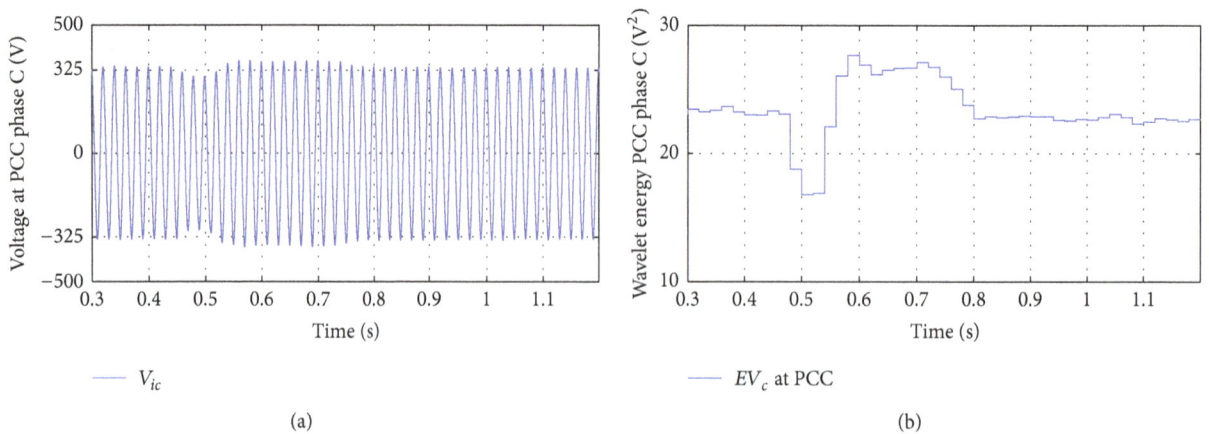

(a)   (b)

FIGURE 21: (a) Voltage signal and (b) wavelet energy plot at PCC for phase C.

7. Conclusions

The proposed method has been implemented for a 10 kW wind energy conversion system with rectifier-inverter interface which can work in grid connected mode as well as in stand-alone mode. The benefit of the used control scheme is that switching between the two operating modes happens automatically on the basis of output of energy function. The most important feature of the system is its adaptability to work in both of the operating modes properly. PLL is used for synchronization in grid connected mode. Simulation results demonstrate the working and transition between the modes of WECS in only 3 ms time and no transients appear during transition of modes.

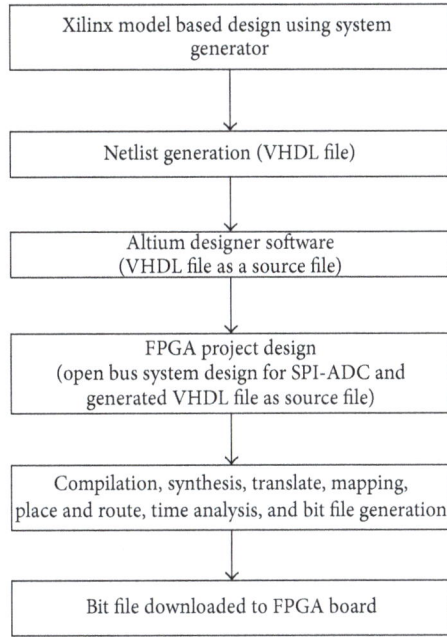

FIGURE 22: Step-by-step procedure for inverter gate pulse generation using FPGA.

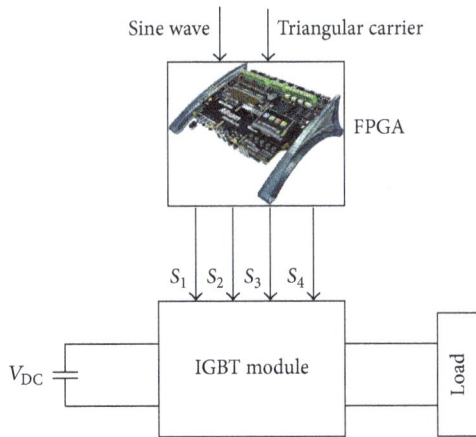

FIGURE 23: Schematic diagram for hardware setup.

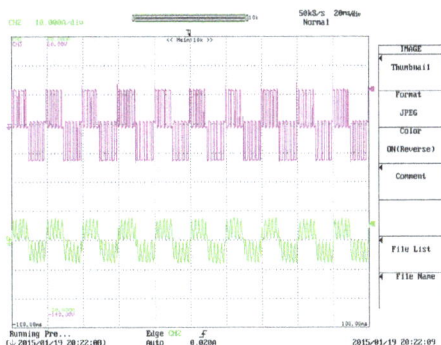

FIGURE 24: Load voltage and load current through RL load.

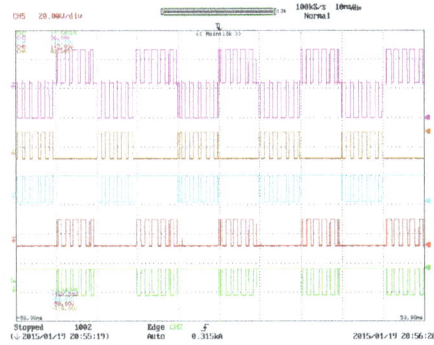

FIGURE 25: Inverter output voltage and output gate pulses for single phase bridge inverter.

The tests concluded that detection method has the following properties.

(i) Detect grid disturbances event in just 3 ms.

(ii) Accurate detection and quick transition maintain the power quality and supply uninterrupted power to critical load in case of grid outage.

(iii) It is suitable for detection of steady state and transient state disturbances both.

Conflict of Interests

The authors declare that there is no conflict of interests regarding the publication of this paper.

References

[1] R. Teodorescu, F. Iov, and F. Blaabjerg, "Flexible development and test system for 11kW wind turbine," in *Proceedings of the IEEE 34th Annual Power Electronics Specialists Conference*, pp. 67–72, June 2003.

[2] N. A. Orlando, M. Liserre, R. A. Mastromauro, and A. Dell'Aquila, "A survey of control issues in pmsg-based small wind-turbine systems," *IEEE Transactions on Industrial Informatics*, vol. 9, no. 3, pp. 1211–1221, 2013.

[3] R. Teodorescu and F. Blaabjerg, "Flexible control of small wind turbines with grid failure detection operating in stand-alone and grid-connected mode," *IEEE Transactions on Power Electronics*, vol. 19, no. 5, pp. 1323–1332, 2004.

[4] B. Singh and G. K. Kasal, "Solid state voltage and frequency controller for a stand alone wind power generating system," *IEEE Transactions on Power Electronics*, vol. 23, no. 3, pp. 1170–1177, 2008.

[5] A. Milczarek and M. Malinowski, "Monitoring and control algorithms applied to small wind turbine with grid-connected/ stand-alone mode of operation," *Przeglad Elektrotechniczny*, vol. 88, pp. 18–22, 2012.

[6] S. I. Jang and K. H. Kim, "An islanding detection method for distributed generations using voltage unbalance and total harmonic distortion of current," *IEEE Transactions on Power Delivery*, vol. 19, no. 2, pp. 745–752, 2004.

[7] S. I. Jang and K. H. Kim, "A new islanding detection algorithm for distributed generations interconnected with utility networks," in *Proceedings of the 8th IEE International Conference*

on Developments in Power System Protection, vol. 2, pp. 571–574, IET, April 2004.

[8] S.-I. Jang and K.-H. Kim, "Development of a logical rule-based islanding detection method for distributed resources," in *Proceedings of the IEEE Power Engineering Society Winter Meeting*, vol. 2, pp. 800–806, January 2002.

[9] J. W. Resende, M. L. R. Chaves, and C. Penna, "Identification of power quality disturbances using the MATLAB wavelet transform toolbox," in *Proceedings of the 4th International Conference on Power Systems Transients (IPST '01)*, Rio de Janeiro, Brazil, June 2001.

[10] Ç. Kocaman and M. Özdemir, "Comparison of statistical methods and wavelet energy coefficients for determining two common PQ disturbances: sag and well," in *Proceedings of the 6th International Conference on Electrical and Electronics Engineering (ELECO '09)*, pp. 180–184, November 2009.

[11] P. K. Ray, N. Kishor, and S. R. Mohanty, "Islanding and power quality disturbance detection in grid-connected hybrid power system using wavelet and S-transform," *IEEE Transactions on Smart Grid*, vol. 3, no. 3, pp. 1082–1094, 2012.

[12] R. Tirumala, N. Mohan, and C. Henze, "Seamless transfer of grid-connected PWM inverters between utility-interactive and stand-alone modes," in *Proceedings of the 17th Annual IEEE Applied Power Electronics Conference and Expositions (APEC '02)*, pp. 1081–1086, March 2002.

[13] A. Timbus, M. Liserre, R. Teodorescu, P. Rodriguez, and F. Blaabjerg, "Evaluation of current controllers for distributed power generation systems," *IEEE Transactions on Power Electronics*, vol. 24, no. 3, pp. 654–664, 2009.

[14] S. W. Mohod and V. A. Mohan, "Power quality issues and it's mitigation technique in wind energy generation," in *Proceedings of the 13th International Conference on Harmonics and Quality of Power (ICHQP '08)*, Wollongong, Australia, October 2008.

[15] C. N. Bhende, S. Mishra, and S. G. Malla, "Permanent magnet synchronous generator-based standalone wind energy supply system," *IEEE Transactions on Sustainable Energy*, vol. 2, no. 4, pp. 361–373, 2011.

[16] J. M. Carrasco, L. G. Franquelo, J. T. Bialasiewicz et al., "Power-electronic systems for the grid integration of renewable energy sources: a survey," *IEEE Transactions on Industrial Electronics*, vol. 53, no. 4, pp. 1002–1016, 2006.

[17] C. L. Anooja and N. Leena, "Single phase shunt active filter with fuzzy controller for harmonic mitigation," *International Journal of Scientific & Engineering Research*, vol. 4, no. 9, pp. 445–451, 2013.

[18] L. G. B. Rolim, D. R. Da Costa Jr., and M. Aredes, "Analysis and software implementation of a robust synchronizing PLL circuit based on the pq theory," *IEEE Transactions on Industrial Electronics*, vol. 53, no. 6, pp. 1919–1926, 2006.

[19] B. Jain, T. Jain, S. Jain, and R. K. Nema, "Power quality improvement of an isolated wind power generation system," *IOSR Journal of Electrical and Electronics Engineering*, vol. 9, no. 3, pp. 33–50, 2014.

[20] K. Zhou and D. Wang, "Relationship between space-vector modulation and three-phase carrier-based PWM: a comprehensive analysis," *IEEE Transactions on Industrial Electronics*, vol. 49, no. 1, pp. 186–196, 2002.

[21] X. Computation, "Getting started with the Xilinx Virtex-6 FPGA MI-605 evaluation Kit," 2010.

2

Design of a Reliable Hybrid (PV/Diesel) Power System with Energy Storage in Batteries for Remote Residential Home

Vincent Anayochukwu Ani

Department of Electronic Engineering, University of Nigeria, Nsukka 410001, Nigeria

Correspondence should be addressed to Vincent Anayochukwu Ani; vincent_ani@yahoo.com

Academic Editor: Mohamed Benghanem

This paper reports the experience acquired with a photovoltaic (PV) hybrid system simulated as an alternative to diesel system for a residential home located in Southern Nigeria. The hybrid system was designed to overcome the problem of climate change, to ensure a reliable supply without interruption, and to improve the overall system efficiency (by the integration of the battery bank). The system design philosophy was to maximize simplicity; hence, the system was sized using conventional simulation tool and representative insolation data. The system includes a 15 kW PV array, 21.6 kWh (3600 Ah) worth of battery storage, and a 5.4 kW (6.8 kVA) generator. The paper features a detailed analysis of the energy flows through the system and quantifies all losses caused by PV charge controller, battery storage round-trip, rectifier, and inverter conversions. In addition, simulation was run to compare PV/diesel/battery with diesel/battery and the results show that the capital cost of a PV/diesel hybrid solution with batteries is nearly three times higher than that of a generator and battery combination, but the net present cost, representing cost over the lifetime of the system, is less than one-half of the generator and battery combination.

1. Introduction

Energy is essential to economic and social development and improves quality of life. It is very important for the developing society [1]. In Nigeria, most residential homes are connected to the electric grid. However, there still exist several "off-grid" or remote locations, which, for financial and/or environmental reasons related to their distance from an existing power line, are not connected to the utility grid. Most of these residences derive their electricity from gasoline or diesel powered generators, which can be noisy and have the disadvantage of increasing the greenhouse gas emission which has a negative impact on the environment. Amid the environmental problems of using petrol and diesel generators, the cost of running them is quite high. Due to the high cost of running petrol/diesel generators, many Nigerians are willing to shift from using these traditional generators to the use of renewable energy technologies.

Renewable energy technologies (such as solar-photovoltaic systems) can be localized and decentralized unlike the national electricity grid. This allows end-users to generate their own electricity wherever they are located. Also, the technologies do not require any running cost, unlike the traditional petrol/diesel generators.

The installation of a solar power system to replace or offset a portion of the diesel electricity generation is an option to consider for remote residential homes. A complete replacement of diesel generation with solar power is usually not feasible, due to low solar input during the rainy season. However, a solar/diesel combination system known as hybrid system can prove to be very reliable and cost effective given the right conditions (such as optimal sizing). Hybrid energy applications are of increasing interest, and a well-managed hybrid solar-diesel system can achieve lifetime fuel savings, while ensuring reliable electricity supply. Insofar as diesel fuel is reduced, and such systems reduce CO_2 as well as particulate emissions that are harmful to health. They are an economical option in areas isolated from the grid.

This paper describes the way to design the aspects of a hybrid power system, a photovoltaic (PV) generator with energy storage for a residential use. The decision to select a PV generator hybrid system rather than a pure PV system for

TABLE 1: Energy needed for the household use.

Description of item	Item abbreviation	Power rating (watts)	Qty	Total load (watts)	Daily hour of actual utilization (hr. per day)
Medium size deep-freezer	DF	130	1	130	24 h (00:00 h–24:00 h)
Water pumping machine	PM	1000	1	1000	1 h (13:00 h-14:00 h)
Washing machine	WM	280	1	280	1 h (09:00 h-10:00 h)
Electric stove	ES	1000	1	1000	2 h (17:00 h–19:00 h)
Microwave oven	MO	1000	1	1000	2 h (06:00 h-07:00 h; 11:00 h-12:00 h)
Electric pressing iron	PI	1000	1	1000	1 h (12:00 h-13:00 h)
Air-conditioner	AC	1170	1	1170	9 h (08:00 h–17:00 h)
Refrigerator	RF	500	1	500	9 h (08:00 h–17:00 h)
Water bath	WB	1000	1	1000	2 h (03:00 h–04:00 h; 18:00 h–19:00 h)
Ceiling fan	CF	100	14	1400	14 h (08:00 h–22:00 h)
Energy efficient lighting	EL	6	23	138	8 h (04:00 h–08:00 h; 18:00 h–22:00 h)
Lighting-outdoor (security)	LO	9	4	36	13 h (18:00 h–07:00 h)
21″ TV with decoder	21″ TV-D	150	1	150	9 h (08:00 h–17:00 h)
21″ television	21″ TV	100	1	100	11 h (18:00 h–05:00 h)
14″ television	14″ TV	80	8	640	22 h (06:00 h–17:00 h; 18:00 h–05:00 h)
Sony music system	SM	100	1	100	1 h (04:00 h–05:00 h)
DSTV receiver	D-R	50	1	50	22 h (06:00 h–17:00 h; 18:00 h–05:00 h)
DVD player	D-P	50	1	50	2 h (19:00 h–21:00 h)
Computer printer	CP	100	1	100	1 h (15:00 h-16:00 h)
Computer PC	PC	115	1	115	9 h (08:00 h–17:00 h)
Computer laptop	CL	35	1	35	9 h (08:00 h–17:00 h)
Miscellaneous	M	100	1	100	24 h (00:00 h–24:00 h)

the considered location is consistent with its solar irradiation. This system will replace an existing diesel powered electric generator and was sized to meet the residence's known lighting and plug loads, refrigeration, cooking, and heating needs. The residence is located about a km from the utility grid and the location is characterized by a yearly global irradiation of about 2150 kWh/m². Also, this study is to produce a detailed experimental accounting of energy flows through the hybrid system and quantify all system losses. In addition, the hybrid system designed will be compared with the diesel/battery system in terms of costs and environmental impacts.

(1) Description of the Residential Home. The residence is a duplex building and has six rooms, a kitchen, and a sitting room at the ground floor, while it has three masters' rooms, a library, and a small sitting room upstairs. The building is furnished with electric power consumptions such as washing machine, electric stove, electric pressing iron, DVD, stereo cassette, television, decoder/cable, water pumping machine, fans, electric bulbs, water bath, deep-freezer, and microwave. Each room has fan, electric bulb, and television. The sitting room at the upstairs uses air-condition, while the one at the ground floor uses four fans. The residence is not connected to the grid and currently utilizes a diesel power generating system to meet its energy needs.

In this research, load assessment and the pattern of using electricity power within the house were carried out based on data provided by the occupant of the house and a site visit to evaluate the characteristics of the power system, power requirements, and power system management and operation. The daily power demands for the residential home are tabulated in Tables 1 and 2 and shown in Figure 1. These tables show the estimation of each appliance's rated power, its quantity, and the hours of use by the residence in a single day. The miscellaneous load is for unknown loads in the house.

(2) Overview of the Study Area. This research focuses on the design of a hybrid power system with energy storage in batteries for a residential home. The residential home where the study was done is located in a remote setting of Ndiagu-Akpugo. Ogologo-Eji Ndiagu-Akpugo is in Nkanu-West LGA of Enugu State in South-Eastern Nigeria on latitude 6°35′N and longitude 7°51′E. The data for solar resource (used in generating Figure 2) were obtained from the National Aeronautics and Space Administration (NASA) Surface Meteorology and Solar Energy web site [2]. After scaling on this data, the scaled annual average resource of 4.7 kWh/m²/d was obtained for the site. As can be seen in Figure 2, months below 4.5 kWh/m²/d are the months of June, July, August, and September which are the months of raining season in Nigeria, and there are likely to be more cloudy days on these months.

TABLE 2: The electrical load (daily load demands) data for the household.

The appliance abbreviations are given in Table 1

Time	DF	PM	WM	ES	MO	PI	AC	RF	WB	CF	EL	LO	21"TV-D	21"TV	14"TV	SM	D-R	D-P	CP	PC	CL	M	Total (W/Hr)
0.00-1.00	130											36		100	640		50					100	1056
1.00-2.00	130											36		100	640		50					100	1056
2.00-3.00	130											36		100	640		50					100	1056
3.00-4.00	130								1000			36		100	640		50					100	2056
4.00-5.00	130										138	36		100	640	100	50					100	1294
5.00-6.00	130										138	36										100	404
6.00-7.00	130				1000						138	36			640		50					100	2094
7.00-8.00	130										138				640		50					100	1058
8.00-9.00	130						1170	500		1400			150		640		50			115	35	100	4290
9.00-10.00	130		280				1170	500		1400			150		640		50			115	35	100	4570
10.00-11.00	130						1170	500		1400			150		640		50			115	35	100	4290
11.00-12.00	130				1000		1170	500		1400			150		640		50			115	35	100	5290
12.00-13.00	130					1000	1170	500		1400			150		640		50			115	35	100	5290
13.00-14.00	130	1000					1170	500		1400			150		640		50			115	35	100	5290
14.00-15.00	130						1170	500		1400			150		640		50			115	35	100	4290
15.00-16.00	130						1170	500		1400			150		640		50		100	115	35	100	4390
16.00-17.00	130						1170	500		1400			150		640		50			115	35	100	4290
17.00-18.00	130			1000						1400												100	2630
18.00-19.00	130			1000					1000	1400	138	36		100	640		50					100	4594
19.00-20.00	130									1400	138	36		100	640		50	50				100	2644
20.00-21.00	130									1400	138	36		100	640		50	50				100	2644
21.00-22.00	130									1400	138	36		100	640		50					100	2594
22.00-23.00	130											36		100	640		50					100	1056
23.00-24.00	130											36		100	640		50					100	1056
Total	3120	1000	280	2000	2000	1000	10530	4500	2000	19600	1104	468	1350	1100	14080	100	1100	100	100	1035	315	2400	69282

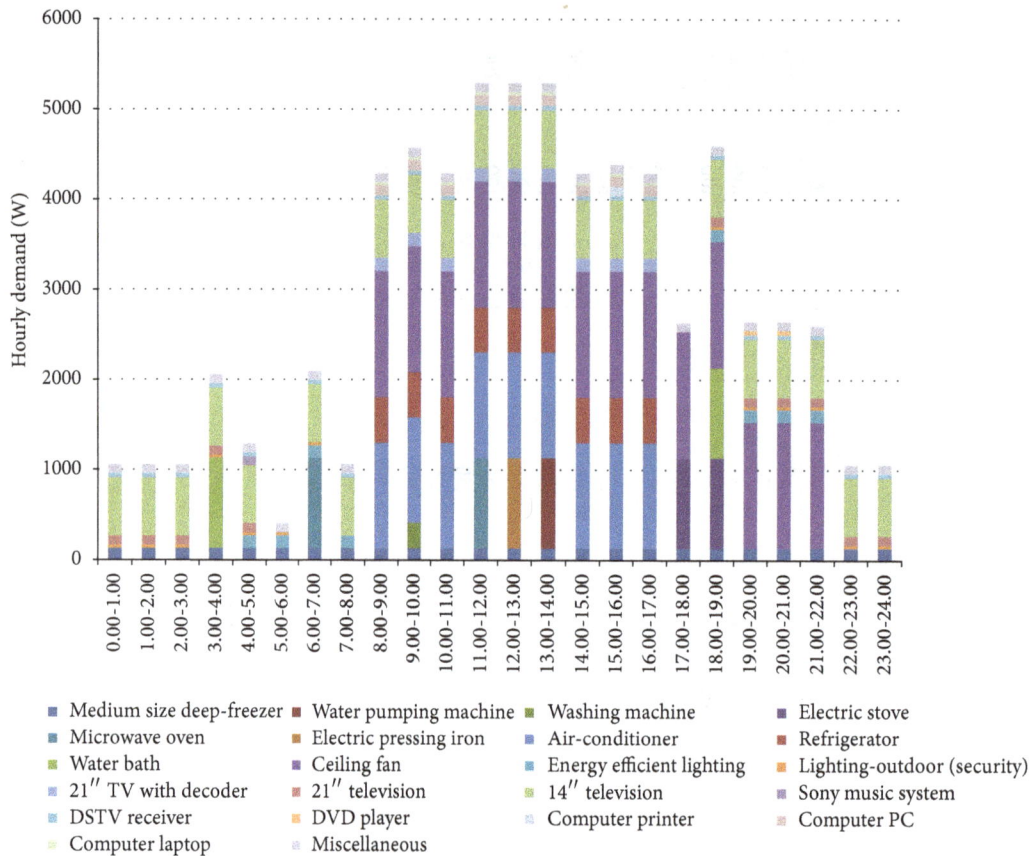

FIGURE 1: Hourly power demand profile of the household.

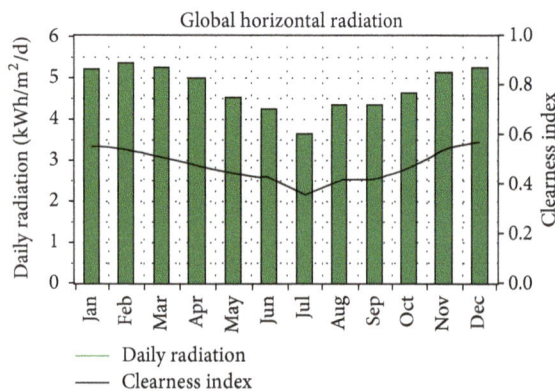

FIGURE 2: Solar daily radiation profile for Ndiagu-Akpugo in Nkanu-West (Enugu State) [2].

2. Energy Models

Energy model depends mainly on the economic feasibility and the proper sizing of the components in order to avoid outages as well as ensuring quality and reliability of supply. Energy design system looks into its sizing and the process of selecting the best components to provide cheap, efficient, reliable, environmentally friendly, and cost effective power supply [3]. The technoeconomic analysis looks at both environmental cost and the cheapest cost of energy produced by the system components. Designing a hybrid system would require correct components selection and sizing, with appropriate operation strategy [4, 5].

In energy systems, the sizing of the individual systems can be made in a variety of ways, depending upon the choice of parameters of interest. Energy models are employed as a supporting tool to develop energy strategies as well as outlining the likely future structure of the system under particular conditions. This helps to provide insights into the technological paths, structural evolution, and policies that should be followed [3]. A lot of research has been conducted on the performance of hybrid power systems and experimental results have been published in many articles [6–13]. The energy output of a hybrid system can be enough for the demands of a house placed in regions where the extension of the already available electricity grid would be financially unadvisable [9]. A method of sizing hybrid PV systems regarding the reliability to satisfy the load demand, economy of components, and discharge depth exploited by the batteries is therefore required.

Several models have been developed, simulating and sizing PV systems using different operation strategies. The estimation of performance of PV systems based on the Loss of Load Probability (LLP) technique is developed by [14–17]. These analytical methods are simple to apply but they are not general. On the other hand, the numerical methods presented by [18–24] present a good solution, but these need a long period solar radiation data record. Other methods estimate the excess of energy provided by PV generators and the storage capacity of the batteries using the utilizability method [25].

The conventional methodology (empiric, analytic, and numeric) for sizing PV systems has been used for a location where the required weather data (irradiation, temperature, humidity, clearness index, etc.) and the information concerning the site where we want to implement the PV system are available. In this case these methods present a good solution for sizing PV systems. However, these techniques could not be used for sizing PV systems in remote areas, in the case where the required data are not available. Moreover, the majority of the above methods need the long term meteorological data such as total solar irradiation and air temperature for their operation. So, when the relevant meteorological data are not available, these methods cannot be used, especially in the isolated areas. In this context, a model was developed, and the methodology aims at finding the configuration, among a set of systems components, which meets the desired system reliability requirements, with the lowest value of levelised cost of energy (LCE). This methodology can be used for determining the optimum number of solar panels and batteries configurations (the storage capacity of the batteries necessary to satisfy a given consumption). Since the investigation of this paper is based on a detailed study of an analysis of the energy flows, the analysis reveals the energy losses (charge controller, rectifier, battery, and inverter) in the system and the storage requirement. In addition, the model developed was used to select the optimal sizing parameters of PV system in which the results obtained have been compared and tested with HOMER software.

2.1. Development of a Model for Energy System Components. Modelisation is an essential step before any phase of component sizing. Various modeling techniques are developed, to model hybrid PV/diesel system components, in previous studies. For a hybrid PV/diesel system with storage battery, three principal subsystems are included, the PV generator, the diesel generator, and the battery storage. A methodology for modeling hybrid PV/diesel system components is described below. The theoretical aspects are given below (Sections 2.1.1, 2.1.2, 2.1.3, 2.1.4, and 2.1.5) and are based on the works of Ani [3], Gupta et al. [26], and Ashok [27].

2.1.1. Modeling of Solar-Photovoltaic Generator. Using the solar radiation available, the hourly energy output of the PV generator (E_{PVG}) can be calculated according to the following equation [3, 27–29]:

$$E_{PVG} = G(t) \times A \times P \times \eta_{PVG}. \quad (1)$$

2.1.2. Modeling of Diesel Generator. Hourly energy generated by diesel generator (E_{DEG}) with rated power output (P_{DEG}) is defined by the following expression [3, 27, 28]:

$$E_{DEG}(t) = P_{DEG}(t) \times \eta_{DEG}. \quad (2)$$

2.1.3. Modeling of Converter. In the proposed scheme, a converter contains both rectifier and inverter. PV energy generator and battery subsystems are connected with DC bus while diesel generating unit subsystem is connected with AC bus. The electric loads connected in this scheme are AC loads.

The rectifier is used to transform the surplus AC power from the diesel electric generator to charge the battery. The diesel electric generator will be powering the load and at the same time charging the battery. The rectifier model is given below:

$$E_{REC\text{-}OUT}(t) = E_{REC\text{-}IN}(t) \times \eta_{REC},$$
$$E_{REC\text{-}IN}(t) = E_{SUR\text{-}AC}(t). \quad (3)$$

At any time t,

$$E_{SUR\text{-}AC}(t) = E_{DEG}(t) - E_{Load}(t). \quad (4)$$

The inverter model for photovoltaic generator and battery bank are given below:

$$E_{PVG\text{-}IN}(t) = E_{PVG}(t) \times \eta_{INV},$$
$$E_{BAT\text{-}INV}(t) = \left[\frac{(E_{BAT}(t-1) - E_{LOAD}(t))}{(\eta_{INV} \times \eta_{DCHG})} \right]. \quad (5)$$

2.1.4. Modeling of Charge Controller. To prevent overcharging of a battery, a charge controller is used to sense when the batteries are fully charged and to stop or reduce the amount of energy flowing from the energy source to the batteries. The model of the charge controller is presented below:

$$E_{CC\text{-}OUT}(t) = E_{CC\text{-}IN}(t) \times \eta_{CC},$$
$$E_{CC\text{-}IN}(t) = E_{REC\text{-}OUT}(t) + E_{SUR\text{-}DC}(t). \quad (6)$$

2.1.5. Modeling of Battery Bank. The battery state of charge (SOC) is the cumulative sum of the daily charge/discharge transfers. The battery serves as an energy source entity when discharging and a load when charging. At any time, t, the state of battery is related to the previous state of charge and to the energy production and consumption situation of the system during the time from $t-1$ to t.

During the charging process, when the total output of all generators exceeds the load demand, the available battery bank capacity at time, t, can be described by [3, 29, 30]

$$E_{BAT}(t) = E_{BAT}(t-1) - E_{CC\text{-}OUT}(t) \times \eta_{CHG}. \quad (7)$$

On the other hand, when the load demand is greater than the available energy generated, the battery bank is in discharging state. Therefore, the available battery bank capacity at time, t, can be expressed as [3, 29]

$$E_{BAT}(t) = E_{BAT}(t-1) - E_{Needed}(t). \quad (8)$$

Let d be the ratio of minimum allowable SOC voltage limit to the maximum SOC voltage across the battery terminals when it is fully charged. So, the depth of discharge (DOD) is

$$DOD = (1-d) \times 100. \quad (9)$$

DOD is a measure of how much energy has been withdrawn from a storage device, expressed as a percentage of full capacity. The maximum value of SOC is 1, and the minimum SOC is determined by maximum depth of discharge (DOD):

$$SOC_{Min} = 1 - \frac{DOD}{100}. \tag{10}$$

2.2. Mathematical Cost Model (Economic and Environmental Costs) of Energy Systems.

This work developed a mathematical model of a system that could represent the integral (total sum) of the minimum economic and environmental (health and safety) costs of the considered options.

2.2.1. The Annualized Cost of a Component. The annualized cost of a component includes annualized capital cost, annualized replacement cost, annual O&M cost, emissions cost, and annual fuel cost (generator). Operation cost is calculated hourly on daily basis [3, 27, 29, 31].

2.2.2. Annualized Capital Cost. The annualized capital cost of a system component is equal to the total initial capital cost multiplied by the capital recovery factor. Annualized capital cost is calculated using [3, 27, 29, 31]

$$C_{acap} = C_{cap} \cdot CRF\left(i, R_{proj}\right). \tag{11}$$

2.2.3. Annualized Replacement Cost. The annualized replacement cost of a system component is the annualized value of all the replacement costs occurring throughout the lifetime of the project minus the salvage value at the end of the project lifetime. Annualized replacement cost is calculated using [3, 27, 29, 31]

$$C_{arep} = C_{rep} \cdot f_{rep} \cdot SFF\left(i, R_{comp}\right) - S \cdot SFF\left(i, R_{proj}\right). \tag{12}$$

f_{rep}, a factor arising because the component lifetime can be different from the project lifetime, is given by

$$f_{rep} = \begin{cases} \dfrac{CRF\left(i, R_{proj}\right)}{CRF\left(i, R_{rep}\right)}, & R_{rep} > 0, \\ 0, & R_{rep} = 0. \end{cases} \tag{13}$$

R_{rep}, the replacement cost duration, is given by

$$R_{rep} = R_{comp} \cdot INT\left(\frac{R_{proj}}{R_{comp}}\right). \tag{14}$$

SFF( ), the sinking fund factor which is a ratio used to calculate the future value of a series of equal annual cash flows, is given by

$$SFF\left(i, N\right) = \frac{i}{(1 + i)^N - 1}. \tag{15}$$

The salvaged value of the component at the end of the project lifetime is proportional to its remaining life. Therefore, the salvage value S is given by

$$S = C_{rep} \cdot \frac{R_{rem}}{R_{comp}}. \tag{16}$$

R_{rem}, the remaining life of the component at the end of the project lifetime, is given by

$$R_{rem} = R_{comp} - \left(R_{proj} - R_{rep}\right). \tag{17}$$

2.2.4. Annualized Operating Cost. The operating cost is the annualized value of all costs and revenues other than initial capital costs and is calculated using [3, 27, 29, 31]

$$C_{aop} = \sum_{t=1}^{365} \left\{ \sum_{t=1}^{24} [C_{oc}(t)] \right\}. \tag{18}$$

2.2.5. Cost of Emissions. The following equation is used to calculate the cost of emissions [3, 27, 29, 31]:

$$C_{emissions} = \frac{c_{CO_2}M_{CO_2} + c_{CO}M_{CO} + c_{UHC}M_{UHC} + c_{PM}M_{PM} + c_{SO_2}M_{SO_2} + c_{NO_x}M_{NO_x}}{1000}. \tag{19}$$

Total cost of a component = economic cost + environmental cost, where economic cost = capital cost + replacement cost + operation and maintenance cost + fuel cost (generator). Also environmental cost = emissions cost.

Annualized Cost of a Component Is Calculated Using [3, 27, 29, 31]

$$C_{ann} = C_{acap} + C_{arep} + C_{aop} + C_{emissions}. \tag{20}$$

Annualized Total Cost of a Component Is Calculated Using [29, 31]

$$C_{ann,tot,c} = \sum_{c=1}^{N_c} \left(C_{acap,c} + C_{arep,c} + C_{aop,c} + C_{emissions}\right). \tag{21}$$

From (21), the economic and environmental cost model through annualized total cost of different configurations of power system results in the hybridizing of the renewable energy generator (PV) with existing energy (diesel) is given below.

Economic and environmental cost model of running *solar + diesel generator + batteries + converter* is calculated as

$$
\begin{aligned}
C_{\text{ann,tot},s+g+b+c} = & \sum_{s=1}^{N_s} \Big(C_{\text{acap},s} + C_{\text{arep},s} + C_{\text{aop},s} \\
& + C_{\text{emissions}} \Big) + \sum_{g=1}^{N_g} \Big(C_{\text{acap},g} + C_{\text{arep},g} + C_{\text{aop},g} \\
& + C_{\text{emissions}} + C_{\text{af},g} \Big) + \sum_{b=1}^{N_b} \Big(C_{\text{acap},b} + C_{\text{arep},b} + C_{\text{aop},b} \\
& + C_{\text{emissions}} \Big) + \sum_{c=1}^{N_c} \Big(C_{\text{acap},c} + C_{\text{arep},c} + C_{\text{aop},c} \Big).
\end{aligned}
\tag{22}
$$

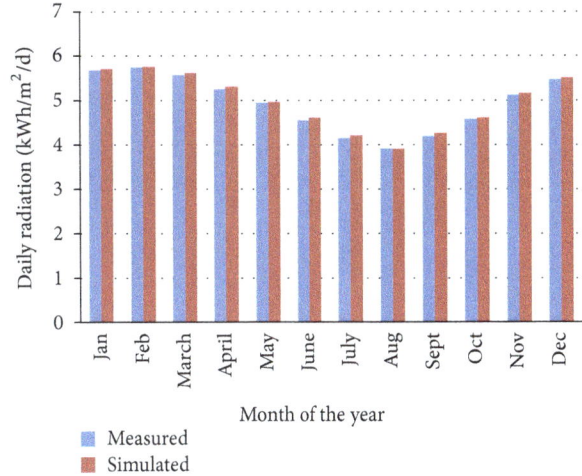

FIGURE 3: Calibrated solar radiation.

2.3. Description of the Computer Simulation. A computer program was developed and used to build the hybrid (PV/diesel) system model. Data inputs to the program are hourly load demand data, latitude, and longitude of the site and reference component cost. The designed software determines as its output the size of system components (sizing parameters) and the performance of the system over the course of the year (see the supplementary data in Supplementary Material available online at http://dx.doi.org/10.1155/2016/6278138) by showing the power supplied by each of the energy systems over the year, given the load conditions and taking into account the technical factors. The designed software can be used to study how the hybrid (PV/diesel) system is being supplied.

2.4. Validation of the Model. The designed software results were carried out followed with HOMER data to validate the analysis. The comparison shows a close agreement between results obtained from the designed software module and results obtained from HOMER setup. In addition, before using the measured data gotten from NASA datasets in simulating the individual components of a PV/diesel hybrid system, the developed program accuracy was established; the simulated data predicted by the software program fall within the bounds of the measured data. The algorithm that the developed program uses to synthesize solar data is based on the work of Graham and Hollands [32]. The realistic nature of synthetic data created by this algorithm is demonstrated and the test shows that synthetic solar data (simulated) produce virtually the same simulation results as real data (measured) as shown in Figure 3.

3. System Description

The designed system considered in this paper is a hybrid system which consists of a renewable (photovoltaic) energy system integrated in a conventional (diesel) power generation system, energy storage in battery, a DC/AC converter (an inverter for the conversion of generated DC power into

required AC power), and an AC/DC converter (a rectifier for the conversion of generated AC power in order to charge the battery) as shown in Figure 4. The inverter used is bidirectional, also known as power converter, which maintains energy flow between AC and DC components, since the flow comes in two different directions (from AC to DC and from DC to AC).

The flow from the solar array passes through the charge controller to charge the battery and at the same time supply electricity to the load through the inverter. The actual AC power obtained after the conversion from a solar array can be seen in Table 3. The charge controller monitors and controls the charging and discharging of the battery in order not to allow the battery to be damaged (due to overcharging or overdischarging).

Another flow comes from diesel generator when the PV and the battery could no longer serve the load; the generator supplies electricity direct to serve the load and at the same time charge the battery through the rectifier. That is how the designed hybrid system is expected to work.

The system design was to be representative of the type of residential systems that were likely to be installed in the foreseeable future. Hence, the system was sized using conventional simulation tool and representative insolation data.

3.1. Cost of Key Components (including Installation and Labour) and Interest Rate for Capital Investments

3.1.1. PV System Cost (US$ 2/Wp). The cost of PV panels on the Nigerian market was estimated as US$ 0.600/Wp based on prices cited by Nigerian suppliers (based on the cost of a module of $1210 \times 808 \times 35$ mm size generating 130 watts of peak power (Wp DC) in controlled conditions) [34]. This was adjusted upward to US$ 2/Wp to account for other support components that are required, also known as balance of system (BOS) parts, such as cables, charge controller with maximum power point tracker, lightening protection, and delivery/labour and installation costs.

TABLE 3: Results of the hybrid electricity production, battery charge, supply, excess, losses, and consumption (kW).

Month	Hybrid PV/diesel electricity generation — Electricity generated and supplied, charging the battery, and excess electricity by the hybrid system (kW)					Rectifier — Energy received by the rectifier to charge the battery (kW)			Battery — Energy received by the battery and supplied to the AC load via inverter (kW)			Inverter — Energy received by the inverter and supplied to the AC load (kW)			AC load — AC load served (kW)
	Electricity generated	*Supplied to the load	*Charging the battery	Losses	Excess electricity generated	Energy in	Energy out	Losses	Charge	Discharge	Losses	Energy in	Energy out	Losses	AC load served (kW)
January	2715.399	1862.277	538.799	0.159	314.164	299.517	254.646	44.871	493.928	-427.515	66.413	1415.143	1273.589	141.554	2148.238
February	2527.428	1695.987	489.700	0.154	341.587	289.982	246.547	43.435	446.265	-379.553	66.712	1351.523	1216.327	135.196	1940.344
March	2802.411	1887.039	533.918	0.154	381.300	301.472	256.311	45.161	488.757	-411.872	76.885	1506.288	1355.615	150.673	2148.238
April	2629.097	1827.413	509.332	0.146	292.206	289.705	246.309	43.396	465.936	-395.701	70.235	1441.311	1297.137	144.174	2078.940
May	2631.846	1876.123	535.826	0.179	219.718	314.101	267.048	47.053	488.773	-419.038	69.735	1468.844	1321.921	146.923	2148.238
June	2522.854	1804.867	528.865	0.177	188.945	313.288	266.355	46.933	481.932	-408.659	73.273	1345.553	1210.967	134.586	2078.940
July	2544.529	1860.281	541.031	0.167	143.050	314.666	267.531	47.135	493.896	-420.576	73.320	1325.833	1193.214	132.619	2148.238
August	2565.565	1849.784	551.749	0.164	163.868	324.882	276.211	48.671	503.078	-425.589	77.489	1270.987	1143.852	127.135	2148.238
September	2568.195	1799.896	529.854	0.171	238.274	312.814	265.950	46.864	482.990	-409.230	73.760	1301.553	1171.367	130.186	2078.940
October	2681.092	1873.098	541.203	0.184	266.607	324.375	275.785	48.590	492.554	-422.554	70.059	1473.828	1326.414	147.414	2148.238
November	2649.461	1812.321	529.743	0.158	307.239	305.717	259.919	45.798	483.945	-409.988	73.957	1433.337	1289.968	143.369	2078.940
December	2668.107	1868.682	541.449	0.185	257.791	314.356	267.268	47.088	494.361	-423.486	70.875	1438.938	1295.008	143.930	2148.238
Total	31505.984	22017.768	6371.469	1.998	3114.749	3704.875	3149.880	554.995	5816.474	-4953.761	862.713	16773.138	15095.379	1677.759	25293.770

*In terms of electricity supply and battery charge, the PV supplies electricity to the AC load via the inverter and charges the battery directly, whereas the diesel generator supplies electricity to the AC load directly and charges the battery through the rectifier, as shown in Table 4. Also, the battery supplies electricity to the AC load through the inverter, as shown in this table.

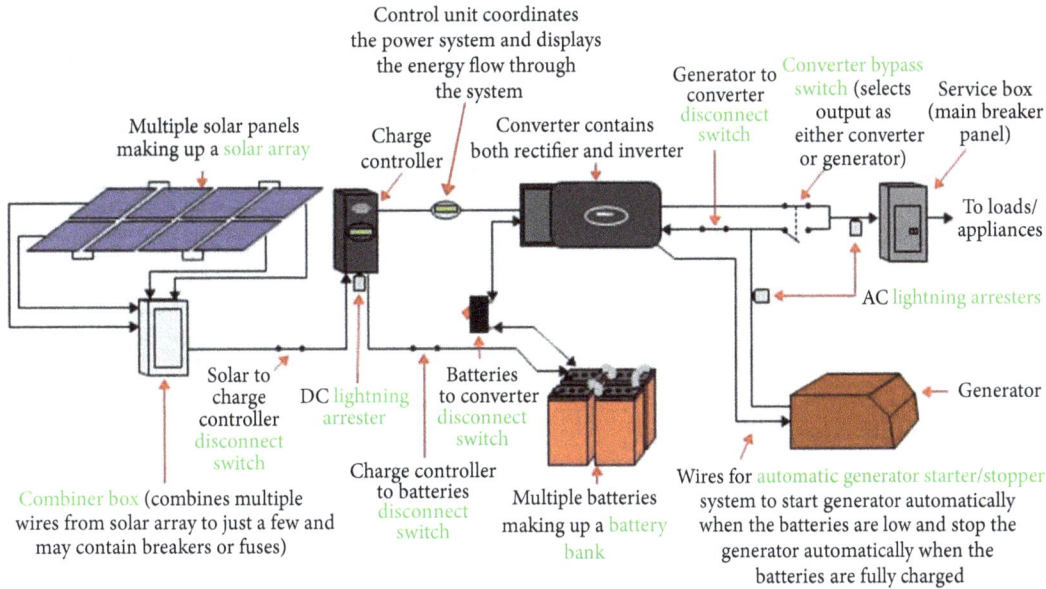

FIGURE 4: Photovoltaic hybrid power system structure [33].

TABLE 4: Results of each of the energy components of the hybrid system (PV and diesel) for electricity production, supply, and battery charging (kW).

Month	Electricity generated and supplied and battery charge by the PV in hybrid system (kW)			Electricity generated and supplied and battery charge by the diesel in hybrid system (kW)		
	Electricity generated	Supplied to the load via inverter	Charging the battery directly	Electricity generated	Supplied to the load directly	Charging the battery via rectifier
January	1538.295	987.628	239.282	1177.104	874.649	299.517
February	1510.492	971.970	199.718	1016.936	724.017	289.982
March	1705.644	1094.416	232.446	1096.767	792.623	301.472
April	1554.395	1045.610	219.627	1074.702	781.803	289.705
May	1488.347	1049.806	221.725	1143.499	826.317	314.101
June	1333.962	936.894	215.577	1188.892	867.973	313.288
July	1271.351	905.257	226.365	1273.178	955.024	314.666
August	1230.539	845.398	226.867	1335.026	1004.386	324.882
September	1343.075	892.323	217.040	1225.120	907.573	312.814
October	1534.355	1051.274	216.828	1146.737	821.824	324.375
November	1552.498	1023.349	224.026	1096.963	788.972	305.717
December	1499.299	1015.452	227.093	1168.808	853.230	314.356
Total	17562.252	11819.377	2666.594	13943.732	10198.391	3704.875

3.1.2. Converter Cost (US$ 0.320/Wp). The cost of a converter, based on prices cited by Nigerian suppliers, was US$ 0.320/Wp [35].

3.1.3. Battery Cost (US$ 180/kWh). The cost of a 6 V/225 Ah lead acid battery on the Nigerian market was found to be in the range of US$ 172 [35]. Including balance of system (BOS) components and labour/installation costs, the capital cost for

the battery arrays was adjusted upward to US$ 180/kWh. The precise number of batteries required for each option is then determined by the simulation.

3.1.4. Generator Cost (US$ 1000/kW). The capital cost of the genset includes the generator itself (usually diesel or gasoline), as well as BOS costs and labour/installation costs. On the Nigerian local market, a generator of smaller range

(2–5 kVA) was priced at about US\$ 991 [36]. Including BOS and labour/installation costs, the total price was estimated at around US\$ 1,000 per kW load.

3.1.5. Fuel Cost (US\$ 1.2/L). The source for this estimate was the Nigerian official market rate as of October 2015.

3.1.6. Interest Rate: 7.5%. Interest rates vary widely and can be particularly high in developing countries, having a profound impact on the cost-benefit assessment. Interest rates on Nigerian commercial bank loans may be between 6% and 7.5%. An estimate of 7.5% was selected for this case study.

4. Energy Losses in Stand-Alone PV/Diesel Hybrid Systems

Stand-alone PV/diesel hybrid systems are designed to be totally self-sufficient in generating, storing, and supplying electricity to the electrical loads in remote areas. Figure 5 shows an energy flow diagram for a typical PV/diesel hybrid system. The following equation (23) shows the energy balance of a PV/diesel hybrid system:

$$E_{IN} \approx E_{OUT}. \tag{23}$$

The energy that has to be supplied from the generator can be determined as

$$E_{MG} = E_{LOAD} + E_{LOSS}R - E_{PV}. \tag{24}$$

The energy that has to be supplied from the photovoltaic can be determined as

$$E_{PV} = E_{LOAD} + E_{LOSS}CC + E_{LOSS}B + E_{LOSS}I - E_{MG}. \tag{25}$$

The objective of this study (efficient energy balance) is to minimize the energy that has to be supplied from auxiliary energy source (diesel generator) by the addition of PV panels. Additionally, the motor generator should be operated near its nominal power to achieve high fuel efficiency by the inclusion of battery bank. As shown in (24) and (25), energy losses are flowing into the energy demand and supply of the system; therefore, it is necessary to identify the energy losses in the system. A classification of all relevant energy losses in a stand-alone PV hybrid system is given as capture losses and system losses [37]. Capture losses account for the part of the incident radiation energy that remains uncaptured and which is therefore lost within a global energy balance. Capture or irradiation losses translate the fact that only part of the incoming irradiation is used for energy conversion. System losses define systematic energy losses that are due to the physical properties of the system components or the entire installation. Energy conversion losses constitute important contributions to this category [38].

System losses cover all energy losses which occur during the conversion of generated energy into usable AC electricity. In this study, only the energy conversion losses were considered, to assess the potential of the designed hybrid system. The losses are indicated in Figure 5.

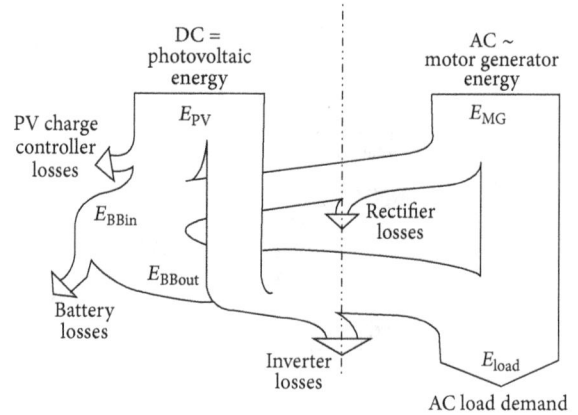

FIGURE 5: Energy flow diagram for a typical PV/diesel hybrid system [37].

5. Results and Discussion

The design provides an interesting example of how optimal combinations of photovoltaic and diesel generation with appropriate energy storage yielded multiple gains: a shift to renewable energy, a reliable supply for household energy needs, and lower overall cost of energy.

5.1. Results

5.1.1. Designed Hybrid System. To overcome the problem of the climatic changes, to ensure a reliable supply without interruption, and to improve the overall system efficiency, a hybrid system (that comprised a PV system, the diesel power system, and storage battery as backup sources) is essential as shown in Figure 4. The reasons for the inclusion of battery bank in this design are due to fluctuations in solar radiation and also for the generator to operate at optimum efficiency, because continued operation of generator at lower loads or severe variation in the load results in an inefficient engine performance and one of the options for the load management is to integrate battery bank (which becomes a load when charging to improve the generator efficiency) to improve the overall system efficiency. Considering various types and capacities of system devices (PV array, diesel generator, and battery size), the configurations which can meet the desired system reliability are obtained by changing the type and size of the devices systems. The configuration with the lowest LCE gives the optimal choice. Therefore, the optimal sizing of the hybrid system (PV-diesel generator-battery system) in terms of reliability, economy, and environment is shown in Tables 3, 5, and 6, respectively. This was determined through rigorous mathematical computations.

From the design results, the PV power supply is between 8:00 h and 19:00 h while the radiation peak is between 12:00 h and 14:00 h as can be seen in the supplementary data. Between 12:00 h and 14:00 h there is no deficit in the system and the PV energy supplies the load and charges the battery, thereby reducing the operational hours of the diesel generator and the running cost of the hybrid energy

TABLE 5: Comparative costs of hybrid power and stand-alone generator supply systems.

Configuration	PV capacity (kW)	Generator capacity (kW)	Number of batteries (6 V/225 Ah)	Converter capacity (kW)	Initial capital (US$)	Annual generator usage (hours)	Annual quantity of diesel (L)	Total net present cost (US$) for 20 years	Cost of energy (US$/kWh)	Renewable fraction
PV + generator + battery	15	5.4	16	5.5	41,048	5,011	5,716	192,231	0.745	0.59
Generator + battery	—	5.4	30	5.5	14,450	5,298	9,183	210,146	0.815	0.00

TABLE 6: Comparative emissions of hybrid power and stand-alone generator supply systems.

Configuration	Pollutant emission (kg/yr)						Fuel consumption (L/yr)	Operational hour of diesel generator (hr/yr)
	CO_2	CO	UHC	PM	SO_2	NO_x		
PV + generator + battery	15,052	37.2	4.12	2.8	30.2	332	5,716	5,011
Generator + battery	24,183	59.7	6.61	4.5	48.6	533	9,183	5,298

Note: PM refers to total particulate matter. UHC refers to unburned hydrocarbons.

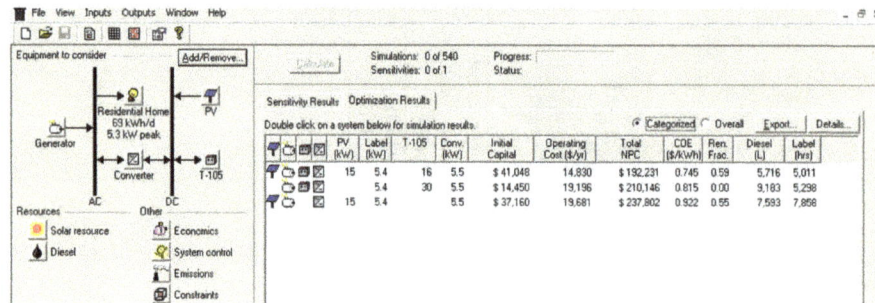

FIGURE 6: The optimization results from HOMER for hybrid PV/diesel energy system.

system as well as the pollutant emissions. There is likely to be deficit in other remaining hours due to poor radiation, and the deficit is being completed by either the battery or the diesel generator. The result of the demand met by the hybrid energy system (PV/diesel) over the course of the year is shown in the supplementary data; it shows how the sources were allocated according to the load demand and availability. It was observed that the variation is not only in the demand but also in the availability of solar resources. The battery or the diesel generator compensates the shortage depending on the decision mode.

5.1.2. Results from HOMER. The derivations from the developed software were compared to HOMER optimization method, and the same inputs used in calculations by the developed software were used by HOMER which produced the same results with the developed software as shown in Figure 6 (Figure 6 compared with Table 5). Therefore, the results from the software can be used as comparison and point of reference.

5.2. Discussion

5.2.1. Overall Energy Production and Utilization. From the design, solar power will not replace the need for diesel generator for this remote residential home but could offset a portion of the diesel fuel used. Although the residential loads provide the best possible match with PV output (since these loads typically peak during daytime and afternoon hours), there is still need for a backup with diesel generator (during the raining season and cloudy days).

In the solar resources, apart from the month of February that has 28 days, the month of March has the highest global and incident solar (207.568 kWh/m²; 213.213 kWh/m²), while the month of August has the least global and incident solar (159.232 kWh/m²; 153.817 kWh/m²) as shown in Table 7.

TABLE 7: Solar resources for the studied zone.

Month	Global solar (kWh/m²)	Incident solar (kWh/m²)	Power generated with 15 kW PV array (kW)
January	173.783	192.285	1538.295
February	176.292	188.814	1510.492
March	207.568	213.213	1705.644
April	198.460	194.312	1554.395
May	197.020	186.037	1488.347
June	178.982	166.744	1333.962
July	168.215	158.916	1271.351
August	159.232	153.817	1230.539
September	166.994	167.880	1343.075
October	182.472	191.792	1534.355
November	175.089	194.054	1552.498
December	165.744	187.409	1499.299
Total	2149.851	2195.273	17562.252

In the hybrid system configuration, the sizing was done in favour of PV system (to overcome the problem of the climatic changes), and in order to accommodate the load demand for all the months, excess electricity was generated by the PV system. The excess electricity generated differs from month to month and depends on the incident solar. The highest excess electricity is observed in March (381.30 kW), while the least is in the months of July (143.05 kW) and August (163.868 kW), the two months most affected by raining season.

In the month of March, the PV generated the highest electricity (1705.644 kW) and supplied to the load via inverter the highest electricity (1094.416 kW). This was because the month of March has the highest global and incident solar (207.568 kWh/m²; 213.213 kWh/m²), while in the month of August, the PV generated the least electricity (1230.539 kW) and supplied to the load via inverter the least electricity (845.398 kW) and this was due to least global and incident solar (159.232 kWh/m^2; 153.817 kWh/m^2). In this month of August, to ensure a reliable supply without interruption, the diesel due to the low electricity generated by the PV (caused by low incident solar) supplies to the load the highest electricity (1004.386 kW) and charges the battery via rectifier (to improve the overall system efficiency). In the month of August, battery charging (503.078 kW) and discharging (−425.589 kW) are highest due to low supply coming from the PV. The generator becomes ON often to serve the AC load and at the same time charge the battery (which is a DC load; battery becomes a load when charging). It is worthwhile noting from Table 3 that the PV-diesel hybrid solution supported by battery storage produces 17,562 kWh/yr (59%) from solar PV array and 13,944 kWh/yr (41%) from diesel generator making a total of 31,506 kWh/yr (100%).

5.2.2. Energy Flows. One of the main objectives of this study was to produce a detailed experimental accounting of energy flows through the hybrid system. In particular, my interest is in quantifying all system losses.

Hybrid PV/Diesel System with Battery. For the PV part of the hybrid system, device losses include PV charge controller losses, DC-AC conversion losses, both for energy flowing directly to the load and for energy transiting through the battery, and storage round-trip losses. On the generator side, the AC-DC conversion losses affect electrical energy that does not flow directly to the load. The reason for these losses on the generator side is that the hybrid system was designed to be cycle charging meaning that the diesel generator is allowed to charge the battery.

All losses through the hybrid system are classified as follows:

(i) PV charge controller losses.

(ii) Battery storage losses.

(iii) Rectifier (battery charger conversion) losses.

(iv) Inverter losses.

PV charge controller losses are due to the DC/DC conversion efficiency (converting energy generated by the PV to charge the storage battery). DC/DC conversion losses are generated during the control of the flow of current to and from the battery by the PV charge controller. Result shows that the losses are minimal when compared to other component losses (storage losses, inverter, and rectifier losses) as shown in Table 3.

Storage losses comprise all energy losses within a battery. They are described by the charge and discharge efficiencies of the battery as well as the self-discharge characteristics. In the month of August, the battery charging and the discharging as well as its losses (due to charge and discharge efficiencies) are highest due to the fact that diesel becomes ON often to charge the battery; when the battery reaches its maximum point of charge, the diesel stops while the battery starts discharging in order to power the load, and once the battery reaches its minimum point of discharge, it stops discharging and

the diesel comes ON again. The process continues the same way until the PV starts to generate electricity to supply it to the load and charge the battery; otherwise, it returns to the diesel charging the battery. Results of the design show that the storage battery was charged with 227.093 kWh/yr and 314.356 kWh/yr by the PV and diesel system, respectively, making a total charge of 5816.474 kWh/yr, while the battery discharged (supplied) to the load via the inverter a total discharge of −4953.761 kWh/yr, having losses of 862.713 kWh/yr as shown in Tables 3 and 4.

Battery charger conversion losses are due to the rectifier's AC/DC efficiency. AC/DC conversion losses are generated during battery charging from an AC source. In the month of August, the rectifier receives the highest electricity from the diesel generator due to the month's least global and incident solar (159.232 kWh/m^2; 153.817 kWh/m^2) and this affects the production from the PV; at this point the diesel comes ON in order to ensure a reliable supply without interruption. Results of the design show that the rectifier was supplied with 3704.875 kWh/yr and rectified to the battery with 3149.880 kWh/yr, having losses of 554.995 kWh/yr as shown in Tables 3 and 4.

Inverter losses are due to the inverter's DC/AC efficiency. DC/AC inverter losses occur before the initially provided energy can be consumed by an AC load. It means that all electrical energy that does not flow directly to the AC load passes through the inverter such as electricity flowing from the PV system, electricity rectified to the battery, and the one coming from the battery. In the month of August, the inverter receives the least electricity from the PV and battery due to the month's least global and incident solar (159.232 kWh/m^2; 153.817 kWh/m^2). Although the battery receives the highest charging of 503.078 kW from both the PV (226.867 kW) and diesel (rectified to the battery with 276.211 kW), the inverter still receives the least electricity because the diesel comes often to supply the AC load and charge the battery; the charging of the battery by the rectifier shows how often diesel supplies electricity to the load in this month of August as shown in Tables 3 and 4.

In conclusion, while the DC/DC conversion efficiency is generally low, the AC/DC rectifier (battery charger conversion) efficiency is somewhat lower than the DC/AC inverter efficiency as shown in Table 3.

5.2.3. Economic Costs. The capital cost of a PV/diesel hybrid solution with batteries is nearly three times higher than that of a generator and battery combination (US\$ 41,048), but the net present cost, representing cost over the lifetime of the system, is less than one-half of the generator and battery combination (US\$ 192,231), as shown in Table 5. The net present cost (NPC) of the PV/diesel/battery hybrid system is slightly lower than the NPC of the diesel/battery combination as a result of less fuel consumption and because fewer storage batteries are needed, and replacing batteries is a significant factor in system maintenance.

5.2.4. Environmental Pollution. On the environmental impact perspective, an increase in the operational hours of diesel generator brings about increase in the fuel consumption as well as an increase in GHG emission, whereas a reduction in the operational hours of diesel generator brings about reduction in the fuel consumption, thereby a reduction in GHG emission. Diesel system operates for 5,298 h/annum, has a fuel consumption of 9,183 L/annum, and generates in kilogrammes (kg) the pollutant emissions as shown in Table 6, while in the hybrid PV-diesel system, diesel generator operates for 5,011 h/annum, has a fuel consumption of 5,716 L/annum, and emits in kilogrammes the pollutant emissions annually into the atmosphere of the location of the residence as shown in Table 6. Reducing fuel consumption also means less emission from the energy system as shown by the solar PV-diesel system which has the lowest pollutant emissions.

6. Conclusion

This paper investigates the designing of a stand-alone hybrid power system focusing on photovoltaic/diesel energy system with energy storage in batteries. Starting from the analysis of the models of the system components, a complete simulation model is realized. From the designed system, a detailed experimental accounting of energy flows through the hybrid system was produced and all system losses caused by PV charge controller, battery storage round-trip, rectifier, and inverter conversions were quantified and documented. Results show that PV charge controller losses are due to the DC/DC conversion efficiency and are generated during the control of the flow of current to and from the battery by the PV charge controller, while storage losses comprise all energy losses within a battery and are described by the charge and discharge efficiencies of the battery as well as the self-discharge characteristics. In addition, battery charger conversion losses are due to the rectifier's AC/DC efficiency and are generated during battery charging from an AC source, while inverter losses are due to the inverter's DC/AC efficiency and occur before the initially provided energy can be consumed by an AC load. From the results, it has proven that the DC/DC conversion efficiency is generally low, while the AC/DC rectifier efficiency is somehow lower than the DC/AC inverter efficiency. Also, it has been demonstrated that the use of hybrid PV/diesel system with battery (one unit of 15 kW PV array, one unit of 5.4 kW generator, with 16 units of battery) can significantly reduce the dependence on solely available diesel resource. The designed hybrid system minimizes diesel operational hour and thereby reduces the fuel consumption which significantly affects (reduces) the pollution, such as carbon emission, thus reducing the greenhouse effect. Although utilization of hybrid PV/diesel system with battery might not significantly reduce the total NPC and COE, it has been able to cut down the dependence on diesel. On the other hand, it was also proven that the use of hybrid PV/diesel system with battery would be more economical if the price of diesel increased significantly. With a projection period of 20 years and 7.5% annual real interest rate, it was found that the use of hybrid PV/diesel system with battery could achieve significantly lower NPC and COE as compared to a stand-alone diesel system. As a conclusion, the hybrid PV/diesel system has potential use in replacing or upgrading existing stand-alone diesel systems in Nigeria.

Nomenclature

A:	The surface area in m^2
$C_{acap,c}$:	Annualized capital cost of a component
$C_{arep,c}$:	Annualized replacement cost of a component
$C_{aop,c}$:	Annualized operating cost of a component
$C_{acap,s}$:	Annualized capital cost of solar power
$C_{arep,s}$:	Annualized replacement cost of solar power
$C_{aop,s}$:	Annualized operating cost of solar power
$C_{acap,g}$:	Annualized capital cost of diesel generator
$C_{arep,g}$:	Annualized replacement cost of diesel generator
$C_{aop,g}$:	Annualized operating cost of diesel generator
$C_{af,g}$:	Annualized fuel cost for diesel generator
$C_{acap,b}$:	Annualized capital cost of batteries power
$C_{arep,b}$:	Annualized replacement cost of batteries power
$C_{aop,b}$:	Annualized operating cost of batteries power
$C_{acap,c}$:	Annualized capital cost of converter power
$C_{arep,c}$:	Annualized replacement cost of converter power
$C_{aop,c}$:	Annualized operating cost of converter power
C_{cap}:	Initial capital cost of the component
c_{CO_2}:	Cost for emissions of carbon dioxide (CO_2) ($/t)
c_{CO}:	Cost for emissions of carbon monoxide (CO) ($/t)
c_{UHC}:	Cost for emissions of unburned hydrocarbons (UHC) ($/t)
c_{PM}:	Cost for emissions of particulate matter (PM) ($/t)
c_{SO_2}:	Cost for emissions of sulfur oxide (SO_2) ($/t)
c_{NO_x}:	Cost for emissions of nitrogen oxide (NO_x) ($/t)
$C_{oc}(t)$:	The cost of operating component
C_{rep}:	Replacement cost of the component
$CRF(i, R_{proj})$:	Capital recovery factor
$E_{BAT}(t-1)$:	The energy stored in battery at hour $t-1$, kWh
$E_{Needed}(t)$:	The hourly load demand or energy needed at a particular period of time
$E_{REC-OUT}(t)$:	The hourly energy output from rectifier, kWh
$E_{REC-IN}(t)$:	The hourly energy input to rectifier, kWh
$E_{SUR-AC}(t)$:	The amount of surplus energy from AC sources, kWh
$E_{DEG}(t)$:	The hourly energy generated by diesel generator
$E_{PVG-IN}(t)$:	The hourly energy output from inverter (in case of SPV), kWh
$E_{PVG}(t)$:	The hourly energy output of the PV generator

$E_{BAT\text{-}INV}(t)$: The hourly energy output from inverter (in case of battery), kWh

$E_{BAT}(t-1)$: The energy stored in battery at hour $t-1$, kWh

$E_{LOAD}(t)$: The hourly energy consumed by the load side, kWh

$E_{CC\text{-}OUT}(t)$: The hourly energy output from charge controller, kWh

$E_{CC\text{-}IN}(t)$: The hourly energy input to charge controller, kWh

$E_{REC\text{-}OUT}(t)$: The hourly energy output from rectifier, kWh

$E_{SUR\text{-}DC}(t)$: The amount of surplus energy from DC source (PV panels), kWh

$E_{BAT}(t)$: The energy stored in battery at hour t, kWh

E_{IN}: Is equal to $E_{PV} + E_{MG}$

E_{OUT}: Is equal to $E_{LOAD} + E_{LOSS}$

E_{PV}: Energy generated by the PV array (kWh)

E_{MG}: Energy generated by the motor generator (kWh)

E_{LOAD}: Energy supplied to the load (kWh)

E_{LOSS}: Energy losses (kWh), which comprise all ($E_{LOSS}CC + E_{LOSS}B + E_{LOSS}R + E_{LOSS}I$)

$E_{LOSS}CC$: Energy losses via charge controller (kWh)

$E_{LOSS}B$: Energy losses via battery (kWh)

$E_{LOSS}R$: Energy losses via rectifier (kWh)

$E_{LOSS}I$: Energy losses via inverter (kWh)

$G(t)$: The hourly irradiance in kWh/m^2

i: Interest rate

$INT(\ )$: The integer function, returning the integer portion of a real value

M_{CO_2}: Annual emissions of CO_2 (kg/yr)

M_{CO}: Annual emissions of CO (kg/yr)

M_{NO_x}: Annual emissions of NO_x (kg/yr)

M_{PM}: Annual emissions of particulate matter (PM) (kg/yr)

M_{SO_2}: Annual emissions of SO_2 (kg/yr)

M_{UHC}: Annual emissions of unburned hydrocarbons (UHC) (kg/yr)

N: Number of years

P: The PV penetration level factor

R_{proj}: Project lifetime

R_{comp}: Lifetime of the component

$SFF(\ )$: Sinking fund factor

η_{PVG}: The efficiency of PV generator

η_{DEG}: The diesel generator efficiency

η_{REC}: The efficiency of rectifier

η_{INV}: The efficiency of inverter

η_{DCHG}: The battery discharging efficiency

η_{CC}: The efficiency of charge controller

η_{CHG}: The battery charging efficiency.

Competing Interests

The author declares no competing interests.

References

[1] M. Jovanović, "An analytical method for the measurement of energy systems sustainability in urban areas," *FME Transactions*, vol. 36, no. 4, pp. 157–166, 2008.

[2] National Aeronautics and Space Administration (NASA) Atmospheric Science Data Center, 2015, http://eosweb.larc.nasa.gov/sse/2012.

[3] V. A. Ani, *Energy optimization at GSM base station sites located in rural areas [Ph.D. thesis]*, 2015, http://www.unn.edu.ng/publications/files/17774_ENERGY_OPTIMIZATION_AT_GSM_BASE_STATION_SITES_LOCATED_IN_RURAL_AREAS.pdf.

[4] B. S. Borowy and Z. M. Salameh, "Optimum photovoltaic array size for a hybrid wind/PV system," *IEEE Transactions on Energy Conversion*, vol. 9, no. 3, pp. 482–488, 1994.

[5] R. Dufo-López and J. L. Bernal-Agustín, "Design and control strategies of PV-diesel systems using genetic algorithms," *Solar Energy*, vol. 79, no. 1, pp. 33–46, 2005.

[6] I. Gross, "The cost of diesel for Africa's mobile operators: 2012 may be the year that this bird comes home to roost," November 2011, http://www.balancingact-africa.com/news/en/issue-no-581.

[7] M. A. Elhadidy, "Performance evaluation of hybrid (wind/solar/diesel) power systems," *Renewable Energy*, vol. 26, no. 3, pp. 401–413, 2002.

[8] W. Kellogg, M. H. Nehrir, G. Venkataramanan, and V. Gerez, "Optimal unit sizing for a hybrid wind/photovoltaic generating system," *Electric Power Systems Research*, vol. 39, no. 1, pp. 35–38, 1996.

[9] M. A. Elhadidy and S. M. Shaahid, "Role of hybrid (wind + diesel) power systems in meeting commercial loads," *Renewable Energy*, vol. 29, no. 1, pp. 109–118, 2004.

[10] M. T. Iqbal, "Simulation of a small wind fuel cell hybrid energy system," *Renewable Energy*, vol. 28, no. 4, pp. 511–522, 2003.

[11] M. H. Nehrir, B. J. LaMeres, G. Venkataramanan, V. Gerez, and L. A. Alvarado, "An approach to evaluate the general performance of stand-alone wind/photovoltaic generating systems," *IEEE Transactions on Energy Conversion*, vol. 15, no. 4, pp. 433–439, 2000.

[12] S. H. Karaki, R. B. Chedid, and R. Ramadan, "Probabilistic performance assessment of autonomous solar-wind energy conversion systems," *IEEE Transactions on Energy Conversion*, vol. 14, no. 3, pp. 766–772, 1999.

[13] C. Protogeropoulos, B. J. Brinkworth, and R. H. Marshall, "Sizing and techno-economical optimization for hybrid solar photovoltaic/wind power systems with battery storage," *International Journal of Energy Research*, vol. 21, no. 6, pp. 465–479, 1997.

[14] L. L. Bucciarelli Jr., "Estimating loss-of-power probabilities of stand-alone photovoltaic solar energy systems," *Solar Energy*, vol. 32, no. 2, pp. 205–209, 1984.

[15] S. A. Klein and W. A. Beckman, "Loss-of-load probabilities for stand-alone photovoltaic systems," *Solar Energy*, vol. 39, no. 6, pp. 499–512, 1987.

[16] L. Barra, S. Catalanotti, F. Fontana, and F. Lavorante, "An analytical method to determine the optimal size of a photovoltaic plant," *Solar Energy*, vol. 33, no. 6, pp. 509–514, 1984.

[17] B. Bartoli, V. Cuomo, F. Fontana, C. Serio, and V. Silvestrini, "The design of photovoltaic plants: an optimization procedure," *Applied Energy*, vol. 18, no. 1, pp. 37–47, 1984.

[18] L. L. Bucciarelli Jr., "The effect of day-to-day correlation in solar radiation on the probability of loss-of-power in a stand-alone photovoltaic energy system," *Solar Energy*, vol. 36, no. 1, pp. 11–14, 1986.

[19] P. P. Groumpos and G. Papageorgiou, "An optimal sizing method for stand-alone photovoltaic power systems," *Solar Energy*, vol. 38, no. 5, pp. 341–351, 1987.

[20] V. A. Graham, K. G. T. Hollands, and T. E. Unny, "A time series model for K_t with application to global synthetic weather generation," *Solar Energy*, vol. 40, no. 2, pp. 83–92, 1988.

[21] R. J. Aguiar, M. Collares-Pereira, and J. P. Conde, "Simple procedure for generating sequences of daily radiation values using a library of Markov transition matrices," *Solar Energy*, vol. 40, no. 3, pp. 269–279, 1988.

[22] R. N. Chapman, "Development of sizing nomograms for stand-alone photovoltaic/storage systems," *Solar Energy*, vol. 43, no. 2, pp. 71–76, 1989.

[23] R. N. Chapman, "The synthesis of solar radiation data for sizing stand-alone photovoltaic systems," in *Proceedings of the 21st IEEE Photovoltaic Specialists Conference*, pp. 965–970, Orlando, Fla, USA, May 1990.

[24] I. Abouzahr and R. Ramakumar, "Loss of power supply probability of stand-alone photovoltaic systems: a closed form solution approach," *IEEE Transactions on Energy Conversion*, vol. 6, no. 1, pp. 1–11, 1991.

[25] A. Mellit, "Sizing of photovoltaic systems: a review," *Revue des Energies Renouvelables*, vol. 10, no. 4, pp. 463–472, 2007.

[26] A. Gupta, R. P. Saini, and M. P. Sharma, "Steady-state modelling of hybrid energy system for off grid electrification of cluster of villages," *Renewable Energy*, vol. 35, no. 2, pp. 520–535, 2010.

[27] S. Ashok, "Optimised model for community-based hybrid energy system," *Renewable Energy*, vol. 32, no. 7, pp. 1155–1164, 2007.

[28] D. K. Lal, B. B. Dash, and A. K. Akella, "Optimization of PV/Wind/Micro-Hydro/diesel hybrid power system in homer for the study area," *International Journal on Electrical Engineering and Informatics*, vol. 3, no. 3, pp. 307–325, 2011.

[29] K. Sopian, A. Zaharim, Y. Ali, Z. M. Nopiah, J. A. B. Razak, and N. S. Muhammad, "Optimal operational strategy for hybrid renewable energy system using genetic algorithms," *WSEAS Transactions on Mathematics*, vol. 7, no. 4, pp. 130–140, 2008.

[30] H. Abdolrahimi and H. K. Karegar, "Optimization and sensitivity analysis of a hybrid system for a reliable load supply in KISH_IRAN," *International Journal of Advanced Renewable Energy Research*, vol. 1, no. 4, pp. 33–41, 2012.

[31] T. Lambert, P. Gilman, and P. Lilienthal, "Micropower system modeling with HOMER," in *Integration of Alternative Sources of Energy*, F. A. Farret and M. G. Simões, Eds., chapter 15, John Wiley & Sons, New York, NY, USA, 2006.

[32] V. A. Graham and K. G. T. Hollands, "A method to generate synthetic hourly solar radiation globally," *Solar Energy*, vol. 44, no. 6, pp. 333–341, 1990.

[33] Off grid solar power system, 2016, http://rimstar.org/renewnrg/off_grid_solar_power_systems.htm.

[34] Solar Power Systems Components—Solar Panels Prices in Nigeria, http://www.naijatechguide.com/2008/11/solar.

[35] The Solar Shop Ltd, 2016, http://www.solarshopnigeria.com/.

[36] Tiger Generators Model Prices, Nigeria Technology Guide, 2015, https://www.naijatechguide.com/2008/03/tiger-generators-nigeria.html.

[37] S. Beverngen, *Mini Grid Kit Report EMS Strategy Review*, University of Kassel, Kassel, Germany, 2002.

[38] D. Mayer and M. Heidenreich, "Performance analysis of stand alone PV systems from a rational use of energy point of view," in *Proceedings of the 3rd World Conference on Photovoltaic Energy Conversion*, pp. 2155–2158, IEEE, Osaka, Japan, May 2003.

3

A Thermodynamic Analysis of Two Competing Mid-Sized Oxyfuel Combustion Combined Cycles

Egill Thorbergsson and Tomas Grönstedt

Division of Fluid Dynamics, Department of Applied Mechanics, Chalmers University of Technology, 412 96 Gothenburg, Sweden

Correspondence should be addressed to Egill Thorbergsson; egill@chalmers.se

Academic Editor: Umberto Desideri

A comparative analysis of two mid-sized oxyfuel combustion combined cycles is performed. The two cycles are the semiclosed oxyfuel combustion combined cycle (SCOC-CC) and the Graz cycle. In addition, a reference cycle was established as the basis for the analysis of the oxyfuel combustion cycles. A parametric study was conducted where the pressure ratio and the turbine entry temperature were varied. The layout and the design of the SCOC-CC are considerably simpler than the Graz cycle while it achieves the same net efficiency as the Graz cycle. The fact that the efficiencies for the two cycles are close to identical differs from previously reported work. Earlier studies have reported around a 3% points advantage in efficiency for the Graz cycle, which is attributed to the use of a second bottoming cycle. This additional feature is omitted to make the two cycles more comparable in terms of complexity. The Graz cycle has substantially lower pressure ratio at the optimum efficiency and has much higher power density for the gas turbine than both the reference cycle and the SCOC-CC.

1. Introduction

The evidence that anthropogenically generated greenhouse gases are causing climate change is ever-increasing. The Intergovernmental Panel on Climate Change (IPCC) has stated that [1]:

> "It is extremely likely that more than half of the observed increase in global average surface temperature from 1951 to 2010 was caused by the anthropogenic increase in greenhouse gas concentrations and other types of anthropogenic forcing together. The best estimate of the human induced contribution to warming is similar to the observed warming over this period."

One of the largest point source emitters of greenhouse gases is fossil fuel based power plants. One of the options to mitigate these greenhouse gases is to utilize Carbon Capture and Storage (CCS) in power plants.

There are three main processes being considered for CCS: the post combustion capture, the precombustion capture, and the oxyfuel combustion capture [2]. This paper focuses on the oxyfuel combustion combined cycle. The oxyfuel combustion fires fuel with nearly pure O_2 instead of air and the resulting combustion products are primarily steam and carbon dioxide. This makes it technically more feasible to implement CO_2 capturing solutions.

Two competing oxyfuel combustion combined cycles have shown promising potential [3]. These are the semiclosed oxyfuel combustion combined cycle (SCOC-CC) and the Graz cycle. Numerous studies have been done on both of these cycles, including a number of studies that have compared the performance of the two cycles [3–37].

1.1. SCOC-CC. The SCOC-CC is essentially a combined cycle that uses nearly pure O_2 as an oxidizer. After the heat recovery steam generator (HRSG) there is a condenser that condenses the water from the flue gas. The flue gas leaving the condenser is then primarily composed of CO_2. Part of the CO_2 is then recycled back to the compressor while the rest is compressed and transported to a storage site.

Bolland and Sæther first introduced the SCOC-CC concept in [4] where they compared new concepts for recovering CO_2 from natural gas fired power plants. They compared a standard combined cycle with a cycle using postcombustion both with and without exhaust gas recirculation and also the SCOC-CC, along with a Rankine cycle that incorporates oxy-fuel combustion. Ulizar and Pilidis [5] were first to present a paper that focused exclusively on the performance of the SCOC-CC. They started with cycle optimization and also simulated off-design performance. They did more extensive work exploring the selection of an optimal cycle pressure ratio and turbine inlet temperature [6] and on operational aspects of the cycle [7]. Bolland and Mathieu also published studies on the SCOC-CC concept [8] comparing its merits with a postcombustion removal plant. Amann et al. [9] also compared the SCOC-CC with a combined cycle using a postcombustion plant and made a sensitivity analysis regarding the purity of the O_2 and the corresponding energy cost of the air separation unit. Tak et al. compared the SCOC-CC with a cycle developed by Clean Energy Systems and concluded that the SCOC-CC seemed to be advantageous [10]. Jordal et al. proceeded to develop improved cooling flow prediction models [11] and Ulfsnes et al. studied transient operation [12] and further explored real gas effects and property modelling [13]. Other researchers have started with a conventional natural gas combined cycle as a starting point in the modelling of the SCOC-CC. Riethmann et al. investigated the SCOC-CC using a natural gas combined cycle as a reference case and concluded that the net efficiency of the SCOC-CC was 8.3% points lower compared to the reference cycle [15]. Corchero et al. did a parametric study with regard to the pressure ratio at a fixed turbine entry temperature of 1327°C [14]. Yang et al. [16] modelled the SOCC-CC along with the ASU and the CO_2 compression train at different pressure ratios and two different turbine inlet temperatures, 1200°C and 1600°C. They concluded that the optimal pressure ratio is around 60 and 90 for the turbine inlet temperatures of 1418°C and 1600°C, respectively. With optimal design conditions, the net cycle efficiency is lower than the efficiency of the conventional CC by about 8 percentage points for both of the two turbine inlet temperatures. Dahlquist et al. optimized a mid-sized SCOC-CC [17]. They concluded that although the optimum pressure ratio was 45 with regard to net efficiency, it would be beneficial to choose a lower pressure ratio for the cycle. Choosing a lower pressure ratio would only penalize the efficiency by a small amount but facilitate the design of the compressor by a great deal. Sammak et al. looked at different conceptual designs for the gas turbine for the SCOC-CC [18]. They compared a single and twin-shaft design and concluded that a twin-shaft would be an advantageous design for the SCOC-CC because of the high pressure ratio.

The main results of these papers have been summed together in Table 1. It is clear that there is no consensus in the power requirement of the air separation unit (ASU) as it ranges from 735 to 1440 kW/(kg/s) based on the mass flow of the O_2 generated. This can be explained partly by the fact that the purity of the O_2 stream varies from 90% to 96% and also that compression of the O_2 stream is included in the ASU energy demand in some of the studies. The pressure ratio (PR) varies considerably between studies, where the lowest is 24.5 while the highest is 90. This can partly explain that only some of the studies aimed at optimizing the cycle with respect to pressure ratio, while others only selected a pressure ratio

TABLE 1: Cycle performance for the SCOC-CC from earlier papers.

Paper	Year	ASU kW/(kg/s)	Output MW	TIT °C	PR	η_C %
[4]	1992	1512[†]	514.7		30	41.4
[5]	1997			1200	48	47.26
[6]	1998		193	1376	56	36.7
[8]	1998	900[*]	57.2	1319	30	44.9
[37]	2003	1440[†]	250		35	47.2
[30]	2004	972[*]	400	1328		44.1
[32]	2007	812[*]	400	1328		47
[33]	2008	1225[†]	400	1400	40	49.75
[15]	2009		329.7	1232	40	48.3
[9]	2009	861[†]	396	1427	24.5	51.3
[10]	2010		475.7	1400	40	53.9
[35]	2012	1225[†]	106	1400	37	46
[16]	2012	1021[§]	291.1	1418	60	48.2
[16]	2012		473.8	1600	90	52.2
[17]	2013	735[ſ]	125	1340	34	47.21

[†] Delivery pressure from ASU is the operating pressure.
[*] Delivery pressure from ASU is 1 bar.
[§] Delivery pressure from ASU is 27 bar.
[ſ] Delivery pressure from ASU is 1.2 bar.

based on, for example, experience. Corchero et al. [14] and Yang et al. [16] looked at different pressure ratios and show similar trends as has been found in the current study. The resulting efficiencies from these studies have a very large spread and ranges from 36.7% to 53.9%. An important fact to take into consideration is that the oldest reference [4] is from 1992 and the state-of-the-art cycle efficiency for a combined cycle at that time was much lower than is today. Bolland and Sæther found that the combined cycle efficiency, without carbon capture, was around 52% and that the SCOC-CC was around 10% lower than the reference cycle. The studies that are presented in Table 1 are not based on the same assumptions, such as condenser cooling methods and turbine cooling which of course influence the net efficiencies reported.

1.2. Graz Cycle . The basic principle of the Graz cycle was developed by Jericha in 1985 [19] and was aimed at solar generated O_2-hydrogen fuel. Jericha et al. modified the cycle in 1995 [20] to handle fossil fuels.

The Graz cycle, similar to the SCOC-CC, uses nearly pure O_2 as oxidizer. In the Graz cycle a major part of the flue gas is recycled back to the compressor; that is, the water is not condensed out of it. Furthermore, the turbine and the combustion chamber are cooled using steam from the steam cycle. This means that the working fluid is mainly steam in the gas turbine. The other part of the flue gas goes through a condenser where the water is condensed from the flue gas. The flue gas is then in major parts CO_2 which can be compressed and transported to a storage site. The full design of the Graz cycle incorporates a second bottoming cycle that uses the heat from the condensation of the flue gas. The second bottoming

TABLE 2: Cycle performance for the Graz cycle from earlier papers.

Paper	Year	ASU kW/(kg/s)	Output MW	TIT °C	PR	η_C %
[21]	2002	1080[†]	92.251	1400	40	57.51[‡]
[22]	2003	900[*]	92.251	1400	40	55[‡]
[30]	2004	972[*]	400	1328	40	42.8
[26]	2005	1355[†]	100	1400	40	55.3
[24]	2005	900[*]	92.2	1312	40	52.5
[32]	2007	812[*]	400	1328	40	48.6
[28]	2008		403	1400	40	53.12
[29]	2008	1225[†]	597	1500	50	54.14
[33]	2008	1225[†]	400	1400	40	53.09
[35]	2012	1225[†]	106	1400	44.7	49

[†]Delivery pressure from ASU is the operating pressure.

[*]Delivery pressure from ASU is 1 bar.

[‡]Energy cost for CO_2 compression not taken into account.

cycle is a subatmospheric steam cycle, since the condensation returns low quality heat, that is, low temperature.

There has been extensive research at the Graz University of Technology in designing the cycle, but the main focus has been on design of the turbomachinery components [21–29]. These studies have not been focused on optimizing the cycle performance with respect to pressure ratio. The main results of papers that study the Graz cycle are shown in Table 2. The design of the Graz cycle has been evolving and the most advanced cycle layout is the S-Graz cycle which was presented in paper [26]. Publications published later all study the S-Graz cycle concept.

1.3. Comparison of Cycles . A number of papers have compared the two different cycles along with other carbon capture technologies for natural gas fired power plants [30–36]. The main results of papers that present cycle results are shown in Tables 1 and 2. Kvamsdal et al. compared nine different carbon capture options for natural gas fired power plants [30, 32]. Among them were the SCOC-CC and the Graz cycle; the results for the cycle simulations are shown in Tables 1 and 2. It was concluded that concepts that employed very advanced technologies that have a low technological readiness level and high complexity achieved the highest performance. Franco et al. evaluated the technology feasibility of the components in 18 different novel power cycles with CO_2 capture [31]. One of the conclusions was that the SCOC-CC would be one of the cycles that incorporates gas turbines that would require the least effort to turn into a real power plant. Sanz et al. made a qualitative and quantitative comparison of the SCOC-CC and the Graz cycle [33]. Their thermodynamic analysis showed that the high-temperature turbine of the SCOC-CC plant needed a much higher cooling flow supply due to the less favourable properties of the working fluid than the Graz cycle turbine. They, in comparison to Franco et al. [31], concluded that all turbomachines of both cycles showed similar technical challenges and that the compressors and high-temperature turbines relied on new designs. Woollatt and

Franco did a preliminary design study for both the compressor and the turbine, in both the SCOC-CC and the Graz cycle [34]. They concluded that the turbomachinery can be designed using conventional levels of Mach number, hub/tip ratio, reaction, and flow and loading coefficients. They furthermore concluded that the efficiencies and the compressor surge margins of the components should be similar to a conventional gas turbine. Thorbergsson et al. examined both the Graz cycle and the SCOC-CC [35]. They conceptually designed the compressor and the turbine for both cycles. They concluded that the Graz cycle, in the original version including the second bottoming cycles, is expected to be able to deliver around 3% points' net efficiency benefit over the semi-closed oxyfuel combustion combined cycle at the expense of a more complex realization of the cycle.

Comparative work on the two cycles has suffered from not having the same technology level in the design of the two cycles. This results in the fact that it is difficult to draw conclusions from the comparisons. The aim of the current study is to assess the two cycles using the same technology level and in addition have comparable complexity levels. The current study goes into more details regarding the optimal pressure ratio and turbine entry temperatures for the oxyfuel combustion cycles then past publications. This is accomplished by establishing a reference cycle, which has a technology level that could enter service around year 2025. The fuel is assumed to be natural gas for all three cycles. The reference cycle is then used as the starting point for the modelling of the oxyfuel combustion combined cycles. In the current study the pressure ratio has been varied to locate the optimal net efficiency with respect to pressure ratio. Previous work has reported around 3% points' benefit for the Graz cycle [32] including work carried out by the authors [35]. It was viewed that a majority of these benefits would be attributed to the use of a second bottoming cycle as included in the original implementation. To make a fair comparison of the two alternatives it was decided to exclude this cycle feature from the original Graz cycle. It should be noted that it is quite feasible to introduce such a bottoming concept also for the SCOC-CC if the target would be to achieve maximum efficiency. The two simpler implementations were preferred in order to keep down complexity and make practical implementation more feasible.

2. Methods

The heat and mass balance program IPSEpro is used to simulate the power cycles [38]. The systems of equations, which are established using a graphical interface, are solved using a Newton-Raphson based algorithm. The simulation program was modified to incorporate the thermodynamic and transport properties program REFPROP to calculate the physical properties of fluids [39].

2.1. Cooled Turbine

2.1.1. Cooling Model. The cooling model is very important when studying the performance of gas turbine based cycles. The cooling model used is the m^* model and is based on the work of Halls [40] and Holland and Thake [41]. The model is based on the standard blade assumption, which assumes that the blade has infinite thermal conductivity and a uniform blade temperature. The model used in this study was originally implemented by Jordal [42].

The main parameters for the cooling model are first the cooling efficiency

$$\eta_c = \frac{T_{ce} - T_{ci}}{T_{bu} - T_{ci}}, \tag{1}$$

where T_{ci} is the temperature of the cooling flow at the inlet, T_{ce} is the temperature of the cooling flow at the exit, and T_{bu} is the uniform blade temperature. The cooling efficiency is set to a moderate limit of $\eta_c = 0.50$.

Second the cooling effectiveness is defined as

$$\varepsilon_c = \frac{T_g - T_{bu}}{T_g - T_{ci}}, \tag{2}$$

where T_g is the hot gas temperature.

The model is a first-law thermodynamic, nondimensional model. The model is based on the dimensionless mass flow cooling

$$\dot{m}^* = \frac{\dot{m}_c C_{p,c}}{\overline{\alpha_g} A_b}, \tag{3}$$

where $\dot{m}_c$ is the cooling mass flow, $C_{p,c}$ is the heat capacity of the cooling fluid, α_g is the convective heat transfer coefficient on the hot gas side, and A_b is the area of the blade. The main parameter of interest is the coolant mass flow ratio

$$\varphi = \frac{\dot{m}_c}{\dot{m}_g} = \dot{m}^* \frac{C_{p,g}}{C_{p,ci}} St_g \frac{A_b}{A_g}, \tag{4}$$

where $C_{p,g}$ is the heat capacity of the hot gas, St_g is the average Stanton number of the hot gas, and A_g is the cross-sectional area of the hot gas. The relations between the cooling mass flow and the temperature differences are

$$\frac{\dot{m}_c C_{p,ci}}{\overline{\alpha_g} A_b} = \frac{T_g - T_{bu}}{T_{bu} - T_{ci}}. \tag{5}$$

The Stanton number is defined as

$$St_g = \frac{\alpha_g}{\rho_g U_g C_{p,g}}, \tag{6}$$

where ρ_g is the density of the hot gas and U_g is the flow velocity.

To estimate the cooling requirements for each cooled turbine blade row, it was assumed that the cooling parameters were constant. The parameters were chosen to represent a cooled turbine that will enter service around 2025.

The uniform blade metal temperature is set to 850°C. This means that the maximum temperature will be around 950°C and the average temperature at the gas side of the blade around 900°C. The uniform blade metal temperature is used as the temperature limit for the cycle simulations.

TABLE 3: Parameters assumed in the cooling model.

T_{bu}	850°C
St_g	0.005
A_b/A_g	5
η_c	0.50
S	0.2

It is assumed, as has been done in other studies [11, 43–46], that the Stanton number is constant, $St = 0.005$, in regard to both the change in the working fluid and the change in the design parameters. The parameters that are assumed to be constant in the cooling model are shown in Table 3. The geometry parameter, A_b/A_g, which is the ratio between the wetted blade and adjacent cooled surface areas over the average gas cross-sectional area, is also held constant between all cases. This parameter is unknown for a thermodynamic analysis where the key dimensions of the turbine have not been designed. El-Masri [44] estimated that this parameter is slightly less than 4.0 for a cascade blade row and around 8.0 for a stage, allowing for a row-to-row spacing. Jordal [47] concluded that when taking into account that rotor disks and the transition piece from the combustion chamber to the first stage nozzles are also subject to cooling, an average value should be around 5.0 for a stage.

The cooling model was used to reproduce the results in [48] and showed good agreement.

2.1.2. Expansion. The expansion in an uncooled turbine is modelled as

$$\eta_p = \frac{(s_2 - s_1) + R \ln (p_2/p_1)}{R \ln (p_2/p_1)}, \tag{7}$$

where R is the gas constant for the working fluid, p, s are the pressure and entropy, respectively, 1 is the inlet, and 2 is the outlet of the turbine stage. This model was evaluated against different models such as Mallen and Saville [49], using numerical integration and the model used gave good agreement with the numerical integration.

For the cooled turbine, the mixing of the coolant and the main stream gas flow result in a loss in stagnation pressure. This irreversibility is taken into account by defining a new polytropic efficiency [42, 50], defined as

$$\eta_{pr} = \eta_p - S \ln \left(\frac{p_{in}}{p_{out}} \right) \frac{p_1}{p_{in} - p_{out}} \frac{\dot{m}_{g,out} - \dot{m}_{g,in}}{\dot{m}_{g,in}}, \tag{8}$$

where p_1 is the stagnation pressure at the inlet of the rotor blade row, in is the inlet to the turbine, and out is the outlet of the turbine. Parameter S is specific to each turbine and models the losses. It is typically in the range of 0.1 for a turbine that has good performance and around 0.5 for a turbine that has poor performance [51]. The polytropic efficiency is set to $\eta_p = 90\%$. The losses are taken into account by assuming that the factor is $S = 0.2$ for all cases. Dahlquist et al. examined the empirical loss models used to design turbomachinery, which are generated using air as the working fluid, and concluded that the loss models generate similar results for the working fluids in oxyfuel cycles [52]. This indicates that it is possible to achieve a similar technology level for the oxyfuel turbines as for state-of-the-art conventional turbines.

2.2. Compressor. The compression is modelled using polytropic efficiency,

$$\eta_p = \frac{R \ln (p_2/p_1)}{(s_2 - s_1) + R \ln (p_2/p_1)}, \tag{9}$$

where R is the gas constant for the working fluid, p and s are the pressure and entropy, respectively, 1 is the inlet, and 2 is the outlet of the compressor.

It is assumed that the polytropic efficiency is constant for all cycles and all cases and is assumed to be $\eta_p = 91\%$. Similar to the turbine, it is assumed that it is possible to achieve a compressor design for the oxyfuel compressor that is on the same level as the state of the art of compressors in conventional gas turbines.

2.3. Combustor. The combustion is a simple energy model based on the assumption that all of the fuel is completed in the combustion, that is, 100% combustion efficiency.

The amount of excess O_2 is calculated as

$$\lambda = \frac{\dot{m}_{O_2,in}}{\dot{m}_{O_2,in} - \dot{m}_{O_2,out}}, \tag{10}$$

where $\lambda = 1.0$ is stoichiometric combustion. For the oxyfuel cycles the combustion is nearly stoichiometric; that is $\lambda = 1.01$. It is preferred that the combustion takes place as close to stoichiometric conditions as possible to reduce the amount of O_2 that the ASU needs to produce. Such a low amount of excess O_2 is very different compared to traditional combustion in gas turbines, which have much larger amount of excess O_2. It is assumed that it is possible to have the combustion under near stoichiometric conditions while the emissions of NO_x, CO and unburned hydrocarbons are within given constraints. Sundkvist et al. found that using excess of 0.5% of O_2, $\lambda = 1.005$, resulted in 400 ppmv of CO at the turbine outlet [53] and increasing the O_2 ratio resulted in reduced levels of CO, while increasing the energy penalty from the ASU, as expected.

The pressure drop in the combustion chamber is assumed to be 4%. And a compressor is used to increase the pressure of the fuel above the pressure in the combustion chamber.

2.4. Air Separation Unit. O_2 is produced with an air separation unit (ASU). The ASU is assumed to be a cryogenic air separation plant. Modelling of the ASU is not within the scope of this current study. ASU power consumption is highly dependent on the purity of the O_2 stream. It is therefore an economic trade-off between purity and cost. Typical state-of-the-art cryogenic ASU can produce O_2 with 99.5%-volume purity at a power consumption of 900 kW/(kg/s) [54]. By decreasing the purity, it is possible to reduce the power consumption of the ASU. At a purity level of 95%, the power

TABLE 4: Oxygen composition.

	Mass fraction
Ar	3.0%
N_2	2.0%
O_2	95.0%

consumption can be assumed to be around 735 kW/(kg/s) [17, 55]. In this study this has been taken into account and a purity level of 95% for the ASU is used. The corresponding O_2 composition is shown in Table 4.

The ASU unit delivers the O_2 stream at a pressure of 1.2 bar and with a temperature of 30°C. An intercooled compressor is used to increase the pressure of the stream to the working pressure in the combustor. The compression process has been modelled in the cycle simulation.

2.5. Flue Gas Condenser. The main purpose of the oxyfuel combustion cycles is to produce CO_2 along with power generation. Because the flue gas consists mostly of CO_2 and steam, the most convenient method is to condense the water from the flue gas to produce the CO_2. The flue gas will also contain small amounts of Ar, N_2, and O_2. There will also be traces of harmful acid gases along with particles such as soot. By using a direct contact condenser, these harmful gases and the particles can be removed from the flue gas when the steam is condensed. The condenser will therefore also act as a scrubber.

The efficiency of the condenser is defined as

$$\eta_{condenser} = \frac{\dot{m}_{condense}}{\dot{m}_{H_2O,in}}, \tag{11}$$

where $\dot{m}_{condense}$ is the amount of water that is condensed from the flue gas and $\dot{m}_{H_2O,in}$ is the amount of water in the flue gas that enters the condenser. The flue gas condenser efficiency is a simple way to evaluate the performance of the condensers [56]. The parameter does not represent an efficiency in its true sense but is a metric commonly used to describe the performance of condensers [56].

2.6. CO_2 Compression. The CO_2 stream from the condenser that will be sent to storage needs to be compressed to a higher pressure and the remaining water vapour and noncondensable gases need to be removed. This process, the CO_2 recovery and compression process, is not within the system boundaries of the current study. It is instead taken into account by assuming a fixed energy cost, 350 kW/(kg/s) of wet CO_2 [25]. This energy cost assumes that the stream is compressed to 100 bar. This value also takes into account the removal of water and other gases that are present in the CO_2 stream.

3. Power Cycles

The fuel is assumed to be natural gas and the composition is shown in Table 5. Common assumptions used in the simulations of the cycles are shown in Table 6.

TABLE 5: Natural gas fuel composition.

	Mass fraction
CH_4	84.7%
N_2	3.3%
CO_2	2.5%
C_2H_6	7.0%
C_3H_8	2.6%

TABLE 6: Assumptions used in cycle simulations.

Compressor polytropic efficiency	0.91
Compressor mechanical efficiency	0.99
Combustor pressure drop	4%
Turbine polytropic efficiency	0.90
Power turbine polytropic efficiency	0.89
Gas turbine mechanical efficiency	0.99
Generator electricity efficiency	0.985
Generator mechanical efficiency	0.994
Lower heating value for fuel	46885 kJ/kg
Fuel temperature	15°C
Fuel compressor isentropic efficiency	0.80
Ambient temperature	15°C
Ambient pressure	1.013 bar
Ambient humidity	60%
Condenser pressure	0.045 bar
HRSG heat exchangers Δp, hot side	0.001 bar
HRSG heat exchangers Δp, cold side	0.9 bar
Steam turbine isentropic efficiency	0.89
Superheater, LP, ΔT_{pinch}	10 K
Superheater, HP, minimum ΔT_{pinch}	25 K
Evaporator ΔT_{pinch}	10 K
HP steam pressure	140 bar
HP steam maximum temperature	560°C
LP steam pressure	7 bar
Pump efficiency	0.7
Pump mechanical efficiency	0.9
Deaerator operating pressure	1.21 bar
Deaerator saturation temperature	105°C
ASU power consumption	735 kW/(kg/s)
O_2 purity	95%
O_2 compressor polytropic efficiency	0.88
ASU delivery pressure	1.2 bar
ASU delivery temperature	30°C
Carbon dioxide compression power	350 kW/(kg/s)
Condenser efficiency, maximum	0.85
Gross power output	100 MW

3.1. Reference Cycle.

A reference cycle was modelled that is in the mid-size range. The mid-size range is from 30 to 150 MW [42]. Here we have aimed at keeping the gross combined power output from the gas turbine and the steam turbine constant at 100 MW. The reference cycle has been modelled as a gas generator and a separate power turbine, that is, a two-shaft gas turbine. The gas generator turbine consists of two cooled stages. The cooling flow is bled from the compressor. The steam cycle for a power plant in this power range usually employs single or double pressure levels and does not use reheat. Here we have used a dual-pressure steam cycle without reheat. The steam turbine is a single-casing nonreheat. The pressure was set to 140 bar and the maximum temperature to 560°C at the inlet to the steam turbine. If the exhaust temperature from the gas turbine goes below 585°C the steam temperature decreases so that the temperature difference is 25°C. A schematic of the cycle is shown in Figure 1.

3.2. SCOC-CC.

A schematic of the SCOC-CC is shown in Figure 2. The SCOC-CC is based on the reference cycle. Now, however, the fuel is combusted with O_2 that is produced in the ASU. The fuel is combusted near to stoichiometric ratio, meaning that nearly no excess O_2 is produced. This minimizes the power demand of the ASU. The combustion chamber is cooled using the recycled flue gas, after most of the steam is condensed from it, in the condenser. The flue gas leaving the combustion chamber is mainly CO_2 and also a small amount of steam. The gas turbine layout is the same as the reference cycle with a gas generator and a power turbine. The turbine in the gas generator has two stages, which are both cooled. The cooling flow is also bled from the compressor, similar to the reference cycle. The layout of the steam cycle is unchanged from the reference cycle. The flue gas goes to the condenser after the heat recovery steam generator, where the major part of the steam is condensed from the exhaust gas. The flue gas is cooled in this process. The CO_2 stream that leaves the condenser has near 100% relative humidity. A small part of the CO_2 stream is sent to the compression and purification process and is then transferred to the storage site. The major part of the CO_2 is recycled back to the compressor. The water in the CO_2 stream can possibly condense at the entry to the compressor, which could have a deteriorating effect for the compressor. The CO_2 stream is therefore heated before it enters the compressor using the heat from the flue condensation.

3.3. Graz Cycle.

The main features are that the gas turbine cooling is implemented with steam and that the flue gas is sent straight to the compressor after the HRSG without condensing the steam from it. Part of the flue gas is sent to a condenser where a major part of the water is condensed from it; after this it is sent to the CO_2 compression and purification process. The CO_2 is afterwards transferred to the storage site.

The most common layout of the Graz cycle incorporates two bottoming cycles. The first one uses a typical HRSG and a steam turbine, which only expands, however, to the pressure of the combustion chamber. This is because the steam is used for cooling both the combustion chamber and the gas turbine blades. The second bottoming cycle uses the enthalpy of the condensation and assumes that the pressure at the outlet of the condenser is 0.021 bar, which is particularly low.

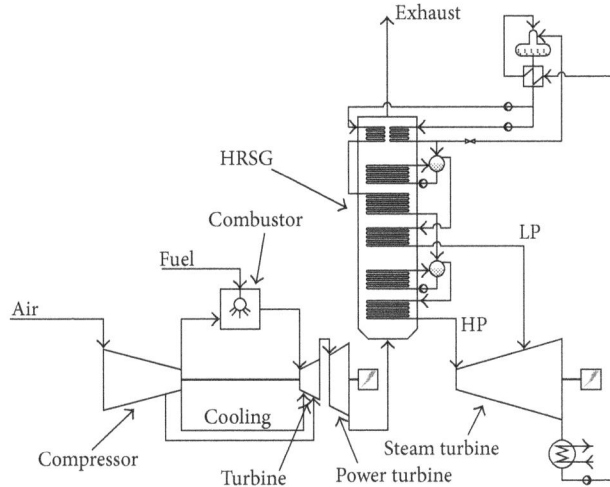

FIGURE 1: Schematic layout of the reference cycle.

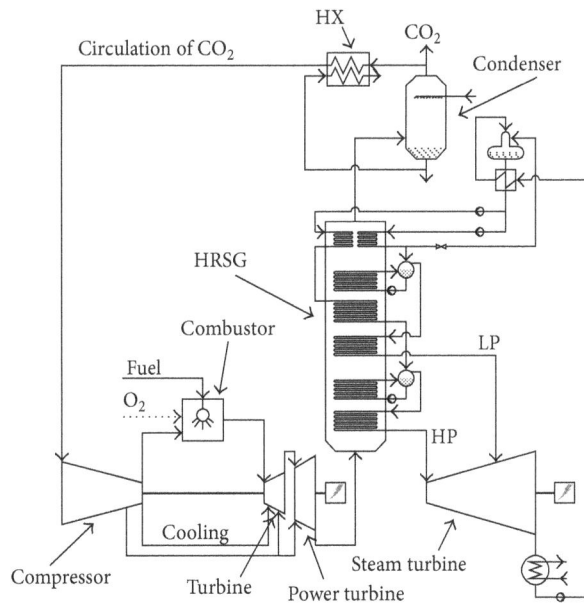

FIGURE 2: Schematic layout of the SCOC-CC.

It is hard to imagine that the first design of the Graz cycle will deviate so greatly from the current layout of the combined cycle. Here we have taken the reference cycle as the basis and implemented the major design features of the Graz cycle. A schematic of the Graz cycle is shown in Figure 3. The cycle incorporates an intercooler to reduce the temperature of the gas at the exit of the compressor as well as steam cooling. This layout, not implementing the second bottoming cycle, is considered more reasonable for the first generation design of the cycle. It also makes the complexity level of the SCOCC-CC and the Graz cycle more comparable, by not including improvements that could be implemented on both cycles.

The cycle illustrated in Figure 3 should therefore be understood as a simplified variant of the Graz cycle.

4. Results

A parametric study of the two oxyfuel cycles was performed by varying the turbine entry temperature (TET) and the pressure ratio (PR) of the gas turbine. The turbine entry temperature is the temperature at the exit of the combustion chamber and is therefore also the temperature at the entry to the first stator in the gas turbine. The temperature has been varied from 1250°C to 1600°C. The pressure ratio was varied freely until the design constraints were attained.

4.1. Reference Cycle. The results for the cycle net efficiency are shown in Figure 4 as a function of pressure ratio. The turbine entry temperature is also shown in Figure 4. The net efficiency

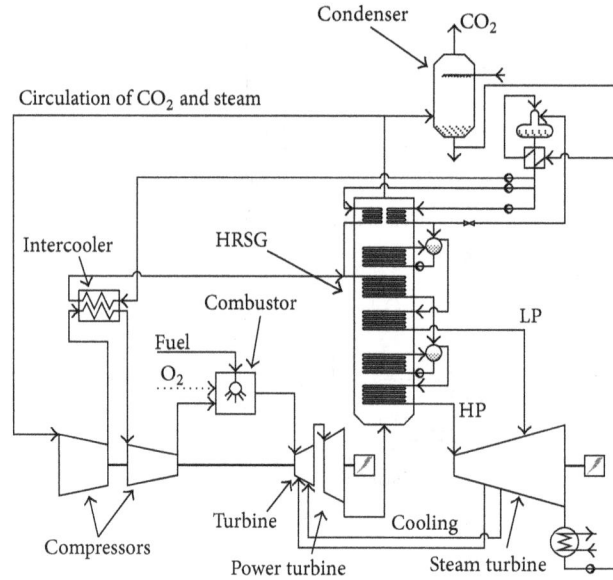

FIGURE 3: Schematic layout of the Graz cycle.

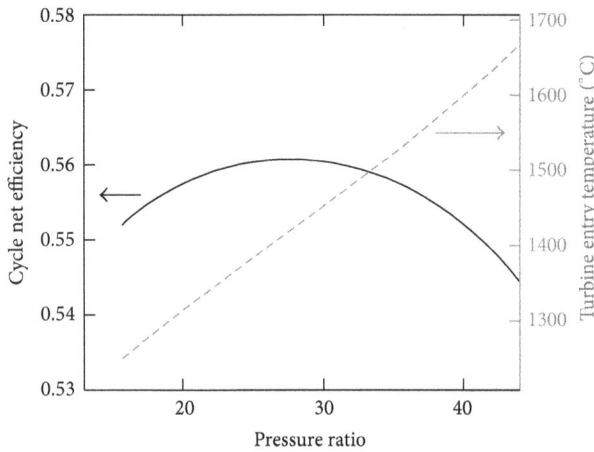

FIGURE 4: Net efficiency and turbine entry temperature as functions of pressure ratio for the reference cycle.

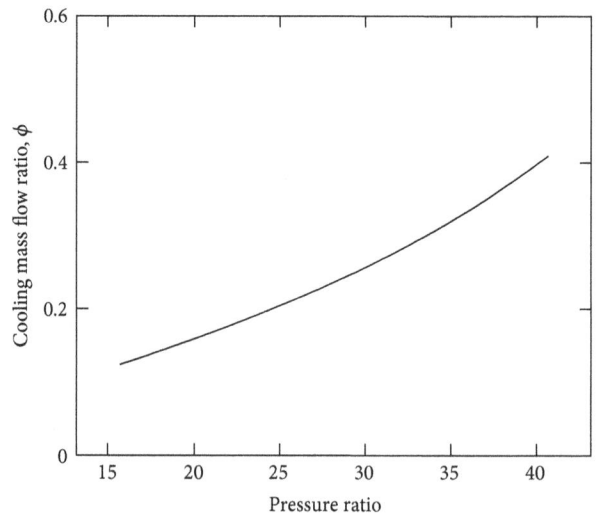

FIGURE 5: Cooling mass flow ratio as function of pressure ratio for the reference cycle.

takes into account the power needed for the pumps in the cycle. The entry temperature for the power turbine has been set to 850°C, to eliminate the need for cooling in the power turbine. If the temperature goes above 850°C, which is the metal temperature limit for the blades, then the first stage in the power turbine would then need to be cooled.

Figure 5 shows the cooling mass flow ratio for the reference cycle. The ratio is defined as the total cooling mass flow divided by the inlet mass flow to the turbine.

4.2. SCOC-CC.

Figure 6 shows the gross efficiency for the SCOC-CC as a function of pressure ratio and turbine entry temperature. The gross efficiency is the total power delivered by the gas turbine and steam turbine generators divided by the energy content of the fuel, based on the lower heating value.

As the pressure ratio decreases, the amount of steam in the low pressure steam is also reduced. The lower limit for the pressure ratio is reached when the mass flow of the low pressure steam approaches zero. The higher pressure ratio limit is reached when the temperature difference for the high pressure steam and the flue gas in the preheater approaches 5°C.

Figure 7 shows the net efficiency for the SCOC-CC as a function of pressure ratio and turbine entry temperature. The net efficiency takes into account the fuel compressor power, the power needed for the pumps, the energy needed for the production of the O_2, the O_2 compressor power, and the power needed to compress the CO_2. The largest decrease in the efficiency comes from the power requirement for the O_2 production and compression. As the pressure ratio is

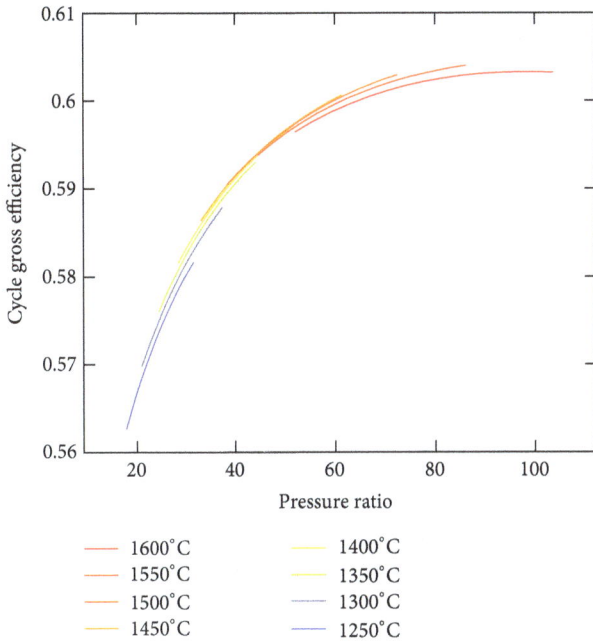

FIGURE 6: Gross efficiency as function of pressure ratio and turbine entry temperature for the SCOC-CC.

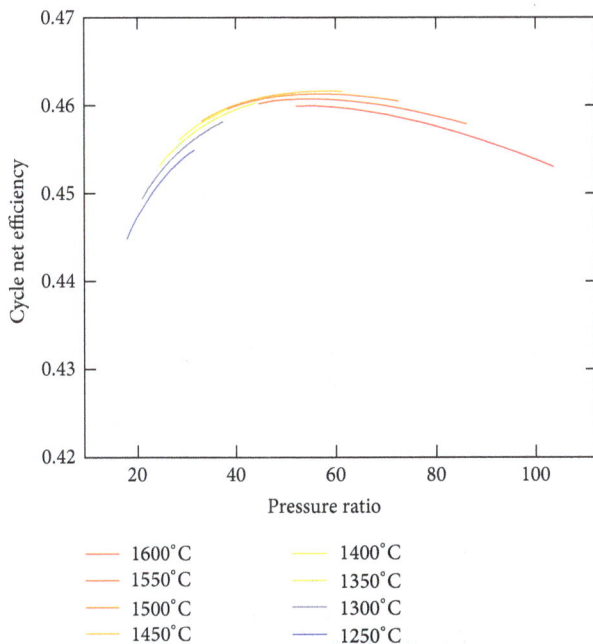

FIGURE 7: Net efficiency as function of pressure ratio and turbine entry temperature for the SCOC-CC.

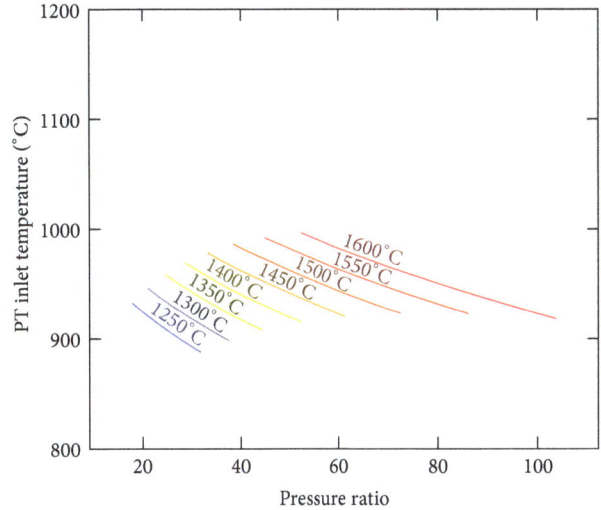

FIGURE 8: Power turbine entry temperature as function of pressure ratio and turbine entry temperature for the SCOC-CC.

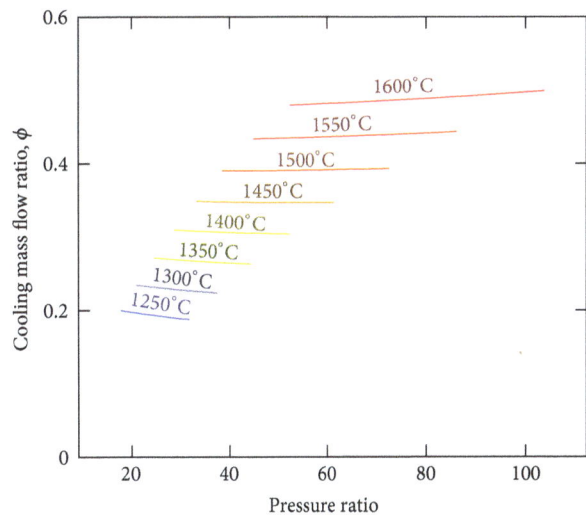

FIGURE 9: Cooling mass flow ratio as function of pressure ratio and the turbine entry temperature for the SCOC-CC.

increased, the O_2 compression power consumption increases very rapidly. This results in there being an optimum pressure ratio.

Figure 8 shows the power entry temperature as a function of pressure ratio and turbine entry temperature for the SCOC-CC. As can be seen in Figure 8 the power turbine entry temperature for all cases is above the blade material temperature limit, 850°C. This means that the first stage in the power turbine needs to be cooled.

Figure 9 shows the cooling mass flow ratio as a function of pressure ratio and turbine entry temperature for the SCOC-CC. The cooling mass flow ratio is higher for the SCOC-CC than the reference cycle, since the heat capacity for the working fluid is lower in the SCOC-CC than in the reference cycle.

4.3. Graz Cycle. The Graz cycle was studied at turbine entry temperatures of 1250°C, 1450°C, and 1600°C. Figure 10 shows the gross efficiency for the Graz cycle as a function of pressure ratio and turbine entry temperature. It can be seen in Figure 10 that there is no global optimum for the gross efficiency.

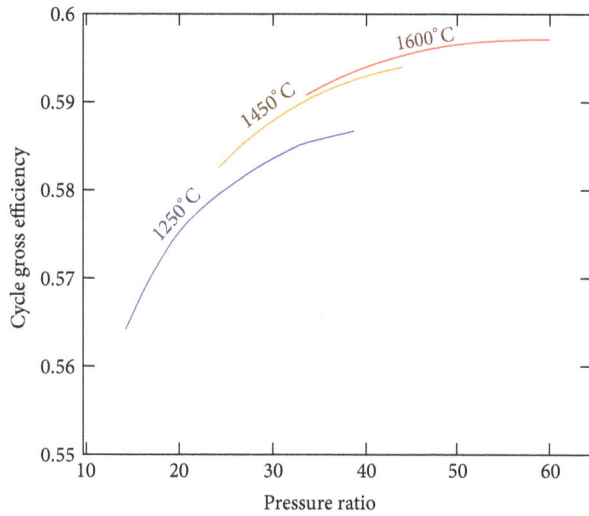

FIGURE 10: Gross efficiency as function of pressure ratio and turbine entry temperature for the Graz cycle.

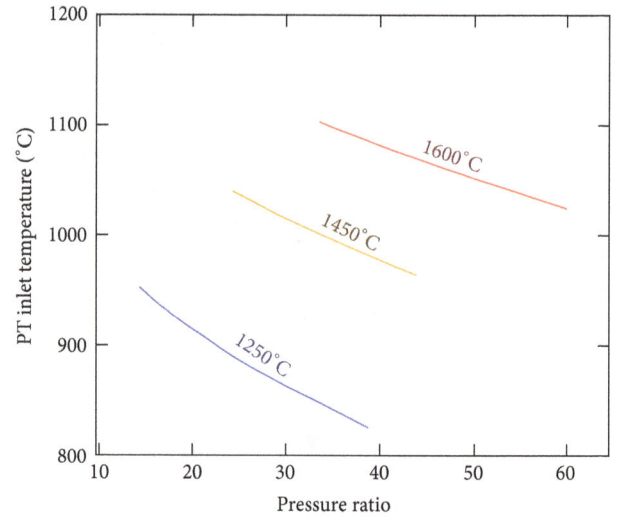

FIGURE 12: Power turbine entry temperature as function of pressure ratio and turbine entry temperature for the Graz cycle.

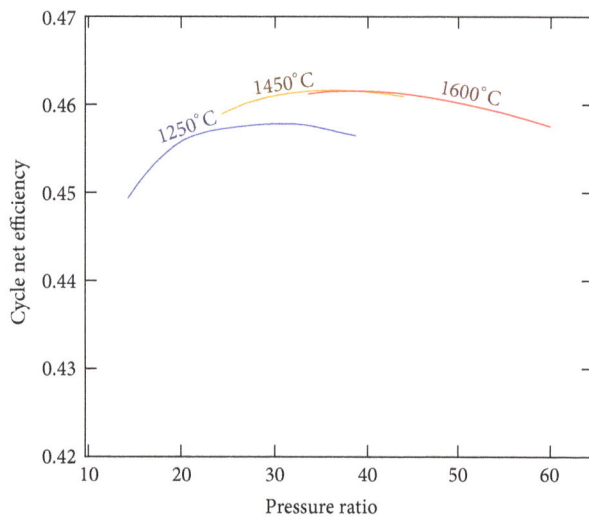

FIGURE 11: Net efficiency as function of pressure ratio and turbine entry temperature for the Graz cycle.

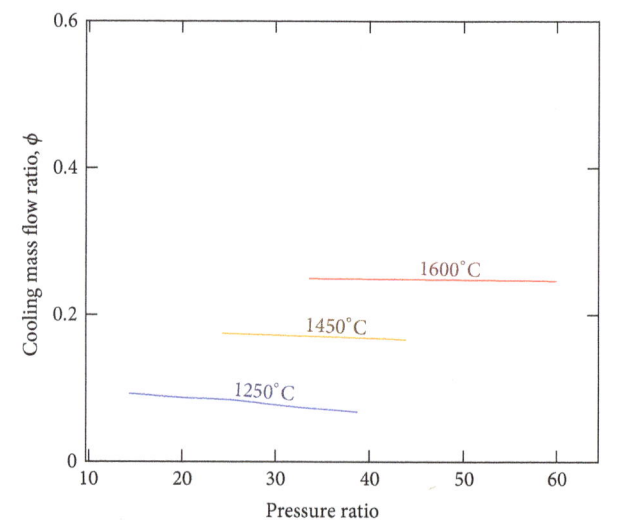

FIGURE 13: Cooling mass flow ratio as function of pressure ratio and the turbine entry temperature for the Graz cycle.

Figure 11 shows the net efficiency for the Graz cycle as a function of pressure ratio and turbine entry temperature. The net efficiency is calculated as is done for the SCOC-CC. The major reduction in the efficiency comes from the power needed for the O_2 production and compression. The relative power consumption of the O_2 compression increases as the pressure ratio increases, which results in an optimum in the net efficiency. Figure 12 shows the power entry temperature as a function of pressure ratio and turbine entry temperature for the Graz cycle. When the power turbine entry temperature is over 850°C, the first stage in the power turbine is cooled.

Figure 13 shows the cooling mass flow ratio as a function of pressure ratio and turbine entry temperature for the Graz cycle. Since steam is used as coolant for the turbine blade cooling for the Graz cycle, the cooling mass flow ratio is considerably lower than for the reference cycle and the SCOC-CC. This is the result of the fact that the steam has a substantially lower temperature than the compressor discharge temperature and that the steam has a higher heat capacity than the working fluid.

4.4. *Optimum Cycles.* The results for the cycles with the optimum performance are shown in Table 7. The optimum reference cycle is determined to be a cycle where there is no need to employ cooling for the power turbine. This means that the power turbine entry temperature is 850°C or lower. The pressure ratio that gives the optimal efficiency is 26.15 and the turbine entry temperature is 1400°C, which results in a net efficiency of 56%. The turbine exhaust temperature is only 526°C,

TABLE 7: Results for the optimal cycles.

		Reference	SCOC-CC	Graz
Heat input	MW	176	167	169
GT power output	MW	69	62	76
GT efficiency	%	39.48	37.08	45.18
ST power output	MW	31	38	24
ST efficiency	%	17.35	22.86	14.00
Gross power output	MW	100	100	100
O_2 production	MW		10.3	10.4
O_2 compression	MW		6.91	5.98
CO_2 compression	MW		3.60	3.79
Net power output	MW	98.6	77.0	78.0
Gross efficiency	%	56.83	59.94	59.18
Net efficiency	%	56.04	46.16	46.16
Compressor pressure ratio		26.2	57.3	36.5
Compressor outlet temp	°C	507	474	605
Compressor mass flow at inlet	kg/s	181	149	73
Cooling mass flow ratio		0.22	0.35	0.17
TET*	°C	1400	1450	1450
TIT†	°C	1251	1208	1274
Power TET	°C	850	927	998
Exhaust gas flow	kg/s	185	166	106
Exhaust temperature	°C	526	618	614
Stack temperature	°C	96	65	100
LP steam pressure	bar	7	7	7
LP steam temperature	°C	337	337	337
LP steam mass	kg/s	7	3	3
HP steam pressure	bar	140	140	140
HP steam temperature	°C	501	560	560
HP steam mass	kg/s	21	27	26
O_2 mass flow	kg/s		14.0	14.2
CO_2 mass flow	kg/s		10.3	10.8

*Same as combustor outlet temperature (COT).
†Turbine inlet temperature based on the ISO definition.

which results in the high pressure steam having a temperature of 501°C since the pinch temperature difference has a minimum value of 25°C in the high pressure (HP) superheater.

The optimum SCOC-CC has a relatively high pressure ratio, or around 57.3, and the turbine entry temperature is 1450°C. Even though the pressure ratio is so high, the compressor outlet temperature is only 474°C, which is below the temperature limit of the blade material. The steam turbine produces more of the power in the SCOC-CC compared to the reference cycle. The exhaust gas is cooled down from 618°C to 65°C in the HRSG. This comes from the fact that the working fluid achieves a better fit to the steam cycle. The main decrease in power comes from the O_2 production and the O_2 compression. The gross efficiency for the cycle is close to 60% but, taking into account the O_2 production and compression, CO_2 compression, and also the pumps, this is lowered to 46%. The SCOC-CC cycle produces 10.3 kg/s of CO_2, which is about 890 tonnes per day. The SCOC-CC also produces about

170 kg/s of water with a temperature of 46°C in the flue gas condenser.

The optimum Graz cycle has a pressure ratio of about 36.5 and a turbine entry temperature of 1450°C. Even though the pressure ratio is lower in the Graz cycle than in the SCOC-CC and the compression is intercooled, the compressor outlet temperature is much higher or around 605°C. The reason for such a high temperature is mainly the fact that the compressor inlet temperature is 100°C. The working fluid saturation temperature is around 95°C so, to avoid condensation at the inlet of the compressor, the temperature needs to be higher than the saturation temperature. To be able to withstand the high temperature at the outlet of the compressor, the blade material will be more expensive than is normally used in compressors. The gas turbine produces a larger share of the power compared to the reference cycle and the SCOC-CC. This is because the cooling in the gas turbine uses steam from the steam cycle. The amount of steam needed for cooling is around 15.4 kg/s, which is about 50% of the steam produced in the steam cycle. This steam will therefore be expanded in the gas turbine and not in the steam turbine. This will have a negative effect on the efficiency since the steam will be expanded to 1 bar instead of 0.045 bar as it is in the steam turbine. One aspect of the Graz cycle is that the power density is much higher compared to both the reference cycle and the SCOC-CC. The compressor inlet mass flow is only 40% of the reference cycle mass flow and 50% of that of the SCOC-CC. The gross efficiency for the Graz cycle is around 59%. The major deduction in efficiency comes from the O_2 production and compression. However, the compression power consumption is lower in the Graz cycle than the SCOC-CC because of the lower pressure ratio. The CO_2 compression is similar to that in the SCOC-CC cycle. This results in nearly the exact same net efficiency as for the SCOC-CC or 46%. The Graz cycle produces slightly more CO_2, or around 10.8 kg/s, which is about 933 tonnes/day. The Graz cycle produces significantly more water than the SCOC-CC, or about 270 kg/s of water with a temperature of 60°C in the flue gas condenser.

5. Discussion and Conclusion

The study compared three combined cycles, a conventional cycle, the SCOC-CC, and the Graz cycle, at the mid-size level power output. The gross power output for all cycles was set to 100 MW. The conventional cycle was used as the basis for the modelling and as a reference for the oxyfuel combustion cycles. A detailed literature review was conducted for the oxyfuel combustion combined cycles. The literature review showed that there is no consensus on the power requirement for the air separation unit. It also showed that the comparison of the SCOC-CC and the Graz cycle has lacked consistent assumptions and agreement on the technology parameters used to model the cycles.

A parametric study was conducted by varying the pressure ratio and the turbine entry temperature for the cycles. A constraint for the conventional cycle was set on the power turbine entry temperature to eliminate the need for cooling in the power turbine. The resulting optimal conventional cycle achieved a 56% net efficiency at a pressure ratio of 26.2 and a turbine entry temperature of 1400°C. The optimal SCOC-CC achieved only a 46% net cycle efficiency at a pressure ratio of 57.3 and a turbine entry temperature of 1450°C. The optimal Graz cycle also achieved a net cycle efficiency of 46% at a pressure ratio of 36.5 and a turbine entry temperature of 1450°C. The main reduction in efficiency for the oxyfuel cycles comes from the O_2 production, which reduced the power output from the cycles by more than 10 MW. An additional reduction of the power output comes from the compression of O_2 to operating pressure. This is about 7 MW and 6 MW for the SCOC-CC and Graz cycle, respectively. The difference comes from the higher pressure ratio of the SCOC-CC.

One of the benefits of the Graz cycle is the high power density of the gas turbine. This results in smaller turbomachinery for the gas turbine in the Graz cycle, which lowers the cost of this machinery. One of the main penalties of the Graz cycle is that the large amount of steam, which is generated in the HRSG, is not expanded in the steam turbine but in the gas

turbine. The result is that the steam does not expand to the condenser pressure of the steam cycle.

The SCOC-CC is considerably simpler than the Graz cycle as it does not implement steam cooling and does not require an intercooler. The optimal SCOC-CC, however, has a much higher pressure ratio than both the reference cycle and the Graz cycle. The efficiency does not vary greatly with the pressure ratio, however, and it is possible to reduce the pressure ratio without significantly penalizing the net efficiency. This would facilitate the compressor design substantially.

Nomenclature

α_g:	Hot gas convective heat transfer coefficient
$\dot{m}^*$:	Dimensionless mass flow cooling
$\dot{m}_c$:	Cooling mass flow
η_c:	Cooling efficiency
η_C:	Cycle net efficiency
η_p:	Polytropic efficiency
$\eta_{condenser}$:	Condenser efficiency
λ:	Ratio of oxygen
ρ_g:	Hot gas density
St_g:	Stanton number of the hot gas
ε_c:	Cooling effectiveness
φ:	Coolant mass flow ratio
A_b:	Blade area
A_g:	Annulus area
$C_{p,c}$:	Heat capacity of the cooling fluid
$C_{p,g}$:	Heat capacity of the hot gas
p:	Pressure
PR:	Pressure ratio
R:	Gas constant
S:	Turbine loss parameter
s:	Entropy
T_{bu}:	Uniform blade temperature
T_{ce}:	Cooling flow exit temperature
T_{ci}:	Cooling flow inlet temperature
T_g:	Hot gas temperature
TET:	Turbine entry temperature
TIT:	Turbine inlet temperature
ASU:	Air separation unit
CCS:	Carbon Capture and Storage
GT:	Gas turbine
SCOC-CC:	Semiclosed oxyfuel combustion combined cycle
ST:	Steam turbine.

Conflict of Interests

The authors declare that there is no conflict of interests regarding the publication of this paper.

Acknowledgments

This research was funded by the Swedish Energy Agency, Siemens Industrial Turbomachinery AB, GKN Aerospace, and the Royal Institute of Technology through the Swedish research program TURBOPOWER. Their support is gratefully acknowledged. The financial grant from Landsvirkjun's Energy Research Fund is gratefully acknowledged by the first author.

References

[1] IPCC, "2013: Summary for policymakers," in *Climate Change 2013: The Physical Science Basis. Contribution of Working Group I to the Fifth Assessment Report of the Intergovernmental Panel on Climate Change*, T. Stocker, D. Qin, G.-K. Plattner et al., Eds., Cambridge University Press, Cambridge, UK, 2013.

[2] B. Metz, O. Davidson, H. de Coninck, M. Loos, and L. Meyer, "IPCC special report on carbon dioxide capture and storage," Tech. Rep., Intergovernmental Panel on Climate Change, Working Group III, Geneva, Switzerland, 2005.

[3] F. Bolland, R. Naqvi, R. Span et al., "Public summary report of ENCAP delivarables D6.1.4 evaluation of technologies and benchmarking based on reference cases and D6.2.1 modelling, design and operational analysis," Tech. Rep., 2007, http://www.encapco2.org/.

[4] O. Bolland and S. Sæther, "New concepts for natural gas fired power plants which simplify the recovery of carbon dioxide," *Energy Conversion and Management*, vol. 33, no. 5–8, pp. 467–475, 1992.

[5] I. Ulizar and P. Pilidis, "A semiclosed-cycle gas turbine with carbon dioxide-argon as working fluid," *Journal of Engineering for Gas Turbines and Power*, vol. 119, no. 3, pp. 612–616, 1997.

[6] I. Ulizar and P. Pilidis, "Design of a semiclosed-cycle gas turbine with carbon dioxide: argon as working fluid," *Journal of Engineering for Gas Turbines and Power*, vol. 120, no. 2, pp. 330–335, 1998.

[7] I. Ulizar and P. Pilidis, "Handling of a semiclosed cycle gas turbine with a carbon dioxide-argon working fluid," *Journal of Engineering for Gas Turbines and Power*, vol. 122, no. 3, pp. 437–441, 2000.

[8] O. Bolland and P. Mathieu, "Comparison of two CO_2 removal options in combined cycle power plants," *Energy Conversion and Management*, vol. 39, no. 16–18, pp. 1653–1663, 1998.

[9] J.-M. Amann, M. Kanniche, and C. Bouallou, "Natural gas combined cycle power plant modified into an O_2/CO_2 cycle for CO_2 capture," *Energy Conversion and Management*, vol. 50, no. 3, pp. 510–521, 2009.

[10] S. H. Tak, S. K. Park, T. S. Kim, J. L. Sohn, and Y. D. Lee, "Performance analyses of oxy-fuel power generation systems including CO_2 capture: comparison of two cycles using different recirculation fluids," *Journal of Mechanical Science and Technology*, vol. 24, no. 9, pp. 1947–1954, 2010.

[11] K. Jordal, O. Bollard, and A. Klang, "Aspects of cooled gas turbine modeling for the semi-closed O_2/CO_2 cycle with CO_2 capture," *Journal of Engineering for Gas Turbines and Power*, vol. 126, no. 3, pp. 507–515, 2004.

[12] R. E. Ulfsnes, O. Bolland, and K. Jordal, "Modelling and simulation of transient performance of the semi-closed O_2/CO_2 gas turbine cycle for CO_2-capture," in *Proceedings of the ASME Turbo Expo: Power for Land, Sea and Air*, GT2003-38068, Atlanta, Ga, USA, June 2003.

[13] R. Ulfsnes, G. Karlsen, K. Jordal, O. Bolland, and H. M. Kvamsdal, "Investigation of physical properties of CO_2/H_2O-mixtures for use in semi-closed O_2/CO_2 gas turbine cycle with CO_2-capture," in *Proceedings of the 16th International Conference on Efficiency, Cost, Optimization, Simulation, and Environmental Impact of Energy Systems (ECOS '03)*, Copenhagen, Denmark, June 2003.

[14] G. Corchero, V. P. Timón, and J. L. Montañés, "A natural gas oxy-fuel semiclosed combined cycle for zero CO_2 emissions: a thermodynamic optimization," *Proceedings of the Institution of Mechanical Engineers Part A: Journal of Power and Energy*, vol. 225, no. 4, pp. 377–388, 2011.

[15] T. Riethmann, F. Sander, and R. Span, "Modelling of a super-charged semi-closed oxyfuel combined cycle with CO_2 capture and analysis of the part-load behavior," *Energy Procedia*, vol. 1, no. 1, pp. 415–422, 2009.

[16] H. J. Yang, D. W. Kang, J. H. Ahn, and T. S. Kim, "Evaluation of design performance of the semi-closed oxy-fuel combustion combined cycle," *Journal of Engineering for Gas Turbines and Power*, vol. 134, Article ID 111702, 2012.

[17] A. Dahlquist, M. Genrup, M. Sjoedin, and K. Jonshagen, "Optimization of an oxyfuel combined cycle regarding performance and complexity level," in *Proceedings of the ASME Turbo Expo 2013: Power for Land, Sea and Air*, GT2013-94755, American Society of Mechanical Engineers, San Antonio, Tex, USA, 2013.

[18] M. Sammak, E. Thorbergsson, T. Grönstedt, and M. Genrup, "Conceptual mean-line design of single and twin-shaft oxy-fuel gas turbine in a semiclosed oxy-fuel combustion combined cycle," *Journal of Engineering for Gas Turbines and Power*, vol. 135, no. 8, Article ID 081502, 2013.

[19] H. Jericha, "Efficient steam cycles with internal combustion of hydrogen and stoichiometric oxygen for turbines and piston engines," *International Journal of Hydrogen Energy*, vol. 12, no. 5, pp. 345–354, 1987, CIMAC Conference Paper, Oslo, Norway.

[20] H. Jericha, W. Sanz, J. Woisetschläger, and M. Fesharaki, "CO_2-retention capability of CH_4/O_2-fired graz cycle," in *Proceedings of the CIMAC World Congress on Combustion Engine Technology*, Paper G07, pp. 1–13, Interlaken, Switzerland, 1995.

[21] H. Jericha and E. Göttlich, "Conceptual design for an industrial prototype Graz cycle power plant," in *Proceedings of the ASME Turbo Expo 2002: Power for Land, Sea and Air*, GT2002-30118, American Society of Mechanical Engineers, Amsterdam, The Netherlands, June 2002.

[22] F. Heitmeir, W. Sanz, E. Göttlich, and H. Jericha, "The graz cycle-a zero emission power plant of highest efficiency," in *XXXV Kraftwerkstechnisches Kolloquium*, Dresden, Germany, September 2003.

[23] H. Jericha, E. Göttlich, W. Sanz, and F. Heitmeir, "Design optimization of the graz cycle prototype plant," *Journal of Engineering for Gas Turbines and Power*, vol. 126, no. 4, pp. 733–740, 2004.

[24] F. Heitmeir and H. Jericha, "Turbomachinery design for the Graz cycle: an optimized power plant concept for CO_2 retention," *Proceedings of the Institution of Mechanical Engineers Part A: Journal of Power and Energy*, vol. 219, no. 2, pp. 147–155, 2005.

[25] W. Sanz, H. Jericha, F. Luckel, E. Göttlich, and F. Heitmeir, "A further step towards a Graz cycle power plant for CO_2 capture," in *Proceedings of the ASME Turbo Expo: Power for Land, Sea and Air*, GT2005-68456, pp. 181–190, American Society of Mechanical Engineers, Reno, Nev, USA, June 2005.

[26] W. Sanz, H. Jericha, M. Moser, and F. Heitmeir, "Thermodynamic and economic investigation of an improved Graz Cycle power plant for CO_2 capture," *Journal of Engineering for Gas Turbines and Power*, vol. 127, no. 4, pp. 765–772, 2005.

[27] M. H. Jericha, M. W. Sanz, and M. E. Goettlich, "Gasturbine with CO_2 retention—400 MW oxyfuel-system Graz cycle," in

Proceedings of the CIMAC World Congress on Combustion Engine Technology, Vienna, Austria, May 2007.

[28] H. Jericha, W. Sanz, and E. Göttlich, "Design concept for large output graz cycle gas turbines," *Journal of Engineering for Gas Turbines and Power*, vol. 130, no. 1, Article ID 011701, 2008.

[29] H. Jericha, W. Sanz, E. Göttlich, and F. Neumayer, "Design details of a 600 mw Graz cycle thermal power plant for CO_2 capture," in *Proceedings of the ASME Turbo Expo: Power for Land, Sea and Air*, GT2008-50515, pp. 507–516, American Society of Mechanical Engineers, Berlin, Germany, June 2008.

[30] H. Kvamsdal, O. Maurstad, K. Jordal, and O. Bolland, "Benchmarking of gas-turbine cycles with CO_2 capture," in *Proceedings of the 7th International Conference on Greenhouse Gas Control Technologies*, vol. 1, pp. 233–242, Vancouver, Canada, September 2004.

[31] F. Franco, T. Mina, G. Woolatt, M. Rost, and O. Bolland, "Characteristics of cycle components for CO_2 capture," in *Proceedings of the 8th International Conference on Greenhouse Gas Control Technologies*, Trondheim, Norway, June 2006.

[32] H. M. Kvamsdal, K. Jordal, and O. Bolland, "A quantitative comparison of gas turbine cycles with CO_2 capture," *Energy*, vol. 32, no. 1, pp. 10–24, 2007.

[33] W. Sanz, H. Jericha, B. Bauer, and E. Göttlich, "Qualitative and quantitative comparison of two promising oxy-fuel power cycles for CO_2 capture," *Journal of Engineering for Gas Turbines and Power*, vol. 130, no. 3, Article ID 031702, 2008.

[34] G. Woollatt and F. Franco, "Natural gas oxy-fuel cycles-part 1: conceptual aerodynamic design of turbo-machinery components," *Energy Procedia*, vol. 1, pp. 573–580, 2009.

[35] E. Thorbergsson, T. Grönstedt, M. Sammak, and M. Genrup, "A comparative analysis of two competing mid-size oxy-fuel combustion cycles," in *Proceedings of the ASME Turbo Expo: Power for Land, Sea and Air*, GT2012-69676, pp. 375–383, American Society of Mechanical Engineers, Copenhagen, Denmark, June 2012.

[36] P. Mathieu and O. Bolland, "Comparison of costs for natural gas power generation with CO_2 capture," *Energy Procedia*, vol. 37, pp. 2406–2419, 2013.

[37] O. Bolland and H. Undrum, "A novel methodology for comparing CO_2 capture options for natural gas-fired combined cycle plants," *Advances in Environmental Research*, vol. 7, no. 4, pp. 901–911, 2003.

[38] SimTech Simulation Technology, *User Documentation: Program Modules and Model Libraries*, IPSEpro Process Simulator, 2003.

[39] E. Thorbergsson, T. Grönstedt, and C. Robinson, "Integration of fluid thermodynamic and transport properties in conceptual turbomachinery design," in *Proceedings of the ASME Turbo Expo: Power for Land, Sea and Air*, GT2013-95833, American Society of Mechanical Engineers, San Antonio, Tex, USA, June 2013.

[40] G. Halls, "Air cooling of turbine blades and vanes: an account of the history and development of gas turbine cooling," *Aircraft Engineering and Aerospace Technology*, vol. 39, no. 8, pp. 4–14, 1967.

[41] M. J. Holland and T. F. Thake, "Rotor blade cooling in high pressure turbines," *Journal of Aircraft*, vol. 17, no. 6, pp. 412–418, 1980.

[42] K. Jordal, *Modeling and performance of gas turbine cycles with various means of blade cooling [Ph.D. thesis]*, Department of Heat and Power Engineering, Lund University, 2001.

[43] O. Bolland and J. F. Stadaas, "Comparative evaluation of combined cycles and gas turbine systems with water injection, steam injection, and recuperation," *Journal of Engineering for Gas Turbines and Power*, vol. 117, no. 1, pp. 138–145, 1995.

[44] M. A. El-Masri, "On thermodynamics of gas-turbine cycles: Part 2—a model for expansion in cooled turbines," *Journal of Engineering for Gas Turbines and Power*, vol. 108, no. 1, pp. 151–159, 1986.

[45] K. Jordal, L. Torbidoni, and A. Massardo, "Convective blade cooling modelling for the analysis of innovative gas turbine cycles," in *Proceedings of the ASME Turbo Expo 2001: Power for Land, Sea and Air*, GT2001-0390, American Society of Mechanical Engineers, New Orleans, La, USA, 2001.

[46] J. H. Horlock, D. T. Watson, and T. V. Jones, "Limitations of gas turbine performance imposed by large turbine cooling flows," *Journal of Engineering for Gas Turbines and Power*, vol. 123, no. 3, pp. 487–494, 2001.

[47] K. Jordal, *Gas turbine cooling modeling—thermodynamic analysis and cycle simulations [Licentiate thesis]*, Department of Heat and Power Engineering, Lund University, Lund, Sweden, 1999.

[48] R. C. Wilcock, J. B. Young, and J. H. Horlock, "The effect of turbine blade cooling on the cycle efficiency of gas turbine power cycles," *Journal of Engineering for Gas Turbines and Power*, vol. 127, no. 1, pp. 109–120, 2005.

[49] M. Mallen and G. Saville, *Polytropic Processes in the Performance Prediction of Centrifugal Compressors*, I. Mech. E. Conference Publications, Institution of Mechanical Engineers, London, UK, 1977.

[50] W. Traupel, *Thermische Turbomaschinen*, Springer, Berlin, Germany, 1977.

[51] F. Di Maria and K. Jordal, "Blade cooling model comparison and thermodynamic analysis," in *Proceedings of the 5th ASME/JSME Joint Thermal Engineering Conference*, AJTE99-6117, American Society of Mechanical Engineers, San Diego, Calif, USA, March 1999.

[52] A. Dahlquist, M. Thern, and M. Genrup, "The inuence from the working medium on the profile loss in compressor and turbine airfoils," in *Proceedings of the ASME Turbo Expo 2014: Power for Land, Sea and Air*, GT2014-25069, American Society of Mechanical Engineers, Düsseldorf, Germany, 2014.

[53] S. G. Sundkvist, A. Dahlquist, J. Janczewski et al., "Concept for a combustion system in oxyfuel gas turbine combined cycles," *Journal of Engineering for Gas Turbines and Power*, vol. 136, no. 10, Article ID 101513, 2014.

[54] I. Pfaff and A. Kather, "Comparative thermodynamic analysis and integration issues of CCS steam power plants based on oxy-combustion with cryogenic or membrane based air separation," *Energy Procedia*, vol. 1, pp. 495–502, 2009.

[55] J. Tranier, N. Perrin, and A. Darde, "Update on advanced developments for ASU and CO_2 purification units for oxy-combustion (air liquide, France)," in *Proceedings of the 3rd Meeting of the Oxy-Fuel Combustion Network*, IEAGHG International Oxy-Combustion Network, Yokohama, Japan, March 2008.

[56] K. Jeong, M. J. Kessen, H. Bilirgen, and E. K. Levy, "Analytical modeling of water condensation in condensing heat exchanger," *International Journal of Heat and Mass Transfer*, vol. 53, no. 11-12, pp. 2361–2368, 2010.

4

Analysis of the Oceanic Wave Dynamics for Generation of Electrical Energy Using a Linear Generator

Omar Farrok,[1,2] Md. Rabiul Islam,[2] and Md. Rafiqul Islam Sheikh[2]

[1]*Department of Electrical and Electronic Engineering, Ahsanullah University of Science & Technology, Dhaka 1208, Bangladesh*
[2]*Department of Electrical and Electronic Engineering, Rajshahi University of Engineering & Technology, Rajshahi 6204, Bangladesh*

Correspondence should be addressed to Omar Farrok; omarruet@gmail.com

Academic Editor: Yongping Li

Electricity generation from oceanic wave depends on the wave dynamics and the behavior of the ocean. In this paper, a permanent magnet linear generator (PMLG) has been designed and analyzed for oceanic wave energy conversion. The proposed PMLG design is suitable for the point absorber type wave energy device. A mathematical model of ocean wave is presented to observe the output characteristics and performance of the PMLG with the variation of ocean waves. The generated voltage, current, power, applied force, magnetic flux linkage, and force components of the proposed PMLG have been presented for different sea wave conditions. The commercially available software package ANSYS/ANSOFT has been used to simulate the proposed PMLG by the finite element method. The magnetic flux lines, flux density, and field intensity of the proposed PMLG that greatly varies with time are presented for transient analysis. The simulation result shows the excellent features of the PMLG for constant and variable speeds related to wave conditions. These analyses help to select proper PMLG parameters for better utilization of sea wave to maximize output power.

1. Introduction

At present, scientists and engineers are facing two major challenges in the world of energy: electrical energy production and environmental issue. These problems can be amicably solved using renewable energy resources (RERs). The vital factors which have stimulated the use of RERs are energy independence, financial viability, and mainly environmental protection [1]. The problem of oil crises (1973–1983) and environmental pollution concerns urged engineers to harvest electrical energy from the available RERs, for example, wind energy [2–5], solar energy [6–9], and hydro power [10]. As the RER is variable and unpredictable, different types of converters and controls are associated with the RER-based power plants that are connected with standalone or grid systems [11, 12]. The traditional RERs have the problem of uncertainty of availability and they require a large land area. On the other hand, the key advantages of oceanic wave energy (OWE) are as follows: (i) it has huge potential compared to the solar and wind energy, (ii) it is easy to forecast, and (iii) it does not need land area. Hence, it has been of great interest to the industrial

field; particularly, in energy generation, the use of wave power has been more attractive compared to other RERs [13]. OWE is a promising environmental pollution-free energy, which would make significant contribution toward saving biochemical resources and reducing carbon emissions [14]. It is estimated that the total wave energy resource in the open sea around the world is 10 TW (10,000 GW), a comparable amount of the total power consumption in the world [15].

Different types of oceanic wave energy converters (WECs) have been invented and examined for successful conversion of wave energy into electrical energy [16–19]. The wave energy device may be of a rotational or a translational type and each of these devices has different features [20]. The power takeoff devices play a vital role in converting the irregular wave motion to a regular motion for energy conversion. Linear generators (LGs) have directly been implemented to the direct drive wave energy conversion without using medium devices which has unique advantages over all other wave energy devices [21]. An LG has two major parts, namely, translator and stator. The translator mounted to a hollow cylinder is sometimes called a float or floater as shown in

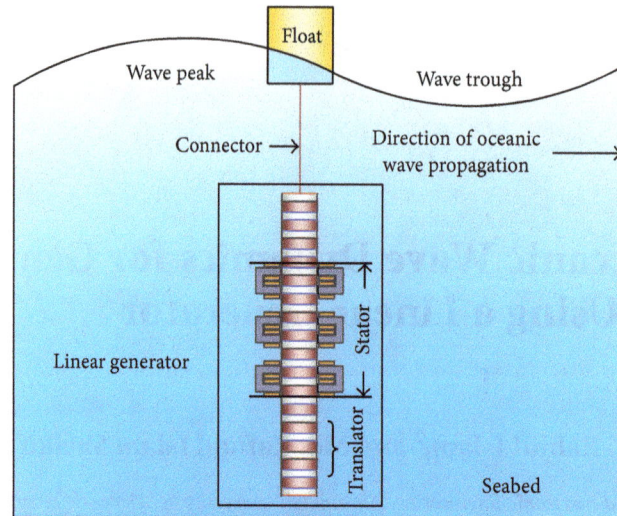

FIGURE 1: Drawing of the LG connected with a float.

Figure 1. The stator is mounted to another mechanical body in order to make it stationary with respect to the sea wave.

The floater tries to float on the surface of the ocean wave with the translator that moves relative to the stator due to wave action. The floater utilizes the rise and fall of the sea wave at a single point for energy conversion. As the translator moves linearly due to the reciprocating motion of the floater, the wave energy so extracted is converted into electrical energy. A lot of mathematicians analyzed and proposed different mathematical models for understanding the nature of oceanic wave [22, 23].

Different types of permanent magnet linear generators (PMLGs) have been designed and analyzed for improved performance. The flat and tubular flux switching permanent magnet linear generators (FSPMLGs) have been proposed [21, 24, 25]. Different analyses have shown that the FSPMLGs made of PMs and steel cores have suffered from the problem of higher leakage flux leading to the reduction of electrical power generation. Tubular PMLGs have been proposed [14, 21, 26, 27] to reduce cogging forces and also increase efficiency. The maintenance of tubular PMLG is difficult due to the presence of the coils inside the periphery. It is seen in the recent works [28] that the linear switched reluctance generator (LSRG) has been proposed for high power generation. Excessive leakage flux and complex control circuit of LSRGs are responsible for the degradation of overall efficiency and reliability.

This paper has presented the mathematical model of the wave motion to analyze the behavior of a PMLG that offers low internal resistance, low loading effect, and high output power. It is essential to consider the nature of wave motion for parameter selection of the PMLG, thus maximizing electricity generation. A relationship between the oceanic wave motions with the parameters of PMLG is established. Different significant parameters, for example, size of PMs, poles, translator length, and stroke length, of the PMLG are found from the relationship and the generated voltage, current, power, applied force, magnetic flux linkage, flux lines, flux

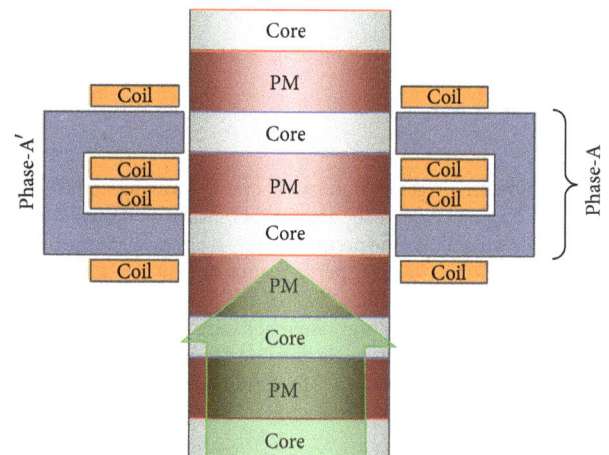

FIGURE 2: Position of the translator for time t_1.

density, and the applied force with force components of the proposed PMLG are shown for different sea wave conditions.

2. Design of the PMLG

2.1. Working Principle. The translator moves vertically with the incident wave; therefore, the direction may be upward or downward. Considering the translator movement in the upward direction as shown in Figures 2 and 3 for a particular time interval, the stator and translator poles are aligned facing each other. The red and green lines are representing the north pole (N) and the south pole (S) of the permanent magnet (PM), respectively. S exists in the upper side and N exists in the lower side of the stator core for a time t_1. The position of the translator varies with time and the direction of magnetic flux changes. The translator position changes in Figure 3 from the position shown in Figure 2. S now exists in the lower side and N exists in the upper side of the stator cores for another time t_2.

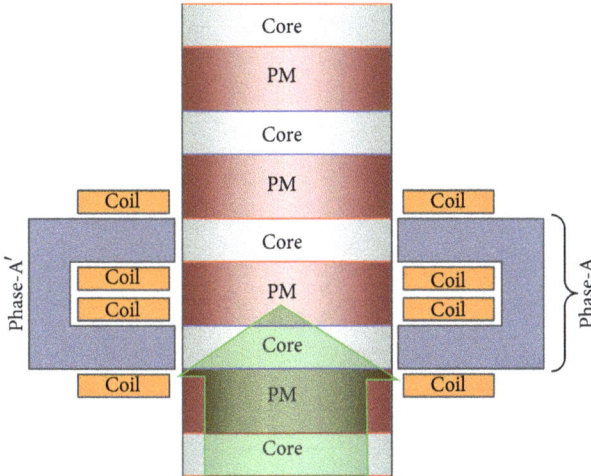

FIGURE 3: Position of the translator for time t_2.

Hence, the direction of magnetic flux according to Figure 2 is opposite for the translator position as in Figure 3 within the time interval $t_2 - t_1$. Therefore, the induced voltage directions across the conductor are opposite to each other and due to this reason the PMLG generates AC power.

2.2. Vector Diagram and Equivalent Circuit. The winding of each phase consists of two coils of opposite phases. The induced voltages are therefore 180° phase shifted to each other. The coils of each phase are connected in series in additive polarity to get higher output voltage. The vector diagrams of induced voltages are represented in Figure 4.

The equivalent circuit diagram of the proposed PMLG for a three-phase load is shown in Figure 5. The equivalent series resistances and inductances of each phase winding coil are considered equal, denoted by R_a and L_S, respectively. The winding consists of two series coils; therefore, the equivalent series resistance and inductance of each coil are denoted by $R_a/2$ and $L_S/2$, respectively. E_a, E_b, and E_c are the induced voltages of phase-A, phase-B, and phase-C, respectively, for simplicity. Similarly, E_a', E_b', and E_c' are the induced voltages of phase-A', phase-B', and phase-C', respectively. The magnetic excitation is fed from the translator's PM array as shown in Figure 6. The terminal voltages, v_a, v_b, and v_c, are measured across the load.

The induced voltage equation may be represented as

$$E_i + E_i' = R_i i_i + L_S \frac{di_i}{dt} + v_i$$
$$= K_m \cos\left(\frac{\pi}{\tau} z + j\frac{2\pi}{3}\right) v_v(t), \quad (1)$$

where i = A, B, and C, j = 0, 1, and −1, v_i is the terminal voltage, i_i is the line current, and L_S is synchronous inductance. K_m is the constant representing the machine construction; $v_v(t)$ is the translator vertical velocity or speed; τ is the pole pitch; and z is the vertical displacement. The terminal voltage is

$$\vec{V}_i = \vec{E}_i + \vec{E}_i' - jX_S\vec{I}_i - R_i\vec{I}_i. \quad (2)$$

2.3. Construction Details. The vertical cross section of the PMLG (front side) is shown in Figure 6. The PMLG basically contains a translator which is made by some PMs with the steel core. The stator contains some copper coils wounded on the stator cores of steel situated on both sides of the translator. Phase-A, phase-B, and phase-C are located on the right side of the translator. Phase-A', phase-B', and phase-C' are located on the left side of the translator which are 180° phase shifted from phase-A, phase-B, and phase-C, respectively. The construction supports the translator to move in the vertical direction with respect to the stator. The orientation of the PM array should be such that N and S can be formed one after another as shown in Figure 6.

3. Model of the Oceanic Wave

In most of the cases, constant speeds or sinusoidal speeds having sinusoidal shapes are common approximations for simulating PMLGs [25–27]. The typical range of vertical velocity of wave is 0–2 m/s with a time period from 4 to 6 s [28–30]. Therefore, a free oceanic wave neither forced nor dissipated on a flat seabed is presented for analysis of the LG as shown in Figure 7 following the mentioned approximations. The amplitude of waveform is A. Hence, the vertical distance between the wave crest and the trough is H which is equal to twice the amplitude A. According to the point of view of oceanographers, it is considered as a linear wave. The sea surface is lying on xy plane, where y components are considered zero as the wave propagates in x-direction, λ is wavelength, d_W is water depth of the ocean with respect to xy plane or water surface, seabed is at $z = -d_W$, and the ocean surface coincides with $z = 0$.

3.1. Generalized Wave. The general description of almost any type of oceanic wave may be considered as, depending only on the interpretation of ζ,

$$\zeta = A\cos(kx - \omega t) \quad (3)$$

$$\omega = \sqrt{gk\tanh(kd_W)}. \quad (4)$$

Here, the number of waves $k = 2\pi/\lambda$ and ω is the frequency that asserts the physics and describes consideration of a water wave relating frequency and wave number. Alternatively, it can be considered as a relation between the phase speed or oceanic wave velocity, v, and the wavelength. Considering gravitational acceleration, $g = 9.8\,\text{m/s}^2$, the phase speed which is a single basic wave that moves along the x-direction can be expressed as

$$v = \sqrt{\frac{g\lambda}{2\pi}\tanh\left(\frac{2\pi}{\lambda}d_W\right)}. \quad (5)$$

According to (3) and (5), the wave velocity is along x-direction only although it has the velocity along z-direction. So, the velocity of oceanic wave is a vector that depends on x, y, and z and time, t. To obtain a complete description of

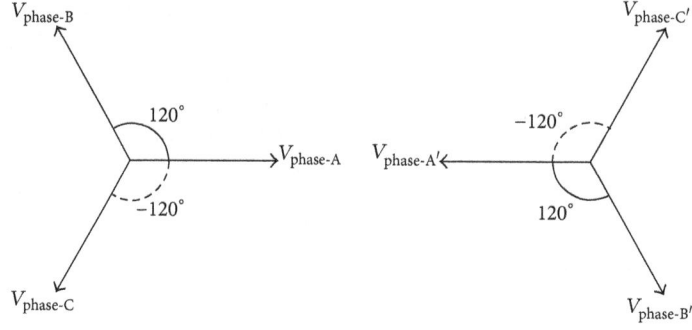

Figure 4: Vector diagram.

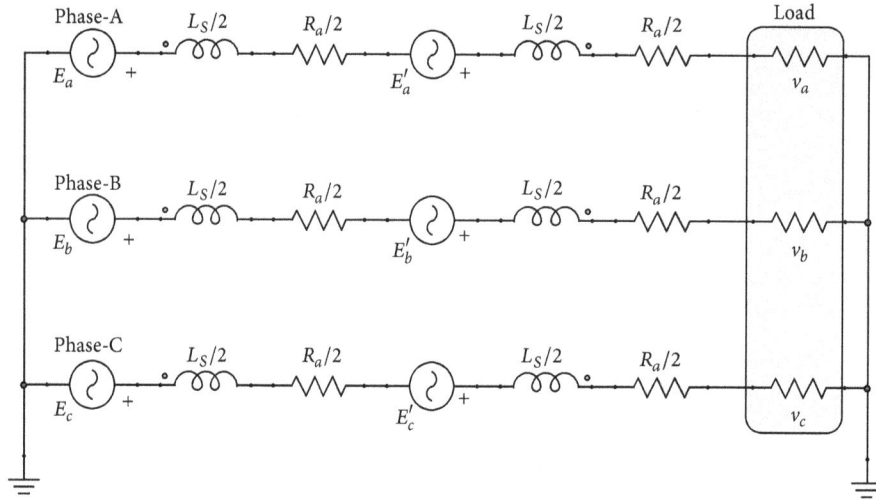

Figure 5: Equivalent circuit diagram of the PMLG.

oceanic wave, the components along x, y, and z have to be calculated. The motion can be expressed as

$$
\begin{aligned}
&V(x, y, z, t) \\
&= (X(x, y, z, t), Y(x, y, z, t), Z(x, y, z, t)),
\end{aligned} \tag{6}
$$

where $V(x, y, z, t)$ represents the velocity of oceanic wave and $X(x, y, z, t), Y(x, y, z, t)$, and $Z(x, y, z, t)$ are the components along x, y, and z. The y components may be considered as zero for the oceanic wave and the x and z components can be expressed as follows:

$$
X = A\omega \frac{\cosh\{k(z + d_W)\}}{\sinh(kd_W)} \cos(kx - \omega t) \tag{7}
$$

$$
Z = A\omega \frac{\sinh\{k(z + d_W)\}}{\sinh(kd_W)} \sin(kx - \omega t). \tag{8}
$$

The x and z components from (7) and (8) can be simplified for individual consideration of shallow water wave, intermediate depth wave, and deep sea wave.

3.2. Shallow Water Wave. In shallow water, water depth is much lower than the wavelength, λ; that is, $d_W \ll \lambda$ or

$kd_W \ll 1$. Another property of shallow water wave is that the amplitude of wave is much smaller than wavelength; so, $A \ll \lambda$ or $Ak \ll 1$. So, (7) and (8) may be written as

$$
X_S = A\omega \frac{e^{k(z+d_W)} + e^{-k(z+d_W)}}{e^{kd_W} - e^{-kd_W}} \cos(kx - \omega t) \tag{9}
$$

$$
Z_S = A\omega \frac{e^{k(z+d_W)} - e^{-k(z+d_W)}}{e^{kd_W} - e^{-kd_W}} \sin(kx - \omega t). \tag{10}
$$

Now, (9) and (10) may be reduced as

$$
X_S \cong \frac{A\omega}{kd_W} \cos(kx - \omega t) \tag{11}
$$

$$
Z_S \cong A\omega \left(1 + \frac{z}{d_W}\right) \sin(kx - \omega t). \tag{12}
$$

Again, if $kd_W \ll 1$, $\omega = \sqrt{gk \tanh(kd_W)}$ can be simplified as $\omega = k\sqrt{gd_W}$ because

$$
\tanh(kd_W) = \frac{e^{kd_W} - e^{-kd_W}}{e^{kd_W} + e^{-kd_W}} = \frac{2kd_W}{2} = kd_W. \tag{13}
$$

As $v = \omega/k$, phase speed can be expressed as $v = \sqrt{gd_W}$.

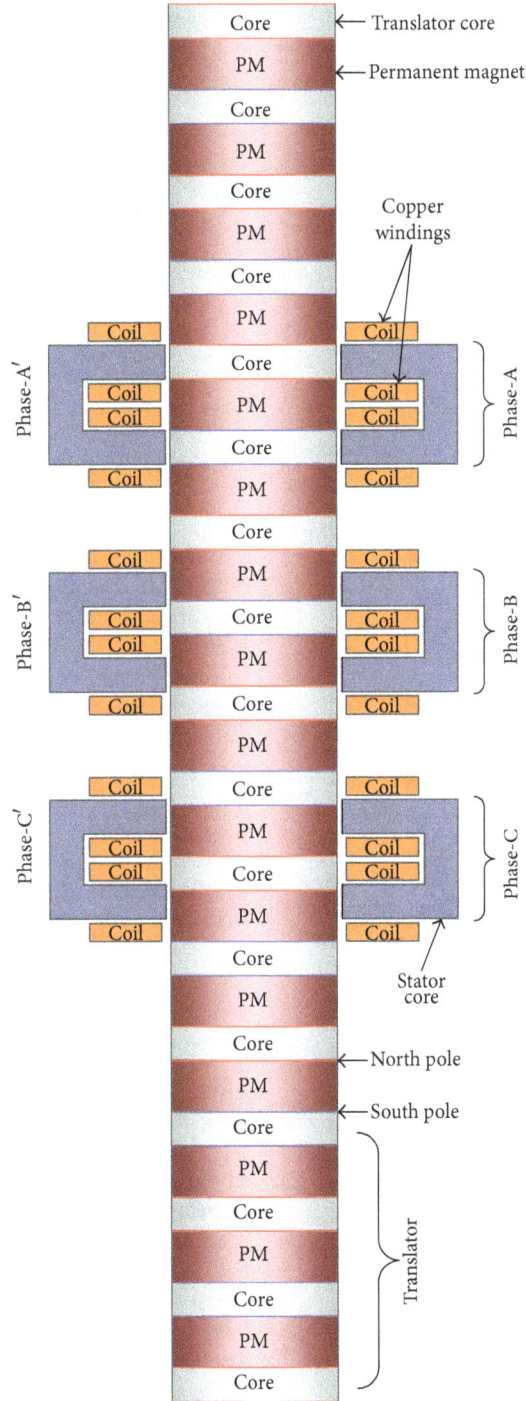

FIGURE 6: Construction of the PMLG.

3.3. *Deep Water Wave.* In deep water, water depth is much higher than wavelength; that is, $d_W \gg \lambda$ or $kd_W \gg 1$. So,

$$\frac{e^{k(z+d_W)} + e^{-k(z+d_W)}}{e^{kd_W} - e^{-kd_W}} \cong \frac{e^{k(z+d_W)}}{e^{kd_W}} = e^{kd_W},$$

$$\frac{e^{k(z+d_W)} - e^{-k(z+d_W)}}{e^{kd_W} - e^{-kd_W}} \cong \frac{e^{k(z+d_W)}}{e^{kd_W}} = e^{kd_W},$$

$$\tanh(kd_W) = \frac{e^{kd_W} - e^{-kd_W}}{e^{kd_W} + e^{-kd_W}} \cong \frac{e^{kd_W}}{e^{kd_W}} = 1. \tag{14}$$

Therefore, (7) and (8) can be expressed as follows:

$$X_D \cong A\omega e^{kd_W} \cos(kx - \omega t) \tag{15}$$

$$Z_D \cong A\omega e^{kd_W} \sin(kx - \omega t). \tag{16}$$

TABLE 1: Comparison of the oceanic wave parameters.

Wave type	Shallow water	Intermediate depth wave	Deep water
Relative depth	$\dfrac{d_W}{\lambda} < 0.05$	$0.05 < \dfrac{d_W}{\lambda} < 0.5$	$\dfrac{d_W}{\lambda} > 0.5$
Wave speed	$\sqrt{gd_W}$	$\sqrt{\dfrac{g\lambda}{2\pi}\tanh\left(2\pi\dfrac{d_W}{\lambda}\right)}$	$\sqrt{\dfrac{g\lambda}{2\pi}}$
Wave length	$\sqrt{gd_W}\,T$	$\dfrac{gT^2}{2\pi}\tanh\left(2\pi\dfrac{d_W}{\lambda}\right)$	$\dfrac{gT^2}{2\pi}$

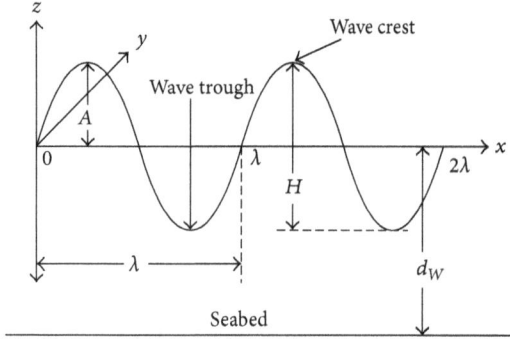

FIGURE 7: Waveform of a free oceanic wave.

Again, if $kd_W \gg 1$, $\omega = \sqrt{gk\tanh(kd_W)}$ can be simplified as $\omega = \sqrt{gk}$. Phase speed, v, can be expressed as $v = \sqrt{g/k}$. The x and z components, both for shallow and for deep water from (6)–(12), (15), and (16), are explained in [31]. Comparisons of relative depth, wave speed, and wavelength of shallow water wave, intermediate depth wave, and deep sea wave are shown in Table 1.

4. The PMLG and Wave Model

4.1. Selection of Translator Length. The translator lengths may be shorter, equal, or longer compared to the stator. The common lengths of the stator and the translator affect the power rating of the PMLG. The power rating increases with increase in the common length. Longer translator has been chosen to obtain the same common length as shown in Figure 8 for different translator positions. The vertical displacement of translator with the incident sea wave is known as stroke length, L_{st}. There is a relationship between the wave amplitude, A_w, and L_{st} during the energy conversion as $L_{st} = 2A_w$. A_w is smaller than the amplitude of free oceanic wave as in (7)–(12), (15), and (16) because of power dissipation.

4.2. Determination of Frequency. The translator pole pitch is the summation of the PM thickness and translator pole thickness of the PMLG as shown in Figure 9. The stator pole pitch and translator pole pitch may be the same or different depending on the design strategy. In the proposed design, the stator pole pitch is the same as the translator pole pitch and the stator pole width is the same as the translator pole width. The pitch and pole width have a vital effect on the generated

voltage, shape, power, frequency, and forces. The frequency, f, of the PMLG is determined by

$$f = \frac{v_v(t)}{2\tau}. \tag{17}$$

4.3. Direction of Forces. The simulation setup of the PMLG design is represented using a 3D Cartesian coordinate system. The applied force is working along z-axis, as shown by bidirectional arrow in Figure 10. The width and thickness of this design are along y-axis and x-axis, respectively. Cogging forces between stator and translator core act along y-axis. Other force components except for applied force and cogging forces work along x-axis. The PMLG converts mechanical energy to electrical energy due to the force along z-axis and the forces along x-axis and y-axis are generated which is the cause of mechanical power loss. In the oceanic wave model, the direction of wave elevation is also along z-axis according to (8), (10), (12), and (16).

4.4. Generated Voltage with Waves. According to Faraday's law of electromagnetic induction, the induced voltage is found as

$$E_i(t) = -N\frac{d\vec{\Phi}}{dt}. \tag{18}$$

In the wave model, the wave elevation is described with respect to x-axis that represents the direction of wave propagation. The induced voltage and related parameters of the PMLG are related to time. The vertical wave displacement is assumed to be sinusoidal and the translator connected to buoy tries to follow the wave elevation; therefore, the vertical displacement, d_{tr}, and velocity, s_{tr}, of the translator can be expressed as given in (19) and (20), respectively [14]. Hence,

$$d_{tr}(t) = A_w \sin\left(\frac{2\pi}{T}t \pm \theta_i\right) \tag{19}$$

$$s_{tr}(t) = A_w \frac{2\pi}{T}\cos\left(\frac{2\pi}{T}t \pm \theta_i\right). \tag{20}$$

Here, A_w, θ_i, and T represent wave amplitude, initial phase angle, and period of oceanic wave, respectively. The translator and buoy move with the wave elevation; therefore, the flux variation with respect to time can be expressed as

$$\Phi(t) = \vec{\Phi}\frac{2\pi}{\lambda}d_{tr}(t). \tag{21}$$

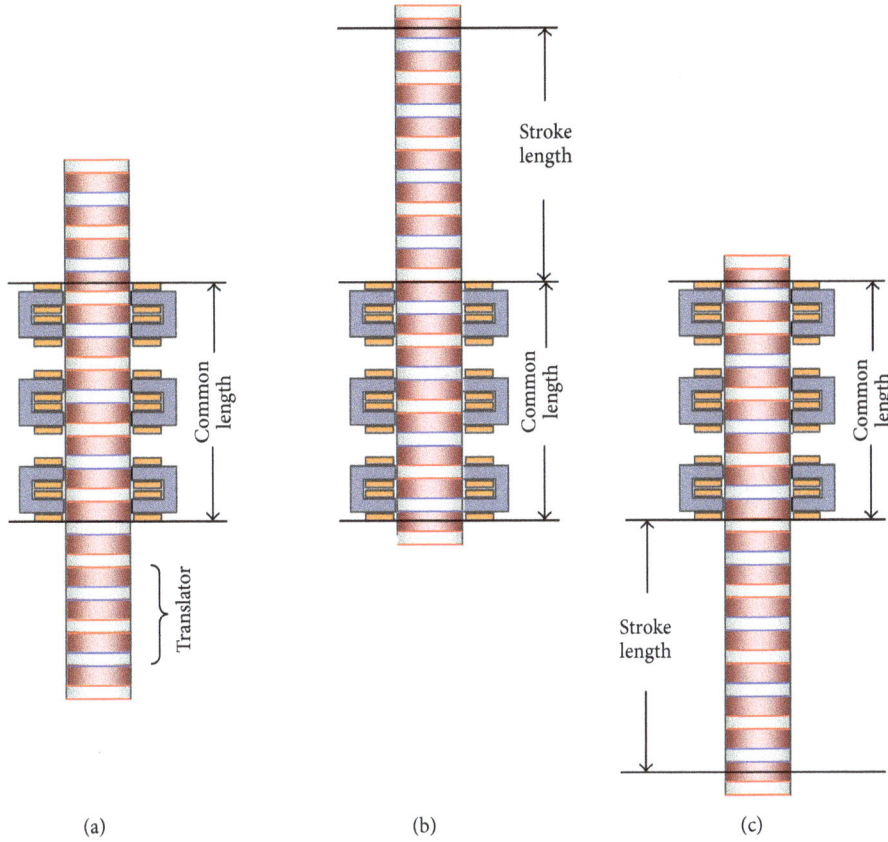

FIGURE 8: Translator position at different time.

If V_m is the peak voltage, combining (18) and (21), the generated EMF per phase and V_m of the PMLG, v_{lg}, are obtained as [32]

$$v_{lg}(t) = \vec{V}_m \cos\left(\frac{2\pi}{T}t\right)\cos\left\{\frac{2\pi A_w}{\lambda}\sin\left(\frac{2\pi}{T}t\right)\right\}$$

$$V_m = N\vec{\Phi}\frac{2\pi}{\lambda}\frac{2\pi}{T}A_w. \tag{22}$$

5. Simulation Results

Two types of speed settings have been used in this simulation; one is consideration of constant translator speed for observation of the performance of the PMLG and the other is translator motion changed with the incident sea wave. The default air gap length is 2 mm, load is 4 Ω, and the translator speed is 1 m/s for the PMLG unless otherwise specified. The input of the PMLG is mechanical thrust, F, in newton. From mechanical power, $P_m = F \times s_{tr}$, and electrical power, P_e, efficiency, η, is calculated by $\eta = P_e/P_m$.

5.1. Constant Translator Speed. The terminal voltage and load current for default condition are shown in Figure 11.

The load voltage and current are in the same phases due to the resistive load. The applied force, F_z, cogging force, F_y, and force component along x-axis, F_x, are shown in Figure 12. The induced voltage, current, and magnetic flux

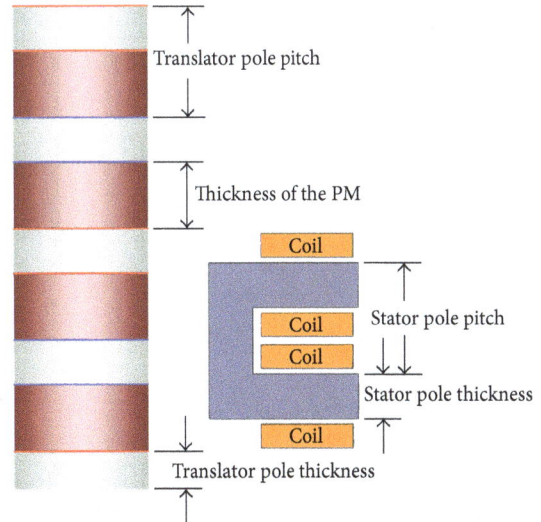

FIGURE 9: Stator and translator pole pitch.

linkage are shown in Figure 13. The generated power is shown in Figure 14.

5.2. Voltage Regulation. The terminal voltage depends on the load for a specific PMLG as in (2), which is shown in Figure 15 to observe the terminal characteristics.

TABLE 2: Numerical values for legend of Figure 16.

Legend name	Load (Ω)	RMS current (A)	DC offset (A)	Period (ms)
I_1	4	8.6656	0.0377	79.9355
I_2	5	7.1327	0.0269	81.7782
I_3	8	4.6401	0.0151	81.2897
I_4	10	3.7586	0.0118	81.1431
I_5	15	2.5455	0.0077	80.85
I_6	20	1.9233	0.0057	80.7523

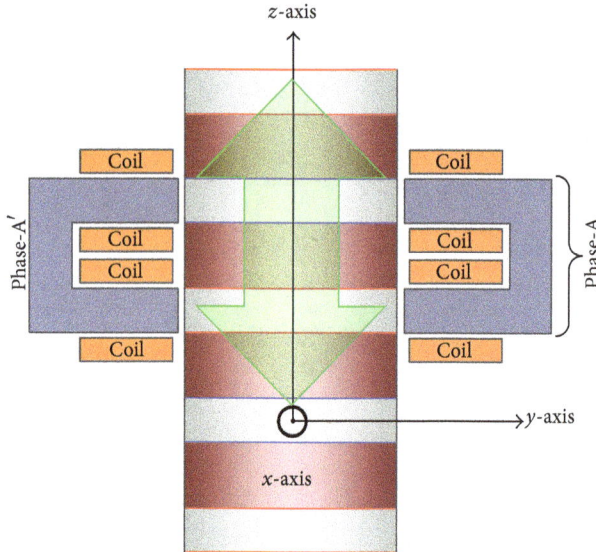

FIGURE 10: Direction of different forces.

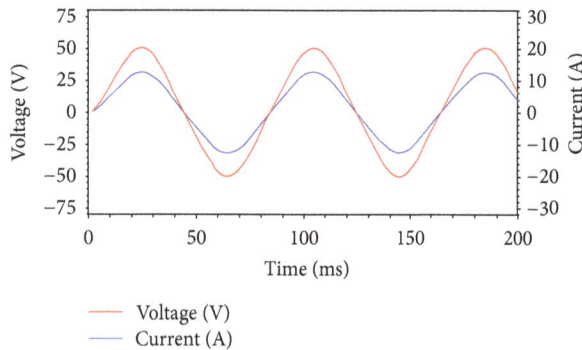

FIGURE 12: Different forces of the PMLG.

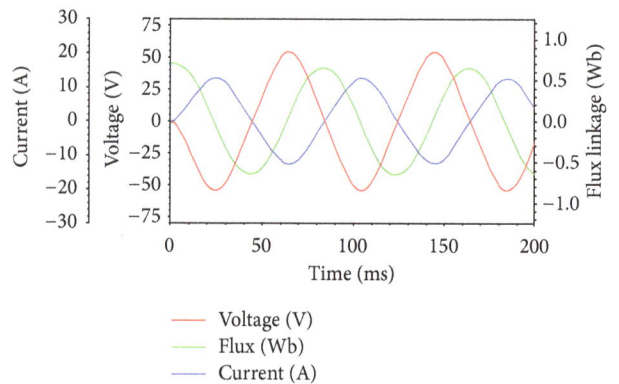

FIGURE 11: Load voltage and current waveforms.

FIGURE 13: Induced voltage, current, and flux linkage.

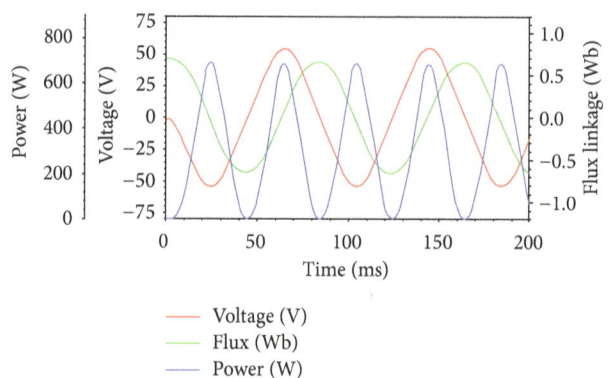

Different loads ranging from $2\,\Omega$ up to $1\,M\Omega$ are shown on a semilog graph. Load voltages measured from any one of the three phases are similar and remain almost unchanged from $25\,\Omega$ to $1\,M\Omega$. The RMS values of currents and power of phase-A for different load conditions are shown in Figures 16 and 17, respectively. The numerical values of the current and power graphs used in the legend are tabulated in Tables 2 and 3, respectively.

5.3. Variable Translator Speed. According to the oceanic wave model, the vertical position of the wave is sinusoidal, so the

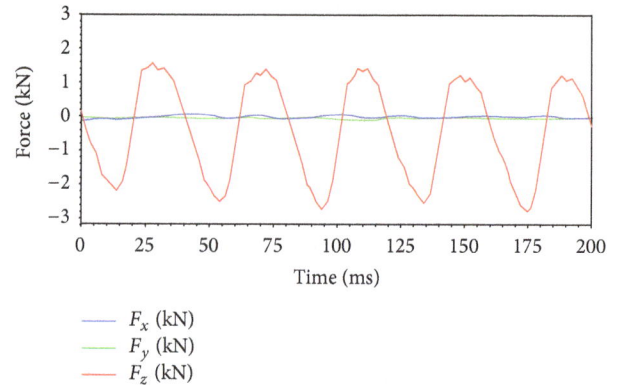

FIGURE 14: Generated power with voltage and flux.

TABLE 3: Numerical values for legend of Figure 17.

Legend name	Load (Ω)	RMS power (W)	Period (ms)	Maximum power (W)
P_1	4	242.0785	40.0733	691.9896
P_2	5	204.3384	40.0488	586.9683
P_3	8	137.9347	40.0488	398.2045
P_4	10	113.0125	40.0366	328.8075
P_5	15	77.636	40.0366	228.3736
P_6	20	59.0638	40.0244	174.9611

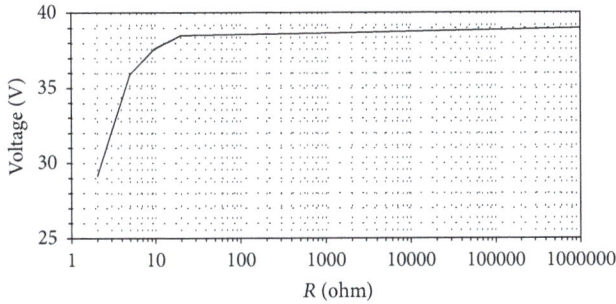

FIGURE 15: Terminal voltage for different loads.

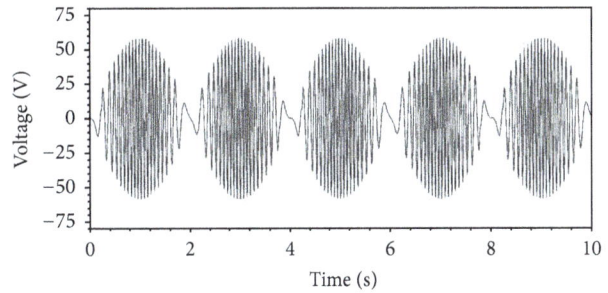

I_1 (A) I_4 (A)
I_2 (A) I_5 (A)
I_3 (A) I_6 (A)

FIGURE 16: Load currents for different loads.

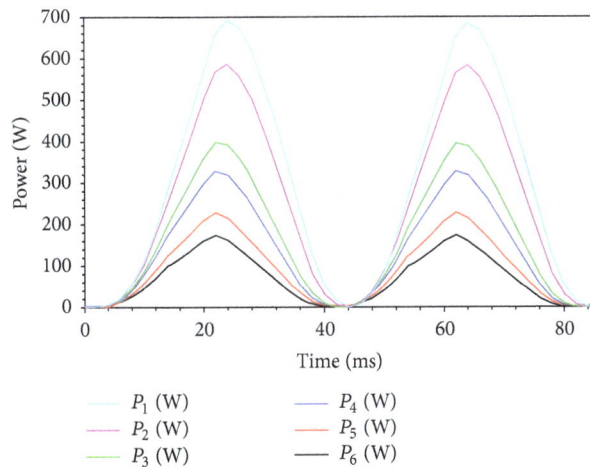

P_1 (W) P_4 (W)
P_2 (W) P_5 (W)
P_3 (W) P_6 (W)

FIGURE 17: Generated power for different loads.

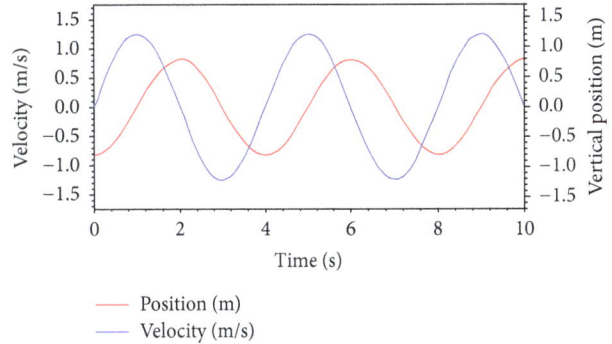

Position (m)
Velocity (m/s)

FIGURE 18: Position and velocity of the translator.

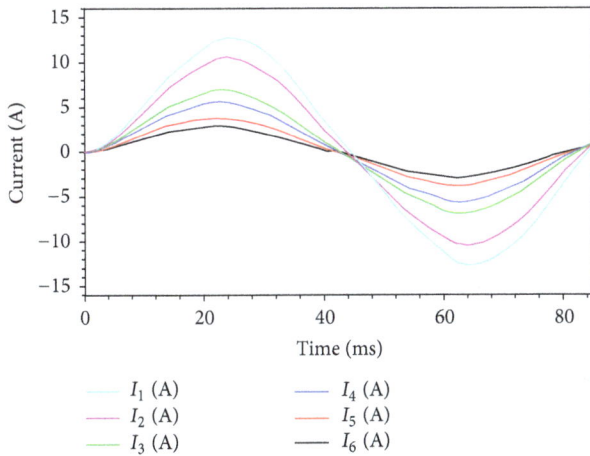

FIGURE 19: Generated voltage.

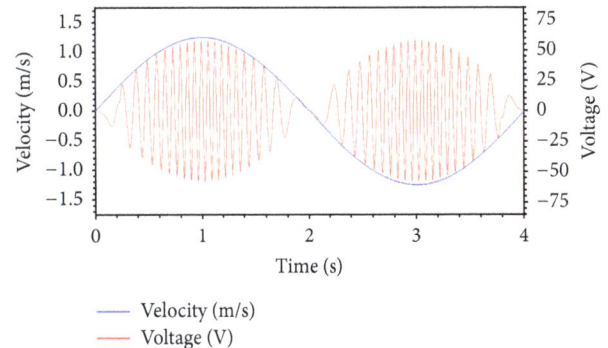

Velocity (m/s)
Voltage (V)

FIGURE 20: Relation between induced voltage and speed.

vertical velocity of the wave elevation is a cosine function. The vertical position and velocity of the translator are shown in Figure 18. The time period $T = 4$ s and $L_S = 1.6$ m which means $A_w = 0.8$ m. The magnitude of the generated voltage of

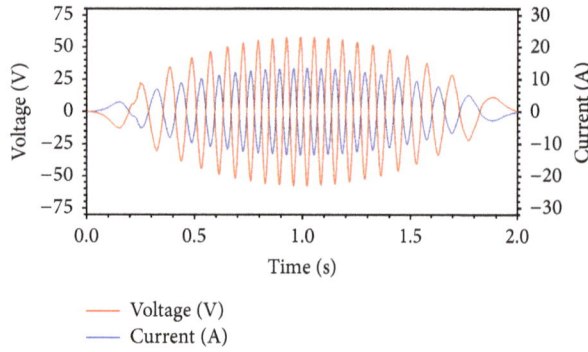

FIGURE 21: The induced voltage and current.

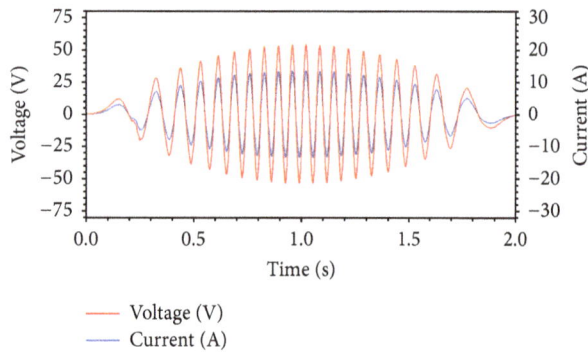

FIGURE 22: Magnetic flux linkage and terminal voltage.

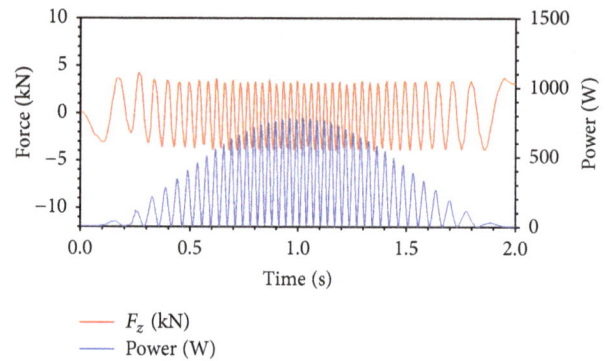

FIGURE 23: Terminal voltage and current.

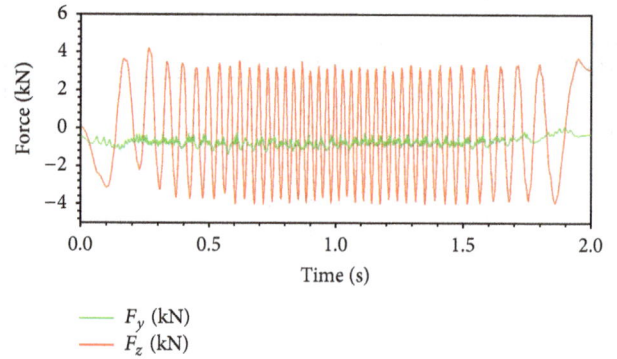

FIGURE 24: Different forces of the PMLG.

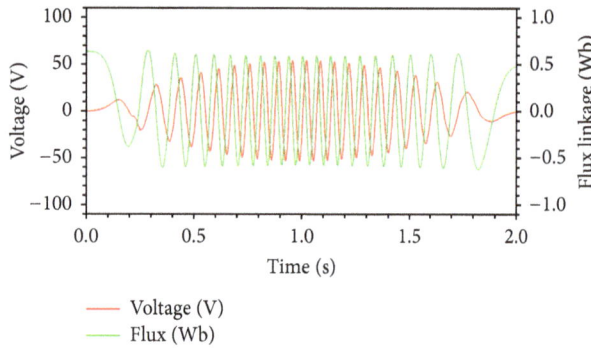

FIGURE 25: Applied force and power.

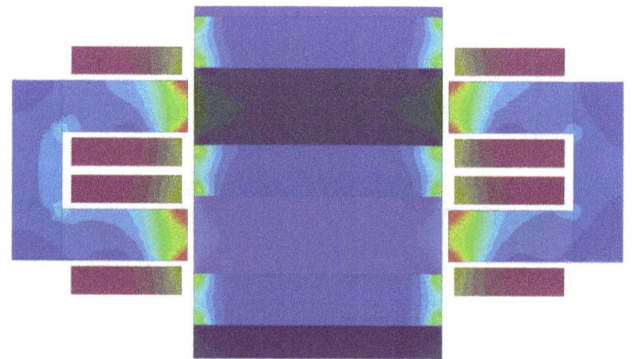

FIGURE 26: Magnetic flux density at $t = 20$ ms.

the PMLG depends on velocity according to (20) and is shown in Figure 19.

As $T = 4$ s, the voltage waveform is repeated in each half cycle or $T/2 = 2$ s as shown in Figure 19. The voltage waveform due to the translator velocity for the time period of one cycle is shown in Figure 20. The induced voltage and current for a half-cycle time interval are shown in Figure 21. The terminal voltage lags by 90° from the magnetic flux linkage as shown in Figure 22.

The voltage magnitude is directly proportional to the translator velocity. According to (17) and (20), the frequency of voltage increases with the increase in velocity. The terminal voltage and current in a winding are shown in Figure 23. The cogging force is the reason of force ripples and is liable for

abnormal operation of the PMLG, so it is maintained at a lower value as shown in Figure 24 with the applied force. The magnitude of power is related to the frequency. The power generation due to the applied force is shown in Figure 25.

Different dimensions of the PMLG active materials of the stator and translator as shown in Figure 6 are given in Table 4. The significant parameters that mainly affect the PMLG performance are given in Table 5.

5.4. Selection of the PM Size. According to Figure 18, the vertical displacement of the translator is 1.6 m owing to the minimum and maximum vertical displacement of −0.8 m and 0.8 m, respectively. On the other hand, the translator

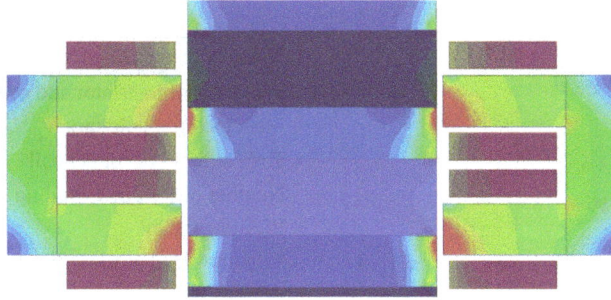

FIGURE 27: Magnetic flux density at $t = 30$ ms.

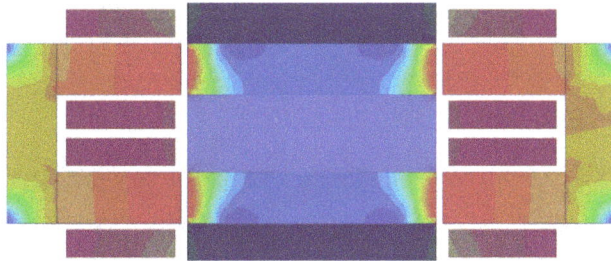

FIGURE 28: Magnetic flux density at $t = 40$ ms.

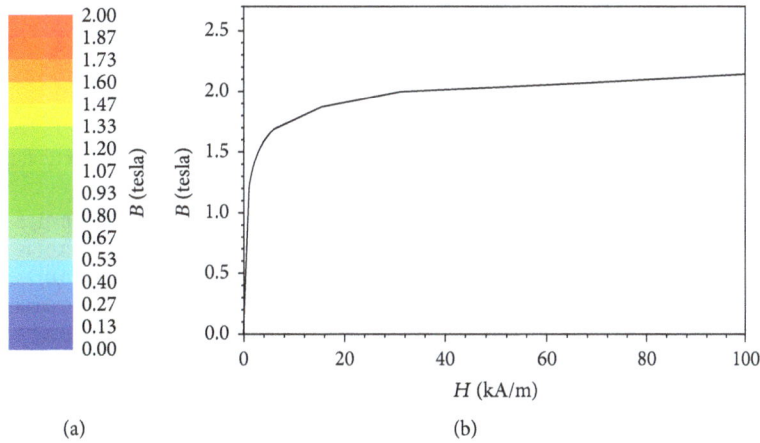

(a)

(b)

FIGURE 29: (a) Scale of B and (b) magnetizing curve.

TABLE 4: Dimensions of the PMLG.

Name of the item	Value	Unit
Width of the translator	8	cm
Width of the stator	5.6	cm
Width of the PMLG	19.6	cm
Thickness of pole shoe	1.6	cm
Depth of the PMLG	10	cm
Thickness of PM	2.4	cm
Cross section of the conductor	2.5	mm^2
Stroke length	1.6	m

length is the summation of stator length and stroke length. Combining the stroke length of 1.6 m and the stator length of 1.4 m, the translator length becomes 3 m. From Table 5, the translator and stator pole pitch is 4 cm which is the summation of the PM and the pole thickness. Therefore, 75 PMs and poles are required to make 3 m length of translator. The dimension of the translator PMs is selected as $8 \times 10 \times 2.4$ cm.

6. Analysis of Magnetic Flux

The magnetic flux density, B, flux lines, and magnetic field intensity of the PMLG are analyzed to observe B in the different portions. The value of B for different time, t, is shown in Figures 26–28. B varies for different positions of the translator which can be realized from the scale given in Figure 29(a). The magnetizing curve of the steel used in the stator and translator core is shown in Figure 29(b).

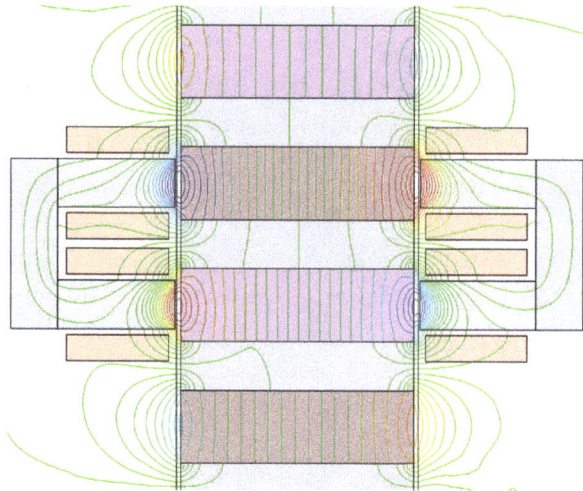

FIGURE 30: Magnetic flux at $t = 20$ ms.

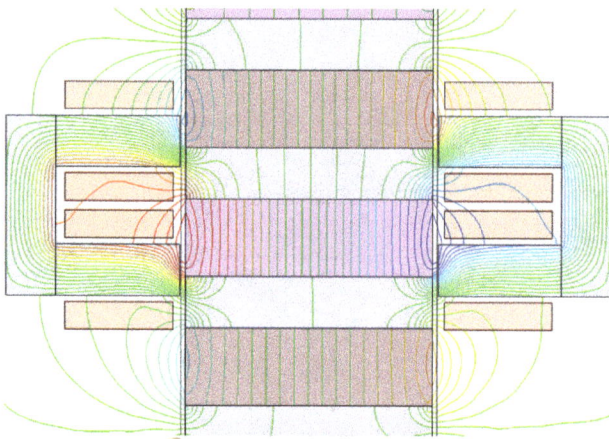

FIGURE 31: Magnetic flux at $t = 30$ ms.

TABLE 5: The parameters of the PMLG.

Name of the item	Value	Unit
Pole pitch of translator and stator	4	cm
Pitch factor of pole shoe	0.4	
Turn number of copper coil	70	Turns
Number of coils in a winding	2	
Winding factor	0.6	
Velocity/speed of the translator	0–1.5	m/s
Air gap length	2	mm
Internal resistance of windings	0.3	Ω
Load resistance	2–25	Ω
Maximum power (depends on some factors)	175–692	W

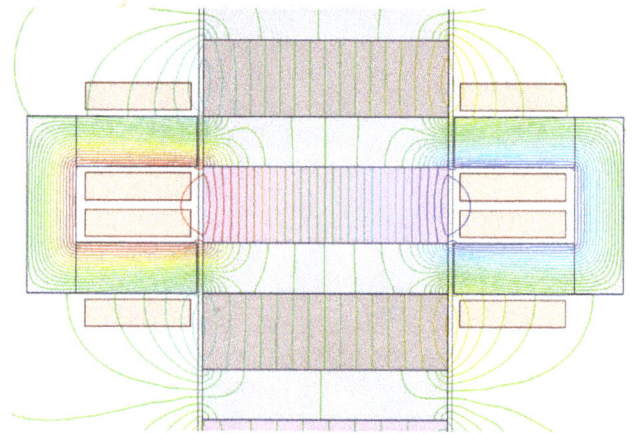

FIGURE 32: Magnetic flux at $t = 40$ ms.

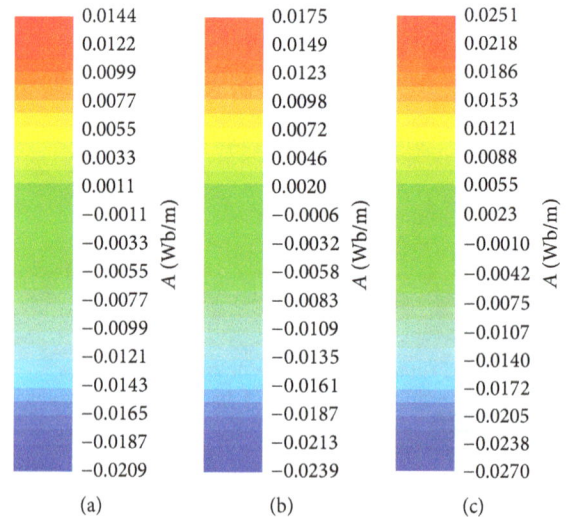

FIGURE 33: Scale of magnetic flux.

The translator position varies with time and the flux lines travel through the low reluctance path. The magnetic flux lines for different times are shown in Figures 30–32.

There is a relation between magnetic flux lines and B. Figures 33(a)–33(c) represent the scale of flux lines for $t = 20$ ms, 30 ms, and 40 ms, respectively. The magnetic properties inside the PMLG are needed for analyses, proper design, and performance check.

The values for the magnetic field intensity at different times are shown in Figures 34–36 and the scale is shown in Figure 37. The magnetomotive force mainly exists in the air gap. When the stator and translator cores come close together, the magnetomotive force reaches a high value. Neodymium iron boron (NdFeB) permanent magnets have been used for magnetic excitation.

7. Conclusions

The simulation results with synchronized translator vertical velocity with wave velocity reflect the model presented in this paper. Different voltages, currents, power, and magnetic flux linkages of phase-A are shown for the three-phase PMLG because the other two phases are just 120° phase shifted from each other and are not necessary to explain the dynamics of energy conversion from wave energy to electrical energy. The cogging force and force ripples are low compared to applied force that helps to prevent mechanical vibration and also

FIGURE 34: Magnetic field intensity at $t = 20$ ms.

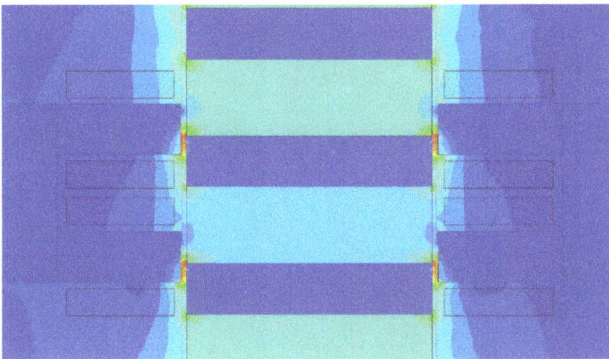

FIGURE 35: Magnetic field intensity at $t = 30$ ms.

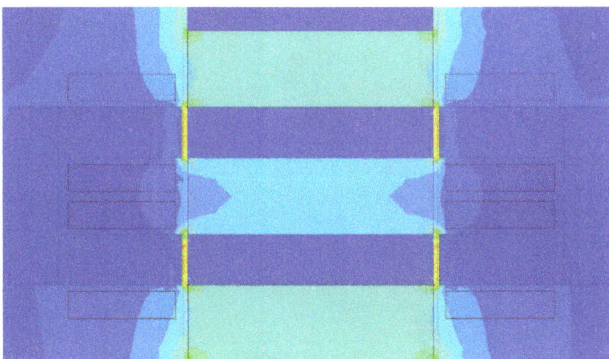

FIGURE 36: Magnetic field intensity at $t = 40$ ms.

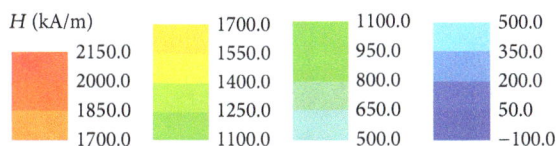

H (kA/m)			
	1700.0	1100.0	500.0
2150.0	1550.0	950.0	350.0
2000.0	1400.0	800.0	200.0
1850.0	1250.0	650.0	50.0
1700.0	1100.0	500.0	−100.0

FIGURE 37: Scale of H.

minimize mechanical power loss. Low core loss is achieved (1.27%) which would produce less heat to prevent overheating of the PMLG and increase service life. The maximum efficiency is calculated as 83% for full-load condition. The loading effect is low and voltage regulation of the PMLG is 13.93% measured from terminal characteristics. Thick copper conductor with smaller turn number is used in the coil to minimize internal resistance.

Competing Interests

The authors declare that there are no competing interests regarding the publication of this paper.

References

[1] H. Polinder, M. E. C. Damen, and F. Gardner, "Linear PM generator system for wave energy conversion in the AWS," *IEEE Transactions on Energy Conversion*, vol. 19, no. 3, pp. 583–589, 2004.

[2] P. Tokat, T. Thiringer, and P. Chen, "Development of an analytically described pitch regulator for a wind turbine to be used for grid disturbance studies," *Journal of Energy*, vol. 2013, Article ID 203174, 9 pages, 2013.

[3] K. Mohammadi, A. Mostafaeipour, Y. Dinpashoh, and N. Pouya, "Electricity generation and energy cost estimation of large-scale wind turbines in Jarandagh, Iran," *Journal of Energy*, vol. 2014, Article ID 613681, 8 pages, 2014.

[4] M. R. Islam, Y. Guo, J. G. Zhu, H. Y. Lu, and J. X. Jin, "High-frequency magnetic-link medium-voltage converter for superconducting generator-based high-power density wind generation systems," *IEEE Transactions on Applied Superconductivity*, vol. 24, no. 5, Article ID 5202605, 2014.

[5] B. Jain, S. Singh, S. Jain, and R. K. Nema, "Flexible mode control of grid connected wind energy conversion system using wavelet," *Journal of Energy*, vol. 2015, Article ID 152898, 12 pages, 2015.

[6] A. O. Adelaja and B. I. Babatope, "Analysis and testing of a natural convection solar dryer for the tropics," *Journal of Energy*, vol. 2013, Article ID 479894, 8 pages, 2013.

[7] M. R. Islam, Y. G. Guo, and J. G. Zhu, "A multilevel medium-voltage inverter for step-up-transformer-less grid connection of photovoltaic power plants," *IEEE Journal of Photovoltaics*, vol. 4, no. 3, pp. 881–889, 2014.

[8] J. Basheer Sheeba and A. Krishnan Rohini, "Structural and thermal analysis of asphalt solar collector using finite element method," *Journal of Energy*, vol. 2014, Article ID 602087, 9 pages, 2014.

[9] O. S. Omogoye, A. B. Ogundare, and I. O. Akanji, "Development of a cost-effective solar/diesel independent power plant for a remote station," *Journal of Energy*, vol. 2015, Article ID 828745, 10 pages, 2015.

[10] D. Borkowski and T. Wegiel, "Small hydropower plant with integrated turbine-generators working at variable speed," *IEEE Transactions on Energy Conversion*, vol. 28, no. 2, pp. 452–459, 2013.

[11] M. Sarvi, I. Soltani, N. NamazyPour, and N. Rabbani, "A new sliding mode controller for DC/DC converters in photovoltaic systems," *Journal of Energy*, vol. 2013, Article ID 871025, 7 pages, 2013.

[12] M. R. Islam, Y. G. Guo, and J. G. Zhu, "A high-frequency link multilevel cascaded medium-voltage converter for direct grid integration of renewable energy systems," *IEEE Transactions on Power Electronics*, vol. 29, no. 8, pp. 4167–4182, 2014.

[13] R. Watanabe, J.-S. Shin, T. Koseki, and H.-J. Kim, "Optimal design for high output power of transverse-flux-type cylindrical linear synchronous generator," *IEEE Transactions on Magnetics*, vol. 50, no. 11, Article ID 8202704, 2014.

[14] C. Liu, H. Yu, M. Hu, Q. Liu, S. Zhou, and L. Huang, "Research on a permanent magnet tubular linear generator for direct drive wave energy conversion," *IET Renewable Power Generation*, vol. 8, no. 3, pp. 281–288, 2014.

[15] F. Wu, X.-P. Zhang, P. Ju, and M. J. H. Sterling, "Modeling and control of AWS-based wave energy conversion system integrated into power grid," *IEEE Transactions on Power Systems*, vol. 23, no. 3, pp. 1196–1204, 2008.

[16] T. Setoguchi, S. Raghunathan, M. Takao, and K. Kaneko, "Air-turbine with self-pitch-controlled blades for wave energy conversion (estimation of performances in periodically oscillating flow)," *International Journal of Rotating Machinery*, vol. 3, no. 4, pp. 233–238, 1997.

[17] M. Takao, Y. Kinoue, T. Setoguchi, T. Obayashi, and K. Kaneko, "Impulse turbine with self-pitch-controlled guide vanes for wave power conversion (effect of guide vane geometry on the performance)," *International Journal of Rotating Machinery*, vol. 6, no. 5, pp. 355–362, 2000.

[18] M. Takao and T. Setoguchi, "Air turbines for wave energy conversion," *International Journal of Rotating Machinery*, vol. 2012, Article ID 717398, 10 pages, 2012.

[19] A. Thakker, J. Jarvis, and A. Sahed, "Quasi-steady analytical model benchmark of an impulse turbine for wave energy extraction," *International Journal of Rotating Machinery*, vol. 2008, Article ID 536079, 12 pages, 2008.

[20] B. Drew, A. R. Plummer, and M. N. Sahinkaya, "A review of wave energy converter technology," *Proceedings of the Institution of Mechanical Engineers, Part A: Journal of Power and Energy*, vol. 223, no. 8, pp. 887–902, 2009.

[21] L. Huang, J. Liu, H. Yu, R. Qu, H. Chen, and H. Fang, "Winding configuration and performance investigations of a tubular superconducting flux-switching linear generator," *IEEE Transactions on Applied Superconductivity*, vol. 25, no. 3, 2015.

[22] A. J. Garrido, E. Otaola, I. Garrido et al., "Mathematical modeling of oscillating water columns wave-structure interaction in ocean energy plants," *Mathematical Problems in Engineering*, vol. 2015, Article ID 727982, 11 pages, 2015.

[23] Z. Liu, Y. Cui, H. Zhao, H. Shi, and B.-S. Hyun, "Effects of damping plate and taut line system on mooring stability of small wave energy converter," *Mathematical Problems in Engineering*, vol. 2015, Article ID 814095, 10 pages, 2015.

[24] M.-J. Jin, C.-F. Wang, J.-X. Shen, and B. Xia, "A modular permanent-magnet flux-switching linear machine with fault-tolerant capability," *IEEE Transactions on Magnetics*, vol. 45, no. 8, pp. 3179–3186, 2009.

[25] L. Huang, H. Yu, M. Hu, J. Zhao, and Z. Cheng, "A novel flux-switching permanent-magnet linear generator for wave energy extraction application," *IEEE Transactions on Magnetics*, vol. 47, no. 5, pp. 1034–1037, 2011.

[26] L. Huang, H. Yu, M. Hu, C. Liu, and B. Yuan, "Research on a tubular primary permanent-magnet linear generator for wave energy conversions," *IEEE Transactions on Magnetics*, vol. 49, no. 5, pp. 1917–1920, 2013.

[27] V. D. Colli, P. Cancelliere, F. Marignetti, R. Di Stefano, and M. Scarano, "A tubular-generator drive for wave energy conversion," *IEEE Transactions on Industrial Electronics*, vol. 53, no. 4, pp. 1152–1159, 2006.

[28] J. F. Pan, Y. Zou, N. Cheung, and G.-Z. Cao, "On the voltage ripple reduction control of the linear switched reluctance generator for wave energy utilization," *IEEE Transactions on Power Electronics*, vol. 29, no. 10, pp. 5298–5307, 2014.

[29] E. S. Vidal, R. H. Hansen, and M. Kramer, "Early performance assessment of the electrical output of avestar's prototype," in *Proceedings of the 4th International Conference on Ocean Energy (ICOE '12)*, Dublin, Ireland, October 2012.

[30] K. Lu and W. Wu, "Electromagnetic lead screw for potential wave energy application," *IEEE Transactions on Magnetics*, vol. 50, no. 11, Article ID 8205004, 2014.

[31] R. Salmon, *Introduction to Ocean Waves*, Scripps Institution of Oceanography, University of California, San Diego, Calif, USA, 2008.

[32] P. R. M. Brooking and M. A. Mueller, "Power conditioning of the output from a linear vernier hybrid permanent magnet generator for use in direct drive wave energy converters," *IEE Proceedings—Generation, Transmission and Distribution*, vol. 152, no. 5, pp. 673–681, 2005.

5

Modelling of Sudan's Energy Supply, Transformation, and Demand

Ali A. Rabah, Hassan B. Nimer, Kamal R. Doud, and Quosay A. Ahmed

Energy Research Centre, Faculty of Engineering, University of Khartoum, P.O. Box 321, Khartoum, Sudan

Correspondence should be addressed to Ali A. Rabah; rabahss@hotmail.com

Academic Editor: Ciro Aprea

The study aimed to develop energy flow diagram (Sankey diagram) of Sudan for the base year 2014. The developed Sankey diagram is the first of its kind in Sudan. The available energy balance for the base year 2012 is a simple line draw and did not count the energy supply by private and mixed sectors such as sugar and oil industries and marine and civil aviation. The private and mixed sectors account for about 7% of the national grid electric power. Four energy modules are developed: resources, transformation, demand, and export and import modules. The data are obtained from relevant Sudanese ministries and directorates and Sudan Central Bank. "e!Sankey 4 pro" software is used to develop the Sankey diagram. The main primary types of energy in Sudan are oil, hydro, biomass, and renewable energy. Sudan has a surplus of gasoline, petroleum coke, and biomass and deficit in electric power, gasoil, jet oil, and LPG. The surplus of gasoline is exported; however, the petroleum coke is kept as reserve. The deficit is covered by import. The overall useful energy is 76% and the loss is 24%. The useful energy is distributed among residential (38%), transportation (33%), industry (12%), services (16%), and agriculture (1%) sectors.

1. Introduction

Figure 1 shows the energy Sankey diagram of USA as an example. The Sankey diagram is an important tool to visualize the energy balance for a system or a country or a region. The Sankey diagram depicts the energy flows from supply to demand taking into account transformation. Sankey diagram was developed over 100 years ago by the Irish engineer Riall Sankey to analyze the thermal efficiency of steam engines. Since then, it has been widely used. Besides visualization, Sankey diagram is a vital tool to identify sources of inefficiency and potential saving in the energy system.

For the preparation of Sankey diagram for a country, four modules are needed. The first is the demand module, which contains the details of the demand for end-use energy (both primary and secondary fuel) for the residential, services, industrial, agricultural, and transportation sectors [1]. The demand for each sector from primary and secondary energy is defined. For example, the demand of residential sector is electricity, oil products, and biomass for lighting, cooking, and HVAC. The second module is the transformation module. It consists of all energy transformation processes, such as

electricity generation, oil refining, and charcoal conversion. In these modules, the energy is divided into useful energy and lost energy. The useful energy is then distributed to demand sectors. The third module deals with the available resources. The main energy sources are coal, crude oil, natural gas, hydroelectricity, biomass, nuclear energy, and renewable energy such as solar, wind, and geothermal energy. The fourth module deals with energy import and export. There exist a number of models used to develop the energy supply and demand modules. Prominent energy models include MARKAL, LEAP, ENERGY 2020, MAPLE C, NEMS, and MAED. Details on these models and their implementation can be found in a number of published studies [2–4].

This study is aimed at providing Sankey diagram for Sudan for base year 2014. What is available now in Sudan is energy balance for 2012 [7] (cf. Table 1). However, the available energy balance data did count only the energy produced by public sectors. The energy produced by private and mixed sectors is not counted. These include the electric energy produced by oil refineries and sugar factories which constitute a significant share of Sudan energy mix. It also did not count for bagasse and bioethanol. Likewise, it does

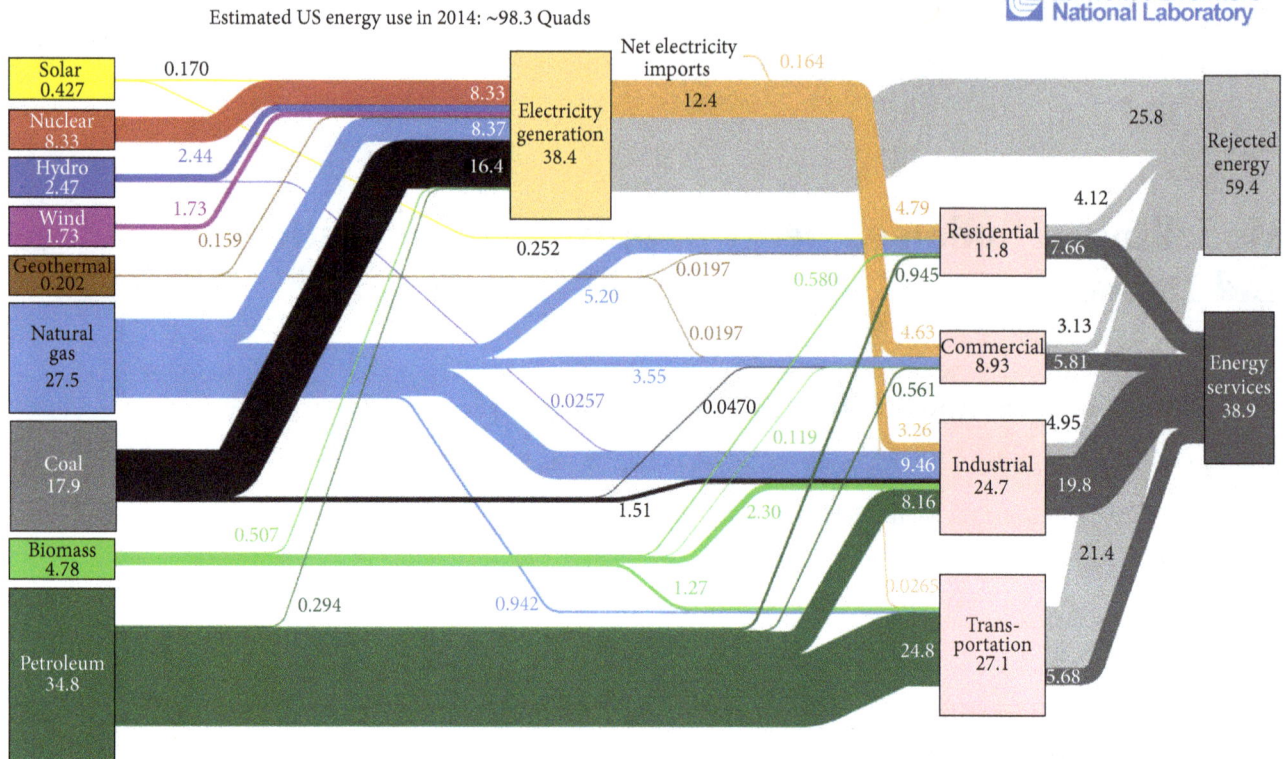

FIGURE 1: Sankey diagram of USA. Source: [5]. Data is based on DOE/EIA-0035(2015-03), March 2014. If this information or a reproduction of it is used, credit must be given to the Lawrence Livermore National Laboratory and the Department of Energy, under whose auspices the work was performed. Distributed electricity represents only retail electricity sales and does not include self-generation. EIA reports consumption of renewable resources (i.e., hydro, wind, geothermal, and solar) for electricity in BTU-equivalent values by assuming a typical fossil fuel plant "heat rate." The efficiency of electricity production is calculated as the total retail electricity delivered divided by the primary energy input into electricity generation. End-use efficiency is estimated as 65% for the residential and commercial sectors, 80% for industrial sector, and 21% for the transportation sector. Totals may not equal sum of components due to independent rounding. LLNL-MI-410527.

TABLE 1: Sudan energy balance 2012 [6].

Demand sectors	Power		Oil		Biomass		Total	
	ktoe	%	ktoe	%	ktoe	%	ktoe	%
Residential	401	54.3	298	7.9	3088	62.2	3911	40.0
Transportation			2994	79.2			3073	31.4
Services	181	24.5	43	1.1	1303	26.2	1579	16.1
Industry	120	16.3	400	10.6	575	11.6	1133	11.6
Agriculture	36	4.9	43	1.1			85	0.9
Total	738	100	3778	100	4966	100	9781	100
Share of energy supply	7.5		38.6		50.8		100.0	

not count the petroleum associated gas (AG) and renewable energy, in particular photovoltaic energy. The most difficult part of this work is the data quality check. There exist a number of reports on Sudan's energy status. These reports include conflicting data, double counts, and inconsistent units. The authors have extensively verified the data from their original sources. Data on renewable energy status in Sudan is scarce as well. Extensive search including field survey was made to collect data on this area. Data on the demand side is not well documented as well and hence

enormous effort was exerted to collect the data from their original sources. The main contribution of this work is to make proper documentation of Sudan energy data on both supply and demand side. The other contribution is to produce Sudan's energy flow diagram for the first time.

2. Methodology

2.1. Data Collection. The energy supply, demand, and transformation data are obtained from the following reports:

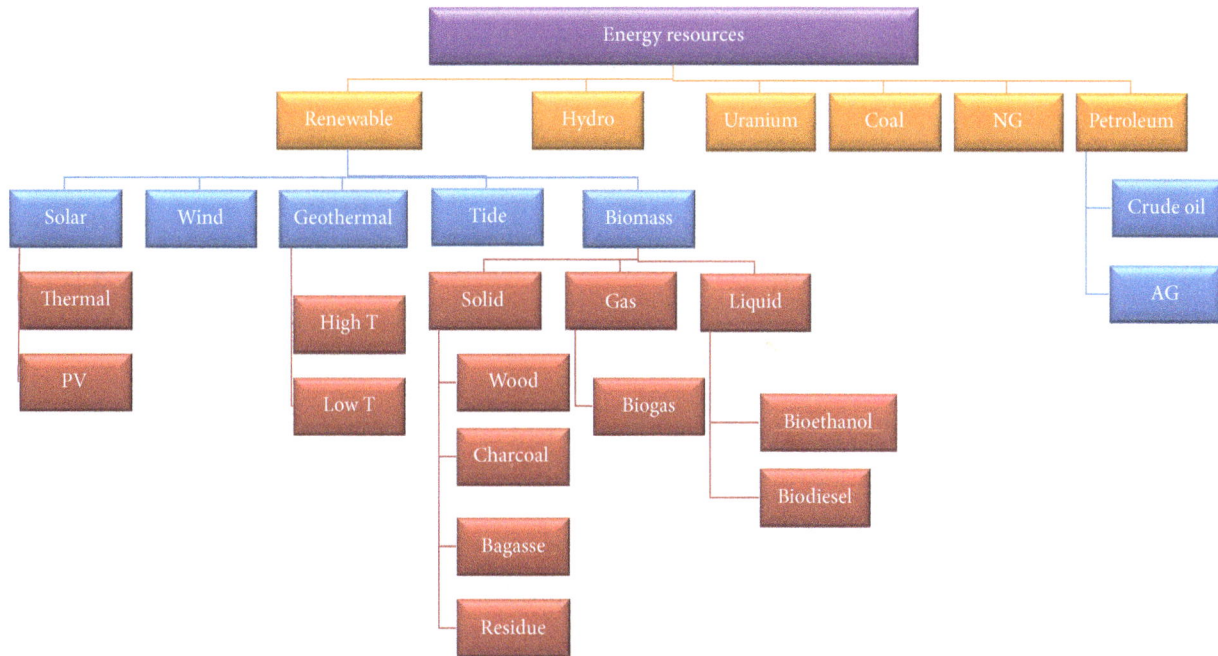

FIGURE 2: Primary energy resources.

(1) Annual report of Sudan Central Bank.

(2) Annual reports from the Ministry of Oil and Gas [6, 7].

(3) Annual report from the Directorate of Agriculture and Forestry [8].

(4) Annual report from Sudanese thermal power generation corporation [9].

The collected data is further subjected to quality check for completeness, avoiding double counting and uncertainty assessment.

2.1.1. Completeness. The available reports on electric power generation in Sudan account only for power produced by the public sector. The power generation of private and mixed sectors is not counted by the published reports. For example, the power generation of petroleum refineries and sugar factory, as mixed sectors, is not accounted for in Sudan energy balance for 2012 [6]. This data gap is filled using field survey. There also exists no inventory of standby generators countrywide.

2.1.2. Avoiding Double Counting. Some data duplicate is found in the available literature. For example, ORC produced Naphtha which is used as feedstock to KRC to produce gasoline and other products; hence, it is double counted in Sudan energy balance for 2012 and the reports of Central Bank and the Ministry of Oil and Gas. The reports of ORC and KRC refineries are considered rather than Sudan Central Bank's reports.

2.1.3. Uncertainty Assessment. There is a lack of proper documentation of bagasse, animal waste, and agriculture waste. However, there are data of production of sugar, cereals, crops, cotton, and count of animals. The waste is estimated from production data. For example, bagasse is estimated from sugar production. The sugar recovery is taken as 9% of the crushed sugar cane and the bagasse is taken as 39.5% of the crushed sugar cane. The estimated bagasse is cross-checked by field survey of the sugar plants and it is found that the assumptions are accurate within 5% for all sugar factories. Similarly, the animal dung is estimated from animal count. The dung production is taken as 10.95, 1.83, 1.83, and 5.48 tons/animal/head for cows, sheep, goats, and camels, respectively [10]. The dung that is potential for energy use is taken as 10%. The agricultural waste is estimated using harvest index (HI) [11]. The harvest index for cereals, oil crops, and cotton is taken as 0.4, 0.52, and 0.16, respectively. The residue is calculated as amount of products multiplied by $[(1 - \text{HI})/\text{HI}]$. The other challenge is that the portion of agriculture residue used as fuel is unknown as the residue is shared by many applications such as animal feed, building material, and pulp and paper industry. This is on the one hand. On the other hand, the residue is mostly left in the farm as a fertilizer or burned during land preparation. Under this circumstance, the ratio of agriculture waste accepted by Sudan energy balance 2012 was considered.

2.2. Energy Modelling. Although there exist a number of energy modelling software programs such Long Range Energy Alternative Planning Systems (LEAP), in this work, a simple Excel worksheet is used for energy modelling.

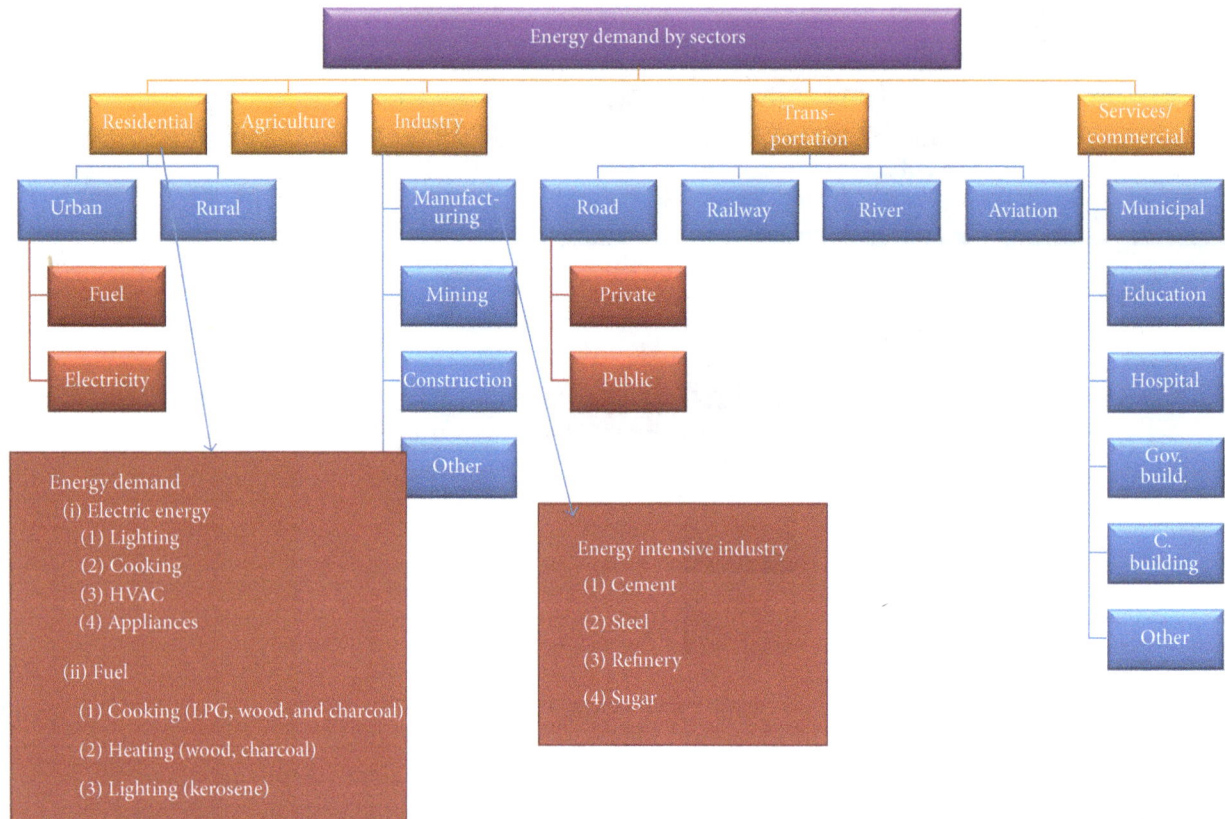

FIGURE 3: Energy demand by sector.

The energy framework for development of Sankey diagram consists of four modules [1].

(1) Energy resource module: this module takes account of all the primary fuel such as petroleum, coal, natural gas, uranium, biomass, and renewable energy (cf. Figure 2).

(2) Transformation module: the transformation module handles data on the conversion of primary fuel into secondary fuel such as power generation and conversion of crude oil in the refinery to secondary fuels.

(3) Demand module: the demand module contains the details regarding the end-use energy demand for both primary and secondary fuel. The consumption sectors include agriculture, industry, services, residential, and transportation sectors (cf. Figure 3).

(4) Import and export module: this module deals with energy import and export. The energy import and export are normally in the end-use form such as electricity and secondary fuel oil.

The data generated by the four modules for the base year 2014 is used to develop the energy Sankey diagram for Sudan. The software "e!Sankey 4 pro" was used. e!Sankey is widely used software for the development of Sankey diagrams for a country or a region.

3. Results and Discussion

The main sources of primary energy in Sudan are oil, hydroelectricity, biomass, and renewable energy. Coal, natural gas, and uranium are nonexistent in Sudan. The main transformation and conversion processes are electric power generation, oil refinery, and wood-to-charcoal conversion.

3.1. Electric Energy Supply and Demand

3.1.1. Hydroelectric Power. Table 2 shows the hydroelectric power generation plants in Sudan. The installed capacity, nominal capacity (MW), expected GWh, and year of establishment are given. The planned and under-construction hydroelectric power plants are also given. The total installed and potential hydroelectric power in Sudan is 4176 MW; the installed capacity is 1585 MW (38%).

3.1.2. Thermal Electric Power Generation. Table 3 shows the installed, under-construction, and planned thermal power plants. The installed power is 1400 MW and the under-construction and planned power are 405 MW and 600 MW, respectively. The installed thermal power generation is about 46% of the total installed public power generation (hydro + thermal). One important point in thermal generation is its diversity in prime movers and fuels sources. The prime movers in thermal generation include steam turbine (ST),

TABLE 2: Hydroelectric power plant (operational, under construction, and planned).

Number	Name	Year	Installed MW	Capacity Nominal MW	Production GWh
			Installed MW	Nominal MW	Production GWh
		Operational plant			
1	Merowe Dam	2009	1250	1240	5580
2	Roseires Dam	1966	280	270	1050
5	Sennar Dam	1962	15	12	49
3	Jebel Aulia Dam	2003	30	19	55
4	Khasm El Girba Dam	1964	10	10	15
	Subtotal A		*1585*	*1551*	*6749*
		Under-construction plant			
5	Upper Atbra and Sitat	2015	323	320	834
6	Sennar upgrading	2015	11	13.7	66
		Planned plant			
7	Shereik		420		2103
8	Kajbar		360		1799
9	Sabaloka		205		866
10	Dal Low		648		2185
11	Dagash		312		1349
12	Mograt		312		1214
	Subtotal B		*2257*		*9515*
	Total (A + B)		*384*		*16264*
	Available power%		*38*		*39*

TABLE 3: Thermal power plant.

Number	Thermal power plant	Year	Prime mover	Fuel	Installed MW	Capacity Nominal MW	Available GWh
					Installed MW	Nominal MW	Available GWh
		Operational thermal power plants					
1	Shahid Mahmoud Sharief 1 + 2 ST	1984	ST[1]	HCGO[2]	60	28	212
2	Shahid Mahmoud Sharief 3 + 4 ST	1994	ST	HFO[3] + HCGO	120	110	840
3	Shahid Mahmoud Sharief 5 + 6 ST	2011	ST	HFO	200	190	1408
4	Shahid Mahmoud Sharief 1 + 2 GT	1992	GT[4]	GO[5]	50	34	265
5	Garri1: CCGT	2003	CCGT[6]	GO + LPG	180	170	1346
6	Garri2: CCGT	2003	CCGT	GO + LPG	180	170	1346
7	Garri4: ST	2010	ST	Pet coke	110	100	675
8	Kusti	2013	ST	Crude oil	500	470	3590
9	Diesel generation		IC	Diesel			
	Available thermal power A				*1400*	*1272*	*9682*
		Under-construction and planned thermal power plant					
9	Al Fula	2016	ST	AG/NG[7]	405	381	2908
10	Red Sea	2016	ST	Coal	600	534	4079
	Planned thermal power cycle B				*1005*	*915*	*6987*
	Total (A + B)				*2405*	*2187*	*16669*

[1]ST: steam turbine.
[2]HCGO: heavy coker gas oil (heavy gasoil).
[3]HFO: heavy fuel oil (equivalent to number 6 fuel oil).
[4]GT: gas turbine.
[5]GO: gasoil.
[6]CCGT: combined cycle gas turbine.
[7]AG: petroleum associated gas; NA: natural gas.

TABLE 4: Electric power profile (GWh).

Year	Hydro	Steam turbine	Gas turbine	Diesel	Combined	Import	Production	Hydro%	Import%
2014	8914	1400	0	202	864	469	11848	75	4
2013	8317	1205	0	183	582	320	10607	78	3
2012	6619	1484	5	182	1145	74	9509	70	1
2011	6452	1631	1	161	210	0	8455	76	0
2010	6199	473	14	192	620	0	7499	83	0
2009	3236	887	94	309	1847	0	6372	51	0
2008	1466	1106	152	442	2340	0	5506	27	0
2007	1457	1057	371	140	1997	0	5021	29	0
2006	1368	1108	501	353	1192	0	4521	30	0
2005	1239	1047	319	385	1135	0	4125	30	0
2004	1107	1037	1256	349	0	0	3749	30	0
2003	1163	1168	210	328	0	0	3354	35	0
2002	1287	116	381	310	0	0	3094	42	0

TABLE 5: Oil industry thermal power generation plant.

Number	Oil industry power plant	Year	Prime mover	Fuel	kW
1	KRC (2×12 MW)	1999	ST	FG[9] + bunker	24000
2	KRC (2×12 MW)	2006	GT	FG + bunker	12000
3	ORC 50 Hz	1999	IC	Diesel	250
4	ORC 60 Hz[8]	1999	IC	Diesel	500
5	Heglig oil field 1	na	IC	Diesel	25.5
6	Heglig oil field 2	na	IC	Diesel	11.5
7	Neem oil field	na	IC	Diesel	11.2
8	Canar oil field	na	IC	Diesel	0.8
9	Diffra oil field	na	IC	Diesel	3.9
10	Other (mainly the power for the oil pipeline)	na	IC	Diesel	1000
	Total				37803

[8]FG: fuel gas mainly CH_4.
[9]Since 2014, it has become standby generator as the refinery is connected to national grid of 50 Hz.

gas turbine (GT), combined cycle of gas turbine (CCGT), and internal combustion (IC) engines. IC engines are used in remote areas that are not covered by the national grid. The fuels used include gasoil (GO), diesel (blend of gasoil and gasoline), heavy fuel oil (HFO) also called furnace and number 6 fuel oil, heavy coker gasoil (HCGO), crude oil, petroleum coke, and liquefied petroleum gas (LPG). Diversity is a merit; however, the use of gasoil in steam turbine is the weak point of thermal generation sector in Sudan. Gasoil is competitive oil as it is used by transportation and agriculture sectors, in addition to its high cost relative to heavy fuel oil and petroleum coke. But it seems that the use of gasoil in ST and GT generation is an issue of "energy security" rather than economic factors; otherwise, the use of gasoil is not a common practice in ST and GT power generation. Table 4 shows the electric power generation in Sudan in the period from 2002 to 2014. The hydroelectric power share increased from 42% in 2002 to more than 75% in 2014. This is mainly due to the inception of Merowe Dam in 2009. The imported electric power from Ethiopia increased to 469 GWh (4%) in 2014.

3.2. Private and Mixed Sectors Power Generation.
There exist a number of private and mixed industries that produce their own power consumption. The major industries are as follows:

(1) Oil refineries.

(2) Oil fields production facilities.

(3) Sugar industry.

(4) Cement industry.

(5) Sea port and civil aviation.

(6) Others.

The power production by these sectors is not accounted for in Sudan energy balance 2012.

3.2.1. Oil Refinery.
Table 5 shows the installed power of oil refinery and oil fields. The total installed power by oil industry is about 38 MW. A large portion of KRC refinery power generation is from waste heat and low grade fuel such as vent gas and oil slope (slurry). The electric power produced by oil

TABLE 6: Sugar industry power plant [12].

Number	Factory	Year	Sugar capacity		Bagasse[10]		Power plant		
			Installed	Current					
			kton	kton	kton	ktoe	MW	D/Y	GWh
1	Kenana	1981	300	307.6	1350	580	20	210	326
2	Sennar	1976	70	73.1	321	138	6.5	210	33
3	Assalaya	1980	70	65.4	287	123	6.5	210	33
4	Guneid	1962	70	73.1	321	138	6.0	210	30
5	New Halfa	1966	70	59.5	261	112	6.0	210	30
6	White Nile[11]	2012	450	73.4	322	138	(104) 6.0[12]	210	33
	Total		*980*	*652.1*	*2862*	*1230*	*(190) 96*		*485*

[10] Bagasse sugar ratio: 4.4.
[11] The design capacity of power plant is 104 MW.
[12] Personnel contact.

TABLE 7: Solar radiation and sunshine duration in Sudan [13].

Station	Mean temperature	Sunshine duration	Solar radiation
	C	h	MJ/m^2/day
Port Sudan	28.4	9.0	20.87
Shambat	29.7	9.9	22.82
Wadi Medani	28.4	9.8	22.84
ElFashir	25.8	9.6	22.80
Abu Na'ama	28.8	8.8	21.90
Ghazala Gawazat	27.2	9.3	21.72
Dongala	27.2	10.5	24.06
Toker	28.8	7.3	17.60
Zalingi	24.5	8.8	22.98
Babanusa	28.2	8.9	21.30
Kadugli	27.5	8.5	21.30

industry is about 1% of the national grid electric power (the hydro and thermal power generation).

3.2.2. Sugar Industry. Sudan ranks high in sugar production among African countries. Sugar industry produces its own power consumption from bagasse, a byproduct of sugar cane. The installed sugar capacity is about 1.0 Mt of sugar/year. Table 6 shows the installed capacity of electric power plants in sugar factory. The installed capacity is about 190 MW, which is about 6% of the national grid electric power. However, the available capacity is 96 MW as the White Nile sugar factory is working at low capacity. Bagasse is sufficient to produce electric power for 210 days/year including cane crushing season of 180 days/year. Besides electric power, the factories also produce their own process heat for sugar cane mills, heating, and evaporation and crystallization process. The potential for high power generation in sugar plant is feasible. If cogeneration is installed, the process heat load will be reduced as the sugar mills and shredders will be electric driven rather than steam driven which in turn saves steam for power generation.

3.2.3. Standby Power Generation. All public utility services such as hospitals, universities, municipals, city water, and commercial buildings have their own standby IC electric power generators. However, there is lack of inventory data on standby generators countrywide. In this work, a field survey was made and about 100 standby IC generators with name plate capacity of about 1.2 kW with diesel as fuel source have been established. However, due to lack of information on the operation hours of the standby power generation, it is excluded in the present study. Besides services, air aviation and sea port also rely heavily on off-grid generation.

3.2.4. Photovoltaic. Table 7 shows the solar radiation in a number of Sudanese cities. The average sunshine duration is about 9 h. Most of the solar installation in Sudan is photovoltaic cell. The total installed capacity is about 2 MW. About 50% of the installed capacity is by telecommunication industry. All remote off-grid antennas and satellites are solar driven.

3.2.5. Summary of Power Supply and Demand. Table 8 shows the power supply by various power producers and consumption by various sectors. The hydropower represents the biggest share of about 70% despite the fact that the installed capacity is about 54% of the total installed power capacity. This is due to the high outage associated with thermal power

TABLE 8: Electric power supply and consumption in base year 2014.

	Power supply (ktoe)					Power demand by sectors (ktoe)					Total
	Hydro	Thermal	IC	Import	PV[13]	Residential	Industrial	Agriculture	Service	Trans.	
Public sector	767	677	17	40	0	534	151	52	283	0	1019
Oil industry		54		0	0	0	29	0	0	0	29
Sugar industry		61		0	0	0	61	0	0	0	61
Renewable		0		0	0.4	0	0	0	0.4	0	0
Total	767	285	17	40	0	534	240	52	283	0	1109
%	69.11	25.65	1.57	3.63	0.03	48.12	21.68	4.67	25.53	0.00	100.00

[13]PV: photovoltaic.

TABLE 9: Oil production.

Year	Oil production	Export
	MMBPD	(000 tonnes) BPD
2014	42.4	11,093
2013	45.1	15,837
2012	37.4	7,210

TABLE 10: Sudan refineries.

Number	Refinery	Capacity (000 tonnes) BPD	
		Installed	Available
1	Khartoum Refinery Corporation (KRC)	100	100
2	Port Sudan Refinery Corporation (PRC)	21.7	0
3	Obeid Refinery Corporation (ORC)	15	15
4	Concorp Refinery	10	0
5	Abu Gabra Refinery	2	0
	Total capacity	143.7	115

TABLE 11: ORC oil products (ton/year) [6].

Year	Fuel oil	Gasoil	Kerosene	Naphtha	Total ktoe
2010	361641.00	134446.00	38011.00	26981.00	405
2011	385330.00	141606.00	40405.00	27299.00	414
2012	385352.00	140899.00	41439.00	24245.00	148
2013	307604.00	114010.00	30937.00	19070.00	328
2014	339663.00	126600.00	32510.00	24547.36	312

and deficit in others. It has surplus in crude oil, gasoline, and pet coke. The surplus of crude and gasoline is for export while the surplus of pet coke is kept as reserve. The deficit of gasoil, LPG, and aviation is met from import. One of the causes of deficit in gasoil is its competitiveness. Gasoil is used by power generation, transportation, and agriculture sectors. As energy conservation measure fuel shift of gasoil is recommended for power generation and agriculture sector and gasoil is to be limited to transportation sector, heavy fuel oil is recommended for thermal power generation and total electrification of agricultural sector via using electric water pump for irrigation system rather than IC motors.

Table 17 shows summary of the oil supply and demand. The total supply is 7594 ktoe (7.594 Mtoe). This oil mix consists of crude oil, AG, and imported oil (gasoil, aviation, and LPG) as shown in Table 17. The imported oils represent about 19% of the total oil mix. 9.6% of the total supply is for thermal power generation including the electric power generated by KRC and ORC. The total energy mix is distributed as follows: 65.6% demand, 20.9% export, 12.9% energy conversion loss, and 1.32% flared gas. The demand is distributed to the following sectors: industry (11%), residential (8%), service (1%), transportation (79%), and agriculture (1%).

generation. The power demand by various sectors is also given. The residential sector has the biggest share of demand followed by the service sector.

3.3. Oil Supply and Demand. Table 9 shows Sudan's oil production in the last 3 years, after the separation of South Sudan. The portion of export is given as well. Besides oil production, a significant amount of associated gas (AG) is also produced. However, the AG is not utilized; it is flared. The historical and projected daily flared amount is about 15–17 MMSCFD from 2006 up to 2030 [14]. For this work, the minimum is considered (15 MMSCFD). Sudan has four oil refineries with a total installed capacity of 143.7 thousand BPD (cf. Tables 9 and 10). However, three of these refineries are out of service representing about 20% of the total installed refinery capacity.

Tables 11–17 show the fuel oil of ORC, KRC, import, total supply, total demand and supply and demand balance, and thermal power generation oil consumption, respectively. The main products include fuel oil, gasoil, kerosene, gasoline (Mogas), LPG, jet fuel, and petroleum coke. The surplus of benzene is for export. Sudan has surplus in some fuel oil

3.4. Biomass. The types of biomass used for energy in Sudan include

(1) firewood,

(2) charcoal,

(3) agriculture residue,

(4) bagasse,

(5) bioethanol,

(6) animal waste.

TABLE 12: KRC products (ton/year) [6].

Year	Crude	Fuel oil	Gasoil	Kerosene	Benzene	LPG	Jet fuel	HCGO	Pet coke	Total ktoe
2010	4265128	0	1860390	0	1241909	327413	133616	379768	322032	4379
2011	3854381	0	1651867	0	1150216	311840	133882	321413	285163	3964
2012	3738191	0	1655445	0	1099800	319690	117924	277230	268102	3850
2013	3403239	0	1515564	0	1009741	289363	79438	253612	255521	3501
2014	3405113	0	1379208	0	1067914	321453	107074	293368	236096	3573

TABLE 13: Fuel import (ton/year) [6].

Year	Gasoil	LPG	Aviation	Total
	Ton			ktoe
2010	337864	37760	136538	540
2011	653661	92282	150459	945
2012	628063	114013	109110	899
2013	898842	195335	82674	1244
2014	1159119	132903	66593	1427

3.4.1. Bagasse and Ethanol. Table 18 shows the sugar produced by 6 Sudanese sugar factories. Bagasse is estimated from the produced sugar. It is taken as 39.5% of crushed sugar cane and the sugar recovery is taken as 9% of the sugar cane. Hence, the bagasse sugar ratio is 4.4-ton bagasse/ton sugar. Table 18 also shows the bioethanol production. Bioethanol is produced from molasses. All bioethanol produced is from Kenana sugar company. It should be remembered that all bioethanol is for export.

3.4.2. Firewood and Charcoal. Tables 19 and 20 show the firewood and charcoal, respectively, obtained from the annual report of the Directorate of Agriculture and Forestry of Sudan. It is clear that there is large data missing/gap. When these data are compared with data in the open literature, a large variation is found. Hence, these data are not considered and literature data is rather considered.

Table 21 shows firewood and charcoal production obtained from the open literature for some years as a cross-check. The data of Sudan energy balance of 2012 is adopted in this work as it is close to that given by other references.

3.4.3. Agriculture Residues. Table 22 shows the cereals, oil crops, and cotton production. The agricultural waste is estimated using harvest index (HI). The harvest index for cereals, oil crops, and cotton is taken as 0.4, 0.52, and 0.16, respectively. The agriculture waste is calculated as the amount of crop multiplied by [(1 − HI)/HI] [11]. The portion of agriculture residue used as fuel is very small as the majority is used as animal feed and building material and in pulp and paper industry, among others. On the other hand, the agriculture waste is mostly burned during land preparation for the next season. The data is cross-checked against Sudan energy balance of 2012 and is found to be consistent.

3.4.4. Animal Waste. Sudan is rich with cows, sheep, goats, and camels, besides donkeys, horses, and chickens. Table 23 shows the animal counts. The animal dung is estimated from average dung production per head. The dung is generally used for biogas production. The dung available for biogas production is estimated by many researchers as 10% [16]. Currently, there are no reports for biogas production in Sudan. There is some biogas production in the 1980s; however, with the inception of crude oil production, biogas production was abandoned nationwide. Hence, the animal waste is not considered in the energy balance for the base year 2014 as there was no information or rates of use in the energy sector. Table 24 shows summary of biomass supply and demand.

3.5. Sankey Diagram. Sudan Sankey diagram has been developed using the energy supply, transformation, and demand data obtained for the base year 2014 (cf. Figure 4). The Sankey diagram gives a detailed flow pattern indicating the proportional consumption of primary and secondary energy in different energy demand sectors. It also shows the energy loss during transformation and conversion. The main features of the Sankey diagram are as follows:

(1) Primary energy: Sudan's primary energy consists of oil (39%), biomass (56%), hydroelectricity (5%), and a very small portion of solar energy (photovoltaic). Besides primary energy, Sudan imports 1427 ktoe of fuel oils (gasoil + aviation + LPG) and 40 ktoe of electricity from the neighbouring country of Ethiopia. The import is about 7% of Sudan's energy mix. However, natural gas, coal, nuclear energy, and other renewable energy sources are nonexistent in the country's primary energy mix, despite the huge potential of solar, wind, and geothermal energy. The high proportion of biomass is due to the large population located in rural areas (70% rural population). They have no access to national grid electricity, fuel gas, and kerosene and therefore are absolutely biomass dependent in meeting their demand for cooking, heating, and lighting.

(2) Transformation and conversion: the transformation processes in Sudan's energy balance are power generation (hydro, oil, and bagasse), refinery of crude oil to secondary fuels, and conversion of wood to charcoal. The transformation and conversion and other losses are about 25% of the end-use energy. These losses are thermodynamic loss and energy management

TABLE 14: Sudan fuel demand/consumption (ton) [6].

Year	Fuel oil	Gasoil	Kerosene	Benzene	LPG	AV	Pet coke	Diesel	Total
2010	555290	2380175	3018	824232	398347	235344	340858	26131	4763395
2011	531991	2358366	3497	838924	231382	289474	211637	22384	4487655
2012	348063	2544450	3119	901671	374968	223797	265292	19887	4681247
2013	337800	2475863	4270	918269	357588	190060	189236	31088	4504174
2014	290696	2616231	3744	895420	397251	144718	245672	27913	4621645

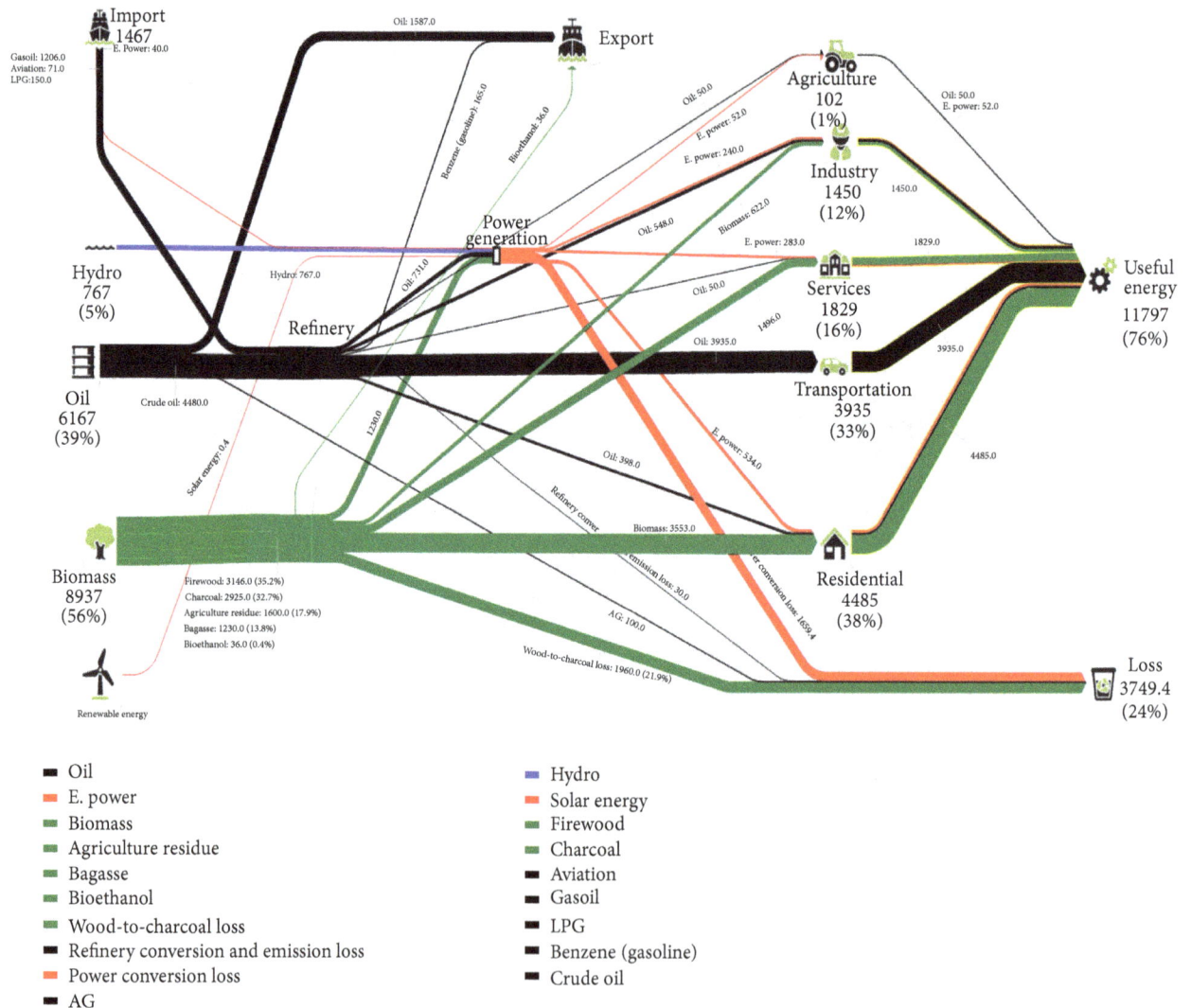

FIGURE 4: Sankey diagram for Sudan. The data used is energy data for the year 2014. The data does not include private power generation. The renewable energy includes only solar photo voltaic power; the other types of energy, such as wind and geothermal, are not included. The hydro power and biomass energy are made as main categories in energy balance. The power generation efficiency and fuel conversion loss and emissions in the refinery are considered. All units are in ktoe (ton of oil equivalent).

loss. It is worth highlighting two examples of energy management loss:

(a) Lack of cogeneration in sugar industry: the potential electric power generation in sugar

industry is double the present level if cogeneration in the sugar industry is introduced [12]. Cogeneration is the standard practice in sugar industry worldwide. Lack of cogeneration made the sugar industry in Sudan burn bagasse

TABLE 15: Supply and demand balance (000 tonnes) [6].

Year	Fuel oil	Gasoil	Kerosene	Benzene	LPG	Naphtha	AV	H. coker GO	Pet coke	Diesel
2010	−295	−79	22	418	−33	20	35	380	−19	−26
2011	−257	47	20	311	173	20	−5	321	74	−22
2012	−248	−227	5	231	59	8	3	277	3	−20
2013	−124	15	15	91	127	18	−28	254	66	−31
2014	−75	64	13	154	50	17	24	292	15	−28

TABLE 16: Fuel demand for thermal electric power base year 2014 [9].

Fuel	Fuel oil	Benzene	H. coker GO	Pet coke	Diesel	Total
tonne	222844	240981	162581	14899	33916	425
ktoe	214	258	156	11	38	677

TABLE 17: Oil supply and demand in 2014.

Item		ktoe	%	Share%
Crude oil production		6067	79.89	
Associated gas		100	1.32	
Gasoil		1206	15.88	
Aviation		71	0.93	
LPG		150	1.98	
Subtotal A		7594	100.00	
Flared AG		100	5.93	1.32
Crude export		1587	94.07	20.90
Subtotal B: export + loss	1687			
Subtotal C = A − B	5907			
Refinery conversion + distribution + emission		30	3.19	
Benzene export		165	17.82	
Power production		677	73.13	
KRC + ORC (internal consumption for power generation)		54	5.85	
Subtotal D = C − oil used in power generation	926		100.00	12.19
Residential		398	8.00	
Transportation		3935	79.00	
[1]Service		50	1.00	
Industry		548	11.00	
Agric.		50	1.00	
Fuel for consumption by sectors	4981			65.59
Total by sectors	5907		100.00	100.00

[1]Service includes public buildings, school, universities, hospitals, municipals, and so forth.

inefficiently; otherwise bagasse poses serious hazard to factory safety.

(b) Flaring of AG: about 15 MMSCFD or 100 ktoe (1.7% of the oil supply) is attributed to flaring of AG. In the oil fields, a significant amount of AG is produced. AG produced is, however, not utilized; it is rather flared. The study conducted by [14] indicated that AG is useful primary energy and if utilized in power generation it can drive steam turbine power plant of 80 MW or can be utilized to produce LPG of annual value of $40 M/year.

(3) Demand: 74% of the total energy demand is useful energy and 26% is loss. 49% of the total loss is due to low efficiency of conversion from wood to charcoal (charcoal conversion efficiency is 33%). The useful energy is distributed among the various demand sectors: agriculture (1%), service (16%), industry (12%), residential (39%), and transportation (32%) sectors.

The developed Sankey diagram is the first of its type in Sudan. The main differences between the present and previous energy balance are the following:

(1) The previous energy balance is presented in simple line draw diagram rather than a professional Sankey diagram as all lines representing different flows are of the same size.

TABLE 18: Bagasse and bioethanol production [15].

| Year | Capacity (000 tonnes/year) | | | | | | | Bagasse | | Bioethanol[14] |
	Kenana	New Halfa	Gunied	Sennar	Assalya	W Nile	Total	Mton	ktoe	(000 m^3)
2002	376	85	94	78	64		697	3059	1315	
2003	398	83	87	85	76		729	3200	1376	
2004	428	88	87	79	74		755	3314	1425	
2005	393	73	87	72	88		713	3127	1344	
2006	400	85	81	81	81		728	3195	1373	
2007	405	83	87	93	90		758	3326	1430	
2008	402	81	85	86	91		745	3268	1405	
2009	382	84	88	87	98		739	3241	1393	
2010	276	57	88	77	75		574	2518	1082	
2011	356	75	98	71	94		693	3040	1307	40
2012	350	66	92	77	90	6	680	2985	1283	55
2013	471	56	77	76	90	73	843	3698	1590	67
2014	308	60	73	73	65	73	652	2862	1230	70

[14] All bioethanol is produced from Kenana sugar factory.

TABLE 19: Firewood by state in m^3 [8].

Year	North	Khartoum	Central	Eastern	Kordofan	Darfur	Total
2005	4267	0	52087	31228	129324	10311	227217
2006	3547	0	37632	16038	577803	138903	773922
2007	5379	0	70113	71116	23792	29079	199479
2008	5153	770	83808	33734	278401	64868	466733
2009	3000	0	39828	15360	0	0	58188
2010	11855	0	75621	29140	127089	33056	276761
2011	6538	340	72992	34307	99828	45955	259960
2012	3410	500	49490	9753	9311	0	72464
2013	3150	0	52079	8400	6515	0	70144
2014	200	0	23757	9627	13837	52363	99784

TABLE 20: Charcoal by state (tons) [8].

Year	North	Khartoum	Central	Eastern	Kordofan	Darfur	Total
2005	895	0	29788	15615	5254	720	52272
2006	912	0	10479	855	13201	2219	27665
2007	1568	0	17239	507	0	1676	20990
2008	1012	158	16016	42479	11820	1015	72501
2009	0	0	310	2995	0	0	3305
2010	8556	0	10004	30545	557026	1512	607643
2011	19276	0	192198	39941	24109	70082	345606
2012	0	0	3200	9875	0	0	13075
2013	0	0	3500	5930	0	0	9430
2014	0	0	1302	155	0	0	1457

TABLE 21: Estimate firewood and charcoal.

Source	Firewood ktoe	Charcoal ktoe	Total as firewood ktoe
[7]	3146	878 (2926)[15]	6072
[16]	2410	795 (2650)	5060
[17][16]	3944	1328 (4427)	8371

[15] Data in brackets is firewood converted into charcoal with conversion efficiency of 33%.
[16] The original data is given as consumption per capita.

TABLE 22: Agriculture waste.

Year	Sorghum	Millet	Wheat	Groundnut HI [11]	Sesame	Sun flower	Cotton	Total residue	
	0.4	0.4	0.4	0.52	0.52	0.52	0.16		
				kton/year				Mton	ktoe
2002	4.394	578	247	990	274	4	72	2.79	2105
2003	2.825	581	330	550	122	18	86	2.46	1852
2004	4.69	769	398	790	399	7	66	3.21	2418
2005	266	280	364	520	277	12	86	2.56	1932
2006	4327	675	416	555	400	44	75	9.44	7117
2007	4.999	796	669	564	242	73	61	3.34	2516
2008	3.869	721	587	716	350	100	24	3.17	2389
2009	4.197	637	641	942	318	247	31	3.48	2620
2010	2.63	471	403	549	248	46	12	2.16	1624
2011	4.605	634	292	1.185	363	124	14	1.92	1445
2012	1.883	378	324	1.032	187	92	55	1.60	1208
2013	4.524	1.09	265	1.767	562	86	25	1.14	857
2014	2.249	359	193	963	205	56	31	2.12	1600

TABLE 23: Animal waste.

Year	Cows	Sheep	Goats	Camels	Dung	
			Dung production (tons/year/head) [10]			
	10.95	1.83	1.83	5.48	Mton	ktoe
			Million head			
2002	39.479	48.136	41.485	3.343	61	25690
2003	39.667	48.44	42.03	3.503	62	25878
2004	39.76	48.91	42.179	3.519	62	25971
2005	40.468	49.797	42.526	3.908	63	26479
2006	40.994	50.39	42.756	4.078	64	26821
2007	41.138	50.651	42.938	4.238	64	26958
2008	41.426	51.067	43.104	4.46	65	27185
2009	41.563	51.555	43.27	4.56	65	27321
2010	41.761	52.079	43.441	4.623	66	27479
2011	29.618	39.296	30.649	4.715	48	19986
2012	29.84	39.483	30.837	4.751	48	20124
2013	30.01	39.568	30.984	4.773	48	20225
2014	30.191	39.846	31.029	4.792	49	20337

(2) The energy balance accounts only for power produced by public sectors. For example, the electric power produced by the refinery and sugar factory is not accounted for in the overall balance. Likewise, renewable energy though small is not accounted for as well.

4. Conclusion

The study provides important information on Sudan's energy sector covering supply and demand sides as well as conversion, distribution, and transmission. For the supply detailed data on electric power generation, oil production and conversion, biomass sources and conversion, and renewable energy was given. Energy import and export were also reported. On the demand side, detailed data of different sectors is reported. The work provided energy flow diagram (Sankey diagram) for the first time. Sankey diagram is an important piece of information for decision-makers. It can be used to develop strategies and identify potential saving, opportunities, and mitigation measures. The weak point of Sudan's energy status is solar energy despite the high intensity of solar radiation and long sunshine hours across the country.

Competing Interests

The authors declare that there are no competing interests regarding the publication of this paper.

TABLE 24: Biomass supply and demand.

Supply	ktoe	%	Demand	ktoe	%
Firewood	3146	28.68	Residential	3553	32.39
Charcoal	2925 (965)[17]	26.66	Services	1496	13.64
Agricultural residue	1600	14.58	Industry	662	6.03
Bagasse	1230	11.21	Transport	0	0.00
Bioethanol	36	0.33	Agriculture	0	0.00
Animal waste	2034[18]	18.54	Electric power	1230	11.21
			Charcoal conversion loss	1960	17.87
			Export[19]	36	0.33
			Not utilized[20]	2034	18.54
Total	10971	100		10971	100.00

[17] Data in brackets is firewood converted into charcoal with conversion efficiency of 33%.
[18] Animal waste is not considered as there is no record for animal waste utilization in the base year 2014.
[19] All bioethanol is exported.
[20] Animal waste is not utilized.

Acknowledgments

The authors acknowledge the research grant by the Ministry of Higher Education and Scientific Research of Sudan.

References

[1] A. Kumar, Subramanyam, and V. R. Kabir, *Development of Energy, Emission and Water Flow Sankey Diagrams for the Province of Alberta Through Modeling. University of Alberta, Department of Mechanical Engineering*, Department of Energy, Government of Alberta, Alberta, Canada, 2011.

[2] IRG, "Energy Planning and Development of Carbon Mitigation Strategies-Using the MARKAL Family of Models," International Resource Group, Washington, DC, USA, 2010.

[3] M. A. Jaccard, *Estimating the Effect of the Canadian Government's 2006-2007 Greenhouse Gas Policies*, C. D. Howe Institute, Toronto, Canada, 2007.

[4] O. Bahn, L. Barreto, B. Büeler, and S. Kypreos, "A multi-regional MARKAL-MACRO model to study an international market of CO_2 emission permits: a detailed analysis of a burden sharing strategy among the Netherlands, Sweden and Switzerland," PSI Technical Paper, Paul Scherrer Institut (PSI), Villigen, Switzerland, 1998.

[5] Quadrennial Technology Review, *An Assessment of Energy Technologies and Research Opportunity*, Department of Energy, Washington, DC, USA, 2015, http://energy.gov/sites/prod/files/2015/09/f26/Quadrennial-Technology-Review-2015_0.pdf.

[6] MOF, *Study of Causes of Demand Increase on Oil Product and Options and Strategy of Subsidy Removal*, Ministry of Finance, Khartoum, Sudan, 2014.

[7] AEC, "Sudan energy status," in *Proceedings of the 10th Arab Energy Conference, Energy and Arab Corporation*, Abudabi, UAE, 2014.

[8] Directorate of Agriculture and Forestry (DAF), *Annual Report*, Directorate of Agriculture and Forestry (DAF), Khartoum, Sudan, 2014.

[9] NEC, *Executive Summary for Year 2014*, National Thermal Electricity Corporation, Khartoum, Sudan, 2014.

[10] M. Niamir, *Report on Animal Husbandry among the Ngok Dinka of the Sudan-Integrated Rural*, Sudan and Harvard Institute for International Elation, Khartoum, Sudan, 1982.

[11] V. Smil, "Crop residues: agriculture's largest harvest," *BioScience*, vol. 49, no. 4, pp. 299–308, 1999.

[12] A. A. Rabah, "Future of sugar industry in Sudan," *Sudanese Engineering Journal*, vol. 5, no. 3, pp. 60–64, 2013.

[13] A. M. Omer, "Energy, environment, and sustainable development in Sudan," *IIOAB Journal*, vol. 2, no. 1, pp. 31–44, 2011.

[14] W. Abdelhamid, *Flared gas uilization [M.S. thesis]*, Chemical Engineering, University of Khartoum, Khartoum, Sudan, 2013.

[15] M. Bashir, B. M. Elhassan, and A. A. Rabah, "Assessment of standby evaporator as an energy conservation measure: case of sudanese sugar industry," *Sugar Tech*, vol. 13, no. 3, pp. 179–184, 2011.

[16] A. M. Omer, "The environmental and economical advantages of agricultural wastes for sustainability development in Sudan," *Journal of Brewing and Distilling*, vol. 1, no. 1, pp. 1–10, 2010.

[17] G. E. Ali, *Studies on Consumption of Forest Products in the Sudan—Woodfuel Consumption in the Household Sector*, Energy Research Institute, Khartoum, Sudan, 1994.

6

Feasibility Analysis and Simulation of Integrated Renewable Energy System for Power Generation: A Hypothetical Study of Rural Health Clinic

Vincent Anayochukwu Ani[1] and Bahijjahtu Abubakar[2]

[1]*Department of Electronic Engineering, University of Nigeria (UNN), Nsukka 410001, Nigeria*
[2]*Renewable Energy Programme, Federal Ministry of Environment, Abuja 900284, Nigeria*

Correspondence should be addressed to Vincent Anayochukwu Ani; vincent_ani@yahoo.com

Academic Editor: Ahmed Al-Salaymeh

This paper presents the feasibility analysis and study of integrated renewable energy (IRE) using solar photovoltaic (PV) and wind turbine (WT) system in a hypothetical study of rural health clinic in Borno State, Nigeria. Electrical power consumption and metrology data (such as solar radiation and wind speed) were used for designing and analyzing the integrated renewable energy system. The health clinic facility energy consumption is 19 kWh/day with a 3.4 kW peak demand load. The metrological data was collected from National Aeronautics and Space Administration (NASA) website and used to analyze the performance of electrical generation system using HOMER program. The simulation and optimization results show that the optimal integrated renewable energy system configuration consists of 5 kW PV array, BWC Excel-R 7.5 kW DC wind turbine, 24 unit Surrette 6CS25P battery cycle charging, and a 19 kW AC/DC converter and that the PV power can generate electricity at 9,138 kWh/year while the wind turbine system can generate electricity at 7,490 kWh/year, giving the total electrical generation of the system as 16,628 kWh/year. This would be suitable for deployment of 100% clean energy for uninterruptable power performance in the health clinic. The economics analysis result found that the integrated renewable system has total NPC of 137,139 US Dollar. The results of this research show that, with a low energy health facility, it is possible to meet the entire annual energy demand of a health clinic solely through a stand-alone integrated renewable PV/wind energy supply.

1. Introduction

The global environmental concerns over the use of fossil fuels for electric power generation have increased the interest in the utilization of renewable energy resources. In particular, rapid advances in wind turbine generator and photovoltaic technologies have brought opportunities for the utilization of wind and solar resources for electric power generation worldwide [1]. Solar and wind energy systems are omnipresent, freely available, and environmental friendly, and they are considered as promising power generating sources due to their availability and topological advantages for local power generations [2, 3]. However, a drawback, common to solar and wind options, is their unpredictable nature and dependence on weather and climatic changes, and the variations of solar and wind energy may not match with the time distribution of load demand. This shortcoming not only affects the system's energy performance, but also results in batteries being discarded too early. Generally, the independent use of both energy resources may result in considerable oversizing, which in turn makes the design costly. It is prudent that neither a stand-alone solar energy system nor a wind energy system can provide a continuous power supply due to seasonal and periodical variations [2, 4] for stand-alone systems. Fortunately, the problems caused by the variable nature of these resources can be partially or wholly overcome by integrating these two energy resources in a proper combination, using the strengths of one source to overcome the weakness of the other. The use of different energy sources allows improving the system efficiency and reliability of the energy supply and reduces the energy storage requirements compared to systems comprising only one single renewable energy source. With the complementary characteristics between solar energy and wind energy for certain locations, the integrated renewable solar/wind power generation systems with storage banks offer a highly reliable source of power [5, 6], which is suitable to electrical loads that need higher reliability such as the health clinic.

Moreover, the economic aspects of these technologies are now sufficiently promising to also justify their use in small-scale stand-alone applications for residential/ranch, communication, and health clinic use; several design scenarios have been proposed for the design of solar/wind power systems for stand-alone applications [7–9].

Integrated renewable (solar/wind) energy systems use two renewable energy sources, allow improving the system efficiency and power reliability, and reduce the energy storage requirements for stand-alone applications. The integrated renewable (solar/wind) systems is becoming popular in remote area power generation applications due to advancements in renewable energy technologies.

This study is on the feasibility analysis of a health facility load data and the renewable resources and evaluates the performance of the designed stand-alone PV/wind generation systems.

2. Hypothetical Study of a Rural Health Clinic in Nigeria

A standard health clinic in rural Nigeria requiring 19 kWh per day to run is considered and was used in the establishment of a hypothetical study for the electrical load data described by [10]. The power consumption rating that was used for the electrical load data and pattern of use are shown in Tables 1 and 2, respectively. Figure 1 shows the daily profile electricity consumption of this health clinic.

2.1. Load Variation. As from 06.00 hrs to 08.59 hrs, load is at the least (462 W). The load increased a little at 09.00 hrs (992 W) and at 10.00 hrs reached a certain level (1255 W) and remains there till 11:59 hr and at 12:00 hr increased again and reached the highest load (3366 W). It comes down at 13.00 hrs (1837 W) and remains there till 14.59 hrs, decreases little at 15.00 hrs (482 W), and remains there till 17.59 hrs where it starts to increase again. As from 10:00 hr to 15:59 hr, most of the energy generated by solar at these times is stored in the battery for use at night with wind energy.

At night between 18.00 hrs and 05.59 hrs, the load is minimal (502 W). Since wind blows much at night than in day, wind energy and the stored energy in the battery can compensate at these hours of time till day time when solar takes up.

2.2. Meteorological Data Generation for Feasibility Study. Climatic conditions determine the availability and magnitude

TABLE 1: Power consumption rating.

Power consumption	Power (Watts)	Qty	Load (watt × qt)
Vaccine refrigerator/freezer	60	1	60
Small refrigerator (nonmedical use)	300	1	300
Centrifuge	575	1	575
Hematology mixer	28	1	28
Microscope	15	1	15
Security light	10	4	40
Lighting	10	2	20
Sterilizer oven (laboratory autoclave)	1,564	1	1,564
Incubator	400	1	400
Water bath	1,000	1	1,000
Communication via VHF radio		1	
Stand-by	2		2
Transmitting	30		30
Desktop computer	200	2	400
Printer	65	1	65

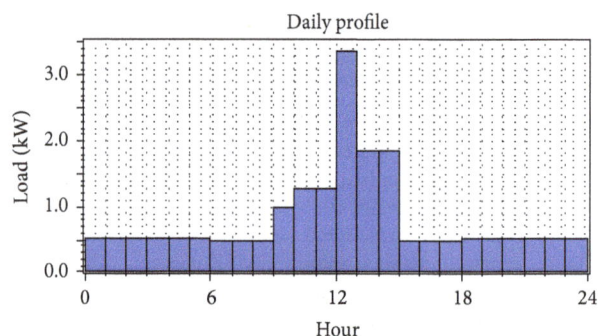

FIGURE 1: Daily load profile of the health clinic (load variation).

of solar and wind energy at a particular location. For different districts and locations, climatic conditions, including solar radiation, wind speed, and air temperature, are always changing. At the potential location, an analysis of the characteristics of solar radiation and wind conditions were made for better utilization of the solar and wind energy resources.

The data for monthly average solar radiation and wind speed for a given year (2013) were obtained from National Aeronautics and Space Administration (NASA) [11]. The specific geographical location (latitude and longitude) of the health clinic in Guzamala (Borno State) is at a location of $11°05'$ N latitude and $13°00'$ E longitude. This location has an annual average solar radiation of 5.90 kWh/m^2/d, whereas its annual average wind is 3.8 m/s. In the solar resource, March was the sunniest month of the year. In this month (March), the solar energy resource is 6.7 kWh/m^2/day, while in August it is only 5.1 kWh/m^2/day, whereas, in the wind resource, September is the least windy month of the year while April

TABLE 2: The electrical load (daily load demands) data for a health facility.

Time	Daily load demands														Total/hr
	1	2	3	4	5	6	7	8	9	10	11	12	13	14	
0.00–0.59	60					40			400		2				502
1.00–1.59	60					40			400		2				502
2.00–2.59	60					40			400		2				502
3.00–3.59	60					40			400		2				502
4.00–4.59	60					40			400		2				502
5.00–5.59	60					40			400		2				502
6.00–6.59	60								400		2				462
7.00–7.59	60								400		2				462
8.00–8.59	60								400		2				462
9.00–9.59	60				15		20		400		2	30	400	65	992
10.00–10.59	60	300		28	15		20		400		2	30	400		1255
11.00–11.59	60	300		28	15		20		400		2	30	400		1255
12.00–12.59	60	300	575		15		20	1564	400		2	30	400		3366
13.00–13.59	60	300	575		15		20		400		2		400	65	1837
14.00–14.59	60	300					20		400	1000	2			65	1847
15.00–15.59	60						20		400		2				482
16.00–16.59	60								400		2				462
17.00–17.59	60								400		2				462
18.00–18.59	60					40			400		2				502
19.00–19.59	60					40			400		2				502
20.00–20.59	60					40			400		2				502
21.00–21.59	60					40			400		2				502
22.00–22.59	60					40			400		2				502
23.00–23.59	60					40			400		2				502
Total	1440	1500	1150	56	75	480	140	1564	9600	1000	48	120	2000	195	19368

1: Vaccine refrigerator/freezer, 2: Small refrigerator (nonmedical use), 3: Centrifuge, 4: Hematology mixer, 5: Microscope, 6: Security light, 7: Lighting, 8: Sterilizer oven (laboratory autoclave), 9: Incubator, 10: Water bath, 11: Communication via VHF radio stand-by, 12: Communication via VHF radio transmitting, 13: Desktop computer, 14: Printer.

and March are the windiest. Figures 2(a) and 2(b) show the solar and wind resource profile of this location.

2.3. Solar Radiation Variation. In the months of September, October, January, February, and March, the solar radiation increases with differences from month to month as (0.43), (0.32), (0.26), (0.69), and (0.40), respectively, whereas, in the months of April, May, June, July, August, November, and December, the solar radiation decreases with differences from month to month as (0.08), (0.26), (0.39), (0.54), (0.29), (0.05), and (0.49), respectively.

2.4. Wind Speed Variation. In the months of January and February, the wind speed remains the same without differences from the months. In the months of March, April, October, November, and December, the wind speed increases with differences from month to month as (0.4), (0.1), (0.3), (0.6), and (0.5), respectively. In the months of May, June, July, August, and September, the wind speed decreases with differences from month to month as (0.4), (0.7), (0.2), (0.2),

and (0.2), respectively. In the months of July, August, and September the wind speed has constant decrease of (0.2) differences from month to month.

The difference in months falls in the range of 0.1–0.7, and these differences are due to earth's rotation.

2.5. System Configuration. The block diagram for a typical stand-alone PV/wind generating system is shown in Figure 3. The system consists of PV panels, wind turbine generator, storage batteries, and dump load. These technologies generate DC current—PV, wind, and battery - and are connected to the DC bus (V_{DC}). An inverter, or a DC-to-AC converter, is used to convert DC current (I_{inv_DC}) to AC current (I_{inv_AC}). The generating system (PV/Wind) supplied to the load and the storage batteries are charged when the renewable energy (wind and solar) generation exceeds the load demand until a specified upper limit for the battery voltage is reached, when it can no longer accept current. At this point the excess available power is diverted to the dump load (a device that sheds excess energy produced by the system), which

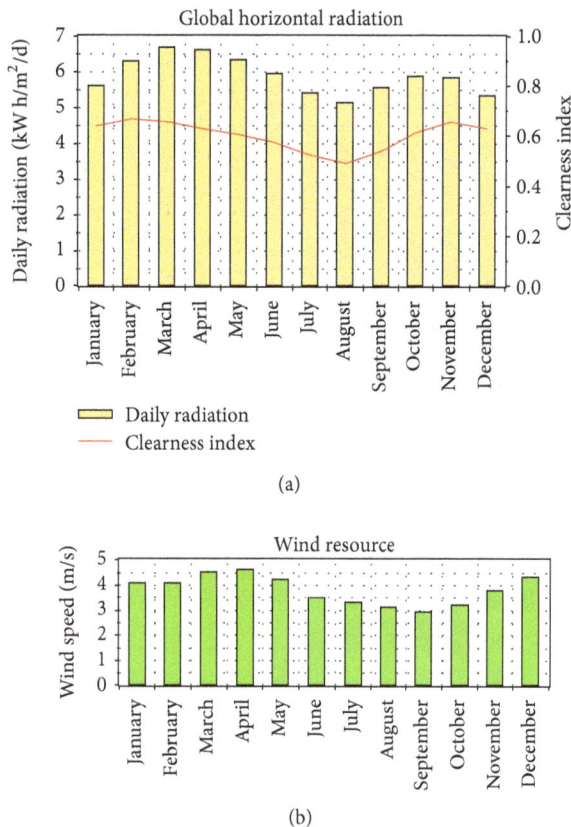

FIGURE 2: (a) HOMER output graphic for solar (clearness index and daily radiation) profile. (b) HOMER output graphic for wind speed profile.

in this study is assumed to be an electric water heater, and the remaining power sold to the community. Batteries will discharge, supplying power to the load when demand exceeds generation. They will continue discharging until a specified lower limit for the battery voltage is reached. At that point, the batteries will stop supplying power to the load, and the renewable energy if available will supply the required power to the load.

3. Modeling and Simulation of Integrated Renewable Solar/Wind System Components

The HOMER software was used to design an optimal integrated renewable power system. The description of HOMER can be found in Ani [12]. Reference to relevant literature provided the design guidelines. The capital costs of all the system components used for the study were gotten from PV system suppliers in Nigeria [13]. The price of PV system is $2000/kW, while a wind turbine (BWC Excel-R) system costs $27,000 and a Surrette 6CS25P battery costs $1,145. The project lifetime is estimated at 20 years and the yearly interest rate applied is 6%. A schematic diagram of an integrated renewable energy system is shown in Figure 4.

4. Results

4.1. Cost of System Component Summary

4.1.1. Solar PV Panels. The proposed PV module is rated at 5 kW. The initial cost of the modules is $10,000, and its operation and maintenance cost is $57, with the total NPC (for the PV module only) of $10,057, as can be seen in Figure 5. The lifetime of the modules will last the project.

4.1.2. Wind Turbine. The BWC Excel-R wind turbine has a capacity of 7.5 kW. Its initial cost is $27,000, and its operation and maintenance cost is $3,441, with the total NPC (for the wind turbine only) of $30,441. The turbine is estimated to last the project.

4.1.3. Battery. The Surrette 6CS25P battery is rated at 6 V and has a capacity 1,156 Ah. Twenty-four batteries initially cost $27,480 and the replacement cost and the operation and maintenance cost add a further $11,927 and $55,056, respectively, and a salvage cost of $-2,494 having the total NPC (for the batteries only) of $91,968.

4.2. Converter. The converter is rated 19 kW. Its initial cost is $3,800, and its operation and maintenance cost is $872, with the total NPC (for the converter only) of $4,672. The converter is estimated to last the project.

4.3. Integrated Renewable Energy System Component. The integrated renewable energy component system has total capital of $68,280; the replacement cost and the operation and maintenance cost add a further $11,927 and $59,426, respectively, and a salvage cost of $-2,494, giving the total NPC (for the coupled (complete) system) of $137,139, as shown in the appendix (Figure 9). This can convince policy makers of the worthiness of the investment in integrated or single renewable energy.

4.4. Electricity Production. The average solar radiation in this location is relatively high. This gives a relatively good possibility and opportunity to engage the photovoltaic (PV) technique and technology as a component of an integrated renewable PV/wind energy system in order to produce clean energy for powering health clinic loads. Although wind speed is relatively moderate with an average of 3.8 m/s throughout the year, it compensates for solar during the months of poor radiation.

It can be noticed that more solar irradiance can be expected from the month of February to June while less solar irradiance is to be expected from December to January. On the other hand, more wind speed can be expected from the month of December to May while less wind speed is to be expected from August to October.

Solar PV compliment wind power during the months (August, September, and October) of poor wind speed, while wind compensates for solar power during the months (December and January) when solar radiation is less as shown in Figure 6.

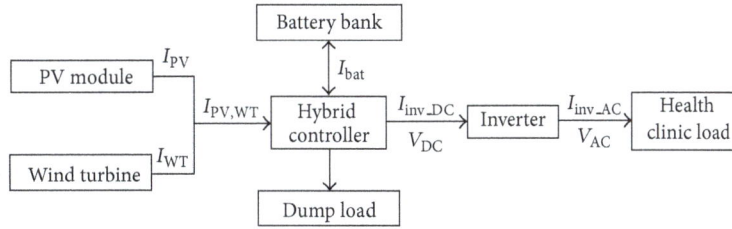

FIGURE 3: Configuration of the integrated renewable (PV/wind) energy system with the energy storage and dump load.

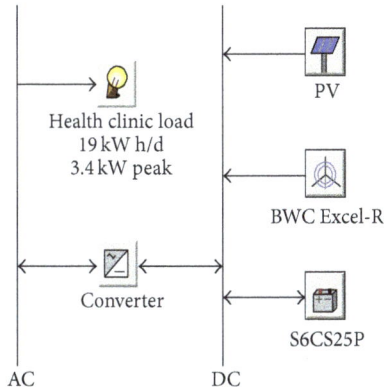

FIGURE 4: The network architecture for the HOMER simulator (proposed PV/wind power system).

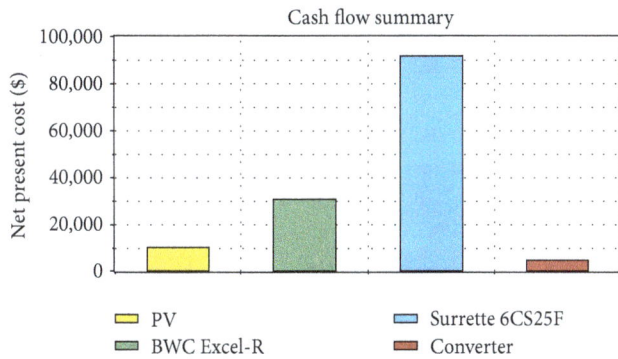

FIGURE 5: Net present cost of components of the integrated renewable PV/wind energy system.

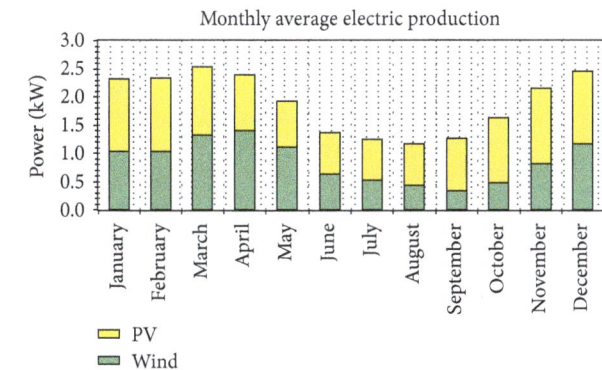

FIGURE 6: Electrical production of integrated renewable PV/wind energy system.

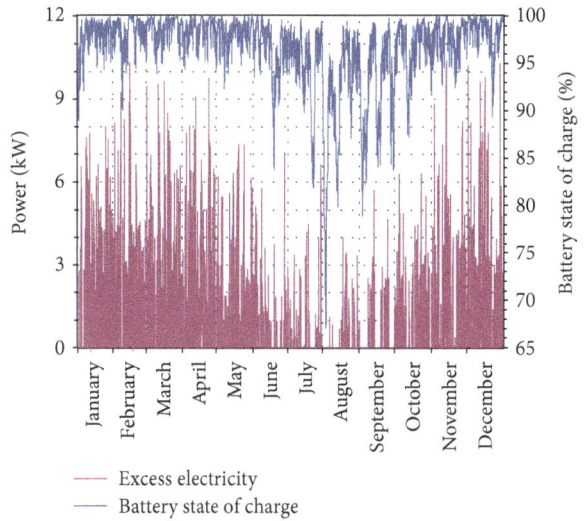

FIGURE 7: Excess electricity versus battery state of charge.

TABLE 3: Simulation results of the electricity production (kWh/yr), battery and inverter losses, and excess energy of the energy system configuration (PV/wind).

System operation	PV/wind system	
Consumption	kWh/yr	%
DC primary load	7,082	100
The total load to be supplied	**7,082**	**100**
Production	kWh/yr	%
PV array	9,138	55
Wind turbine	7,490	45
Total energy generated	**16,628**	**100**
Losses	kWh/yr	
Battery	460	
Inverter	1,250	
Total losses	**1,710**	
Excess energy going to dump load	7,836	
Total energy supplied to the load	7,082	

The integrated renewable energy system (PV/Wind) produces 9,138 kWh/yr (55%) from solar PV array and 7,490 kWh/yr (45%) from wind turbine making a total of 16,628 kWh/yr (100%) as shown in Table 3 and Figure 10 in the appendix. The load demand is 7,082 kWh/yr, while the

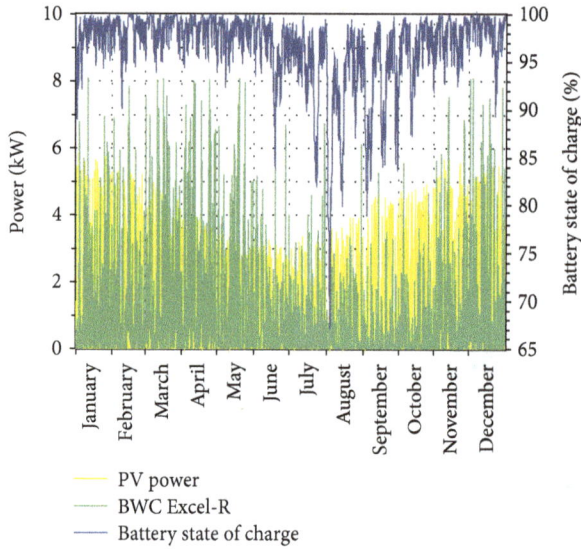

FIGURE 8: Integrated renewable energy (PV/Wind) versus battery state of charge.

Legend:
- PV power
- BWC Excel-R
- Battery state of charge

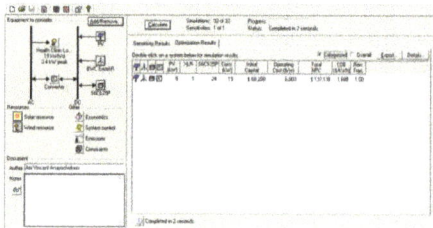

FIGURE 9: The network architecture for the HOMER simulator, proposed PV/wind power system and its optimization results.

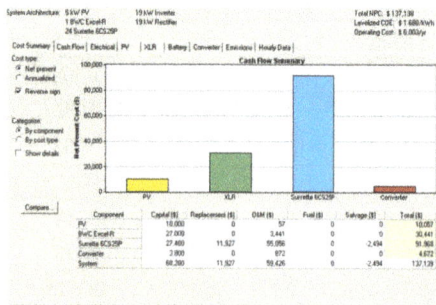

FIGURE 10: Cost summary of the integrated renewable energy system.

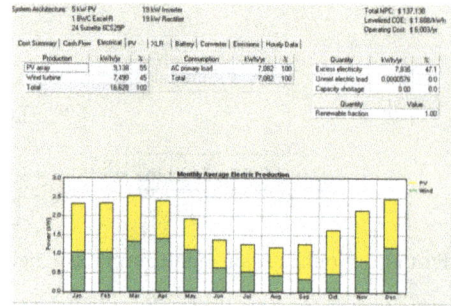

FIGURE 11: Electricity production summary of the integrated renewable energy system.

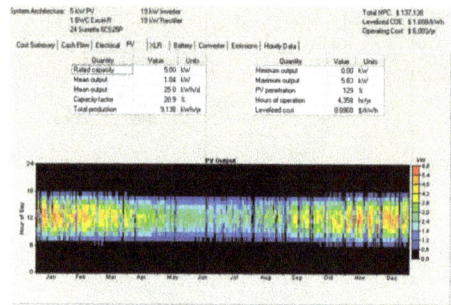

FIGURE 12: PV output summary.

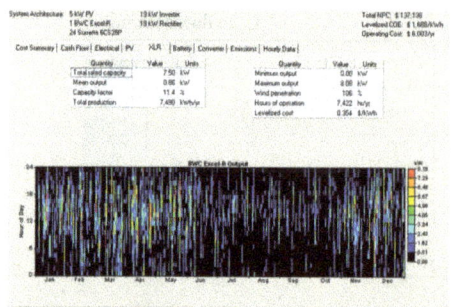

FIGURE 13: Wind turbine (BWC Excel-R) output summary.

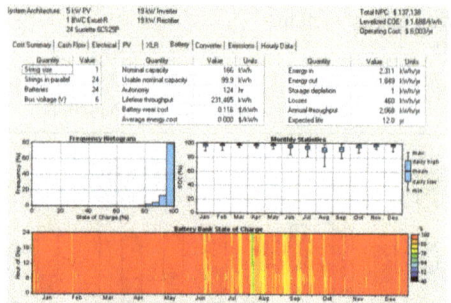

FIGURE 14: Battery bank state of charge summary.

excess electricity from the system is 7,836 kWh/yr also shown in Table 3 and Figure 10 in the appendix.

4.5. Excess Electricity. Excess electricity always occurs when the battery state of charge (SOC) is at 93% upwards and discharges less (at the rate of 1.5 kW to 2.5 kW) and this is between Novembers and Mays. Between Junes and Octobers when the integrated renewable (solar radiation and wind speed) is low, the battery SOC is at 92% downward and

discharges much (at the rate of 3 kW to 10 kW) and there will be no excess electricity from this point downward (due to poor solar radiation and less wind speed) as shown in Figures 7 and 8.

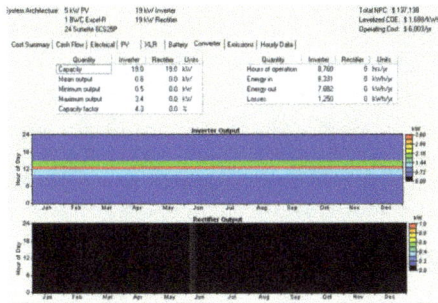

FIGURE 15: Converter output summary.

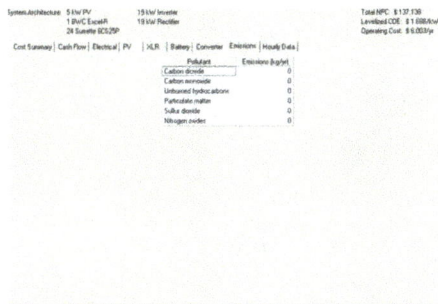

FIGURE 16: Emission output summary.

5. Conclusion

This paper describes the feasibility study of load data and the renewable resources and evaluates the performance of the designed stand-alone PV/wind generation systems. Hourly average wind speed and solar radiation data from the site for the generating unit and the anticipated load data are used to predict the general performance of the generating system. These power systems are very well suited to supply the specific load demand of the rural health clinic that presents a peak in the day time (afternoon) when the solar radiation is maximum and minimal load at night time when the wind blows much (since wind blows much at night than in day). Such performance evaluations are useful in estimating the component sizes needed for generation systems to supply power to loads reliably. It is also helpful in performing a detailed economic analysis (cost benefit study) for the generating unit. Of particular interest is the introduction of dump load model. The excess wind and solar-generated power, when available, are used to heat water in an electric water heater, and the remaining power is sold to the community. This heated water can be used for any purposes in the health clinic such as drinking water. Finally, in the integrated renewable PV/wind energy system, there is no fuel consumption, which means no emission of CO_2, CO, UHC, PM, SO_2, and NO_x from the energy system. This system configuration would be suitable for deployment of 100% clean energy for uninterruptable power performance.

Appendix

See Figures 9, 10, 11, 12, 13, 14, 15, and 16.

Conflict of Interests

The authors declare that there is no conflict of interests regarding the publication of this paper.

References

[1] M. H. Nehrir, B. J. LaMeres, G. Venkataramanan, V. Gerez, and L. A. Alvarado, "An approach to evaluate the general performance of stand-alone wind/photovoltaic generating systems," *IEEE Transactions on Energy Conversion*, vol. 15, no. 4, pp. 433–439, 2000.

[2] V. A. Ani, "Optimal energy system for single household in Nigeria," *International Journal of Energy Optimization and Engineering*, vol. 2, no. 3, pp. 16–41, 2013.

[3] W. Zhou, C. Lou, Z. Li, L. Lu, and H. Yang, "Current status of research on optimum sizing of stand-alone hybrid solar-wind power generation systems," *Applied Energy*, vol. 87, no. 2, pp. 380–389, 2010.

[4] H. Yang, W. Zhou, L. Lu, and Z. Fang, "Optimal sizing method for stand-alone hybrid solar-wind system with LPSP technology by using genetic algorithm," *Solar Energy*, vol. 82, no. 4, pp. 354–367, 2008.

[5] H. Yang, L. Lu, and W. Zhou, "A novel optimization sizing model for hybrid solar-wind power generation system," *Solar Energy*, vol. 81, no. 1, pp. 76–84, 2007.

[6] F. Giraud and Z. M. Salameh, "Steady-state performance of a grid-connected rooftop hybrid wind—photovoltaic power system with battery storage," *IEEE Transactions on Energy Conversion*, vol. 16, no. 1, pp. 1–7, 2001.

[7] W. D. Kellogg, M. H. Nehrir, G. Venkataramanan, and V. Gerez, "Generation unit sizing and cost analysis for stand-alone wind, photovoltaic, and hybrid wind/PV systems," *IEEE Transactions on Energy Conversion*, vol. 13, no. 1, pp. 70–75, 1998.

[8] R. Chedid and S. Rahman, "Unit sizing and control of hybrid wind-solar power systems," *IEEE Transactions on Energy Conversion*, vol. 12, no. 1, pp. 79–85, 1997.

[9] R. Yokoyama, K. Ito, and Y. Yuasa, "Multiobjective optimal unit sizing of hybrid power generation systems utilizing photovoltaic and wind energy," *Journal of Solar Energy Engineering*, vol. 116, no. 4, pp. 167–173, 1994.

[10] V. A. Ani and A. N. Emetu, "Simulation and optimization of photovoltaic/diesel hybrid power generation system for health service facilities in rural environments," *Electronic Journal of Energy and Environment*, vol. 1, no. 1, 2013.

[11] NASA, 2013, http://eosweb.larc.nasa.gov/.

[12] V. A. Ani, "Feasibility assessment of a PV-diesel hybrid power system for an isolated off-grid catholic church," *Electronic Journal of Energy and Environment*, vol. 3, no. 1, 2013.

[13] *Renewable Energy Shop in Nigeria*, http://www.solarshopnigeria.com/.

7

Original Framework for Optimizing Hybrid Energy Supply

Amevi Acakpovi

Department of Electrical/Electronics Engineering, Accra Polytechnic, P.O. Box GP561, Accra, Ghana

Correspondence should be addressed to Amevi Acakpovi; acakpovia@gmail.com

Academic Editor: Antonio Moreno-Munoz

This paper proposes an original framework for optimizing hybrid energy systems. The recent growth of hybrid energy systems in remote areas across the world added to the increasing cost of renewable energy has triggered the inevitable development of hybrid energy systems. Hybrid energy systems always pose a problem of optimization of cost which has been approached with different perspectives in the recent past. This paper proposes a framework to guide the techniques of optimizing hybrid energy systems in general. The proposed framework comprises four stages including identification of input variables for energy generation, establishment of models of energy generation by individual sources, development of artificial intelligence, and finally summation of selected sources. A case study of a solar, wind, and hydro hybrid system was undertaken with a linear programming approach. Substantial results were obtained with regard to how load requests were constantly satisfied while minimizing the cost of electricity. The developed framework gained its originality from the fact that it has included models of individual sources of energy that even make the optimization problem more complex. This paper also has impacts on the development of policies which will encourage the integration and development of renewable energies.

1. Introduction

Recent advances in renewable energy have contributed to the increasing deployment of hybrid energy systems as stand-alone energy systems for providing electricity in remote areas. A hybrid energy system, or hybrid energy, usually consists of two or more renewable energy sources used together to provide increased system efficiency as well as greater balance in energy supply. In other terms, hybrid energy systems are combination of two or more energy conversion devices or two or more fuels for the same device which when integrated overcome limitations that may be inherent in either. Extensive studies have been done to put together different sources of energy in order to design hybrid systems. Moreover, the growing interest for hybrid energy system is being motivated by the identification of a huge deficit of electricity supply in most developing countries.

In Ghana, for instance, based on the Volta River Authority's (VRA) capacity demand and supply balance (2013–2025), and in line with Ghana's power sector reform and major policy objectives, the country's current total installed generating capacity requires to be increased to 5,175 MW by 2023 in order to address the current power shortages, ensure an adequate supply of electricity, meet the country's forecast growth in demand, and improve the quality of service and reliability of the power system.

The shortage of electricity is a crucial problem faced by populations which hinders development in all directions. Ackwa [1] investigated 350 electricity dependent micro and small enterprises and found out that, according to one-third of the samples, electricity supply was insufficient for their business. It was found that the energy deficit can lower a firm's annual sales by 37–48 percent. Considering that about 90 percent of businesses in Ghana are Medium Scale Enterprises (MSE), providing two-thirds of jobs in Ghana, and that only one in every 10 firms operates a back-up generator, the poor electricity supply can be considered as a major constraint to business operations. Reference [1] revealed that presently, on average, the nation is losing about US$ 2.2 million per day or US$ 792 annually, a figure that translates into 2 percent of annual GDP on account of the energy crisis alone. This illustrates the severe impact of electricity shortage on economic development.

One of the most prominent solutions to this problem which is embraced by many countries in our century is the adoption of renewable energy sources to complement the existing deficit in electricity supply. However, the cost of electricity attached to some of these renewable sources especially solar and wind is higher compared to the conventional hydro and this therefore consists of a limiting factor. Instead of fully adopting these renewable energies to cater for load request, they are rather used as supportive energy supplies to the main grid to cater for the deficit of energy only. This brings forth the need of combining existing renewable energy supplies in a manner that minimizes the cost of electricity supply.

In this regard, [2] presented a computational model for optimal sizing of Solar-Wind Hybrid Energy System. They performed mathematical modelling of the individual sources involved, solar and wind, and performed hourly measurement of meteorological and load data. In addition, [3–5] equally worked on optimizing hybrid wind and solar systems by also using their self-developed algorithms that do not respond to a general solution for optimization problems. Their approach can be likened to case study approaches for which the limitation is that the solution is mostly not global; that is, it cannot be generalized or exploited to solve similar problems with some variations in the constraints.

Reference [6] investigated the optimization of a PV/Wind integrated hybrid energy system with battery storage under various loads and unit cost of auxiliary energy sources. The main performance measure was the hybrid energy system cost, and the design parameters were PV size, wind turbine rotor swept area, and battery capacity. The system was simulated with ARENA software, commercial simulation software, and was optimized using the OptQuest tool. Likewise, the studies of [7, 8] were based on software known as HOMER. The lack of evidence in the exact mathematical approach adopted to solve the optimization problem created an uncertainty on the effectiveness and reliability of the aforementioned software. Furthermore [9] developed a computational optimization technique for hybridising solar, wind, and hydro energy. Artificial intelligence was developed to implement the optimization function and this was executed with Matlab software. The system results were satisfactory in terms of reliable combination of available energy sources that feed a particular load at a minimum cost of electricity.

In addition, a computational framework for efficient analysis and optimization of dynamic hybrid energy system was developed by [10]. Their study dealt with a microgrid system having multiple inputs and multiple outputs (MIMO) which was modelled with the Modelica Language in the Dymola Environment. The optimization functions were, however, implemented with Matlab and other tools. The framework was tested over two optimization problems of which one took operating and capital cost into consideration by imposing linear and nonlinear constraints on the decision variables. Reference [11] proposed a hybrid energy systems modelling and simulation framework that targeted a multiple-input and multiple-output system integrating solar power, wind power, hydropower, and biomass. Their study proposed an energy optimization tool that evaluated the optimization cost and showed optimal mix of energy sources considering cost.

However, one of the major limitations to the literature reviewed is the lack of existence of a general framework that covers all aspects of optimizing hybrid energy sources. Many of the optimizations described above are self-developed methods that do not follow any existing framework or guideline which rigorously enforces some common stages to comply with. Even though [10, 11] made an attempt to build such general optimization frameworks for hybrid energy systems, they did not look into the models of hybrid energy system components themselves. They rather resorted to recommending these aspects of the modelling for future studies.

There is a need to fill in the gap identified in order to improve upon hybrid energy system optimization in general. The development of a framework will subsequently guide the implementation of hybrid energy systems in general and provide fair and reliable means of assessing their effectiveness. This paper, therefore, aims at developing a general framework for optimizing hybrid energy systems which encompasses all aspects of the optimization process including the individual modelling of each energy source. The framework will be tested with sample data collected from Accra. The rest of the paper is structured as follows. Section 2 deals with the materials and methods, Section 3 elaborates on the results and discussion, and the last section presents the conclusion and recommendations.

2. Materials and Methods

2.1. Development of the Framework. The proposed framework for hybrid energy optimization is depicted in Figure 1. Figure 1 is divided into four parts. The first part consists of the input variables to the renewable energy sources which directly or indirectly determine the energy produced by each source for a period of a month. For instance, wind speed data will help estimate the energy produced by a wind turbine, solar radiation and average temperature will determine the energy generated by a solar module, and water flow and head can also help determine the available energy produced by a hydroelectric power plant.

In the second stage, the unit model of energy generation will transform the available input variables into energy, based on standard analytical models. This stage mainly deals with a mathematical description of how individual models will transform the input variables into energy and involves deep knowledge of theory of energy conversion for each source of energy. The difficulty at this level lies in the fact that analytical models of renewable energy generation are mostly not linear and do involve many differential and integral equations between input and output variables. Therefore, it is sometimes difficult to establish a straightforward relationship between input and output and this leads to the adoption of computer based modelling like Simulink modelling and others.

The third stage consists of the brain of the system, where the artificial intelligence is being implemented. From Figure 1, the artificial intelligence box takes an input from all the energy generated by the individual sources in stage two.

It also has an estimate of cost of electricity per individual sources which is determined based on the knowledge of

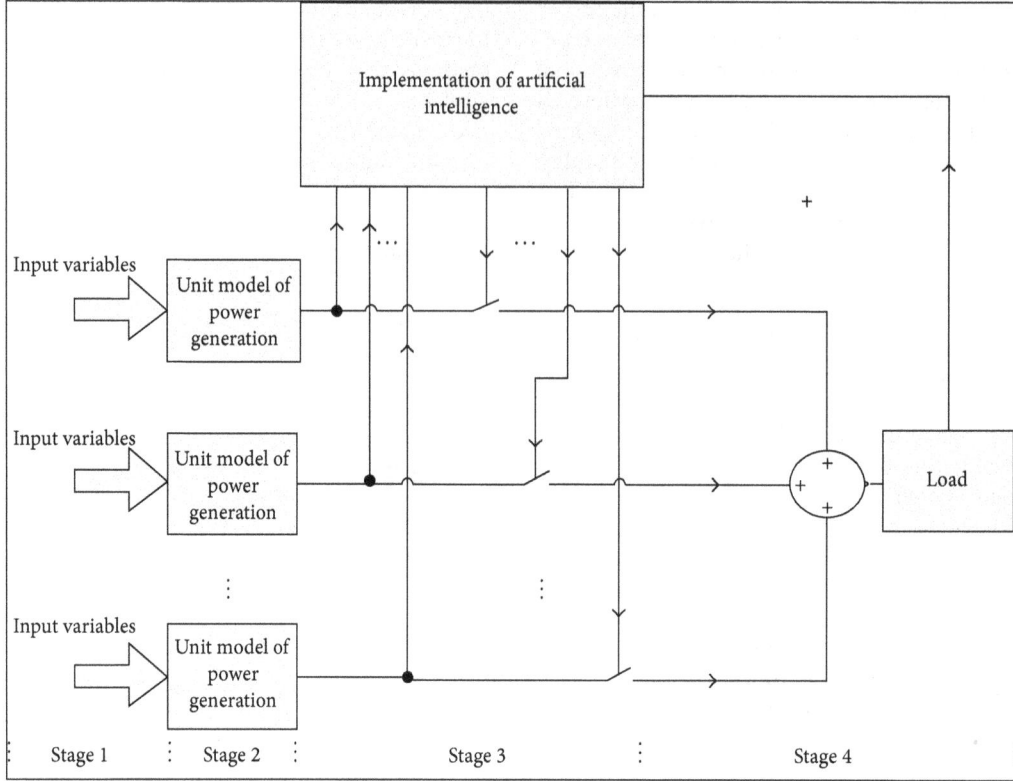

FIGURE 1: Proposed framework for the hybrid energy switching system.

capital cost, capital recovery factor, operation, and maintenance cost. The artificial intelligence box equally monitors the load request and constraints of each variable. It then implements an optimization solution to determine which different sources will be selected to feed the load. The selection is done by a control system applied on the various switches which can be closed or opened to, respectively, select or unselect a particular source of energy generation that will feed the load.

Moreover, the function of the artificial intelligence box is well explained with the flowchart in Figure 2. For a solar, wind, and hydro hybrid energy supply, for instance, the input data to be loaded will consist of wind velocity, solar radiation and temperature, hydro data, and load request. The next step in the flowchart shows that the input energy will be calculated with the unit cost of electricity per each source. The estimated energies with constraints on variables and load satisfactions will be used to solve an optimization problem based on cost, using one method from a set of defined optimization methods. Available optimization solvers include metaheuristic optimization techniques like PSO for nonlinear problems with many inputs. Once the optimization problem is solved, the selected sources and their contribution in terms of energy are derived. This is further used to calculate an average cost of electricity for the hybrid system and the results are saved. The process is repeated for all the available N optimization techniques and the results are finally compared to determine a global best solution which will be implemented.

In the fourth stage, the selected sources are added up together to feed the load and the process is repeated continuously. The paragraph below shows how the cost of electricity is being estimated for a hybrid solar, wind, and hydro hybrid system.

2.2. Estimation of Electricity Cost for a Solar, Wind, and Hydro Hybrid System. Considering a power system made of N_s solar plants, N_w wind power plants and N_h micro-hydro-generating stations located in the same stream, and referring to [8, 9], the total annualized life cycle cost of the system incorporating components of both capital cost and operating cost is given by

$$C_{an} = C_{ans} + C_{anw} + C_{anh} \tag{1}$$

with

$$
\begin{aligned}
C_{ans} &= C_{cs} \cdot CRF_s + C_{os}, \\
C_{anw} &= C_{cw} \cdot CRF_w + C_{ow}, \\
C_{anh} &= C_{ch} \cdot CRF_h + C_{oh},
\end{aligned}
\tag{2}
$$

where (i) C_{ans}, C_{anw}, and C_{anh} represent, respectively, the annualized life cycle cost of energy for the solar, wind, and hydropower generation, (ii) C_{cs}, C_{cw}, and C_{ch} represent, respectively, the capital cost of investment for the solar, wind, and hydropower generation, (iii) C_{os}, C_{ow}, and C_{oh} represent,

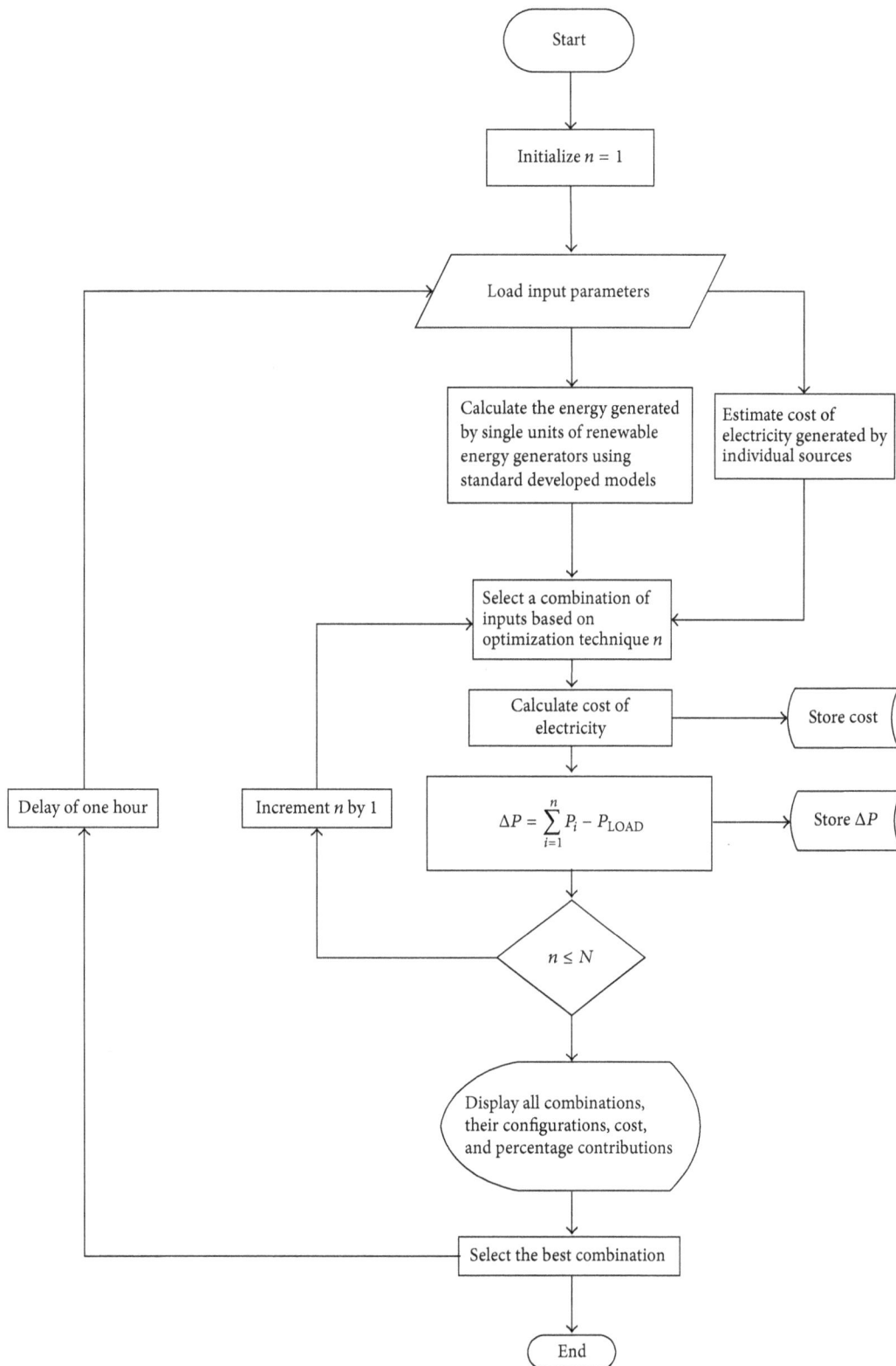

FIGURE 2: Hybrid system optimization framework.

respectively, the operation and maintenance costs for the solar, wind, and hydropower generation, and (iv) CRF_s, CRF_w, and CRF_h represent, respectively, the capital recovery factor for the solar, wind, and hydropower generation.

The unit costs of electricity C_{us}, C_{uw}, and C_{uh}, generated, respectively, by the solar, wind, and hydropower plants, can be expressed as follows:

$$C_{us} = \frac{C_{ans}}{E_s},$$

$$C_{uw} = \frac{C_{anw}}{E_w} \qquad (3)$$

$$C_{uh} = \frac{C_{anh}}{E_h},$$

with E_s, E_w, and E_h being the net energy generated by the solar, wind, and hydropower plants, respectively, over a year.

The cost of electricity generated by the hybrid energy system over a period of time T can be expressed as follows:

$$C_E = C_{us} \cdot \int_0^T a_s \cdot P_s(t)\, dt + C_{uw} \cdot \int_0^T a_w \cdot P_w(t)\, dt$$
$$+ C_{uh} \cdot \int_0^T a_h \cdot P_h(t)\, dt, \qquad (4)$$

where a_s, a_w, and a_h represent, respectively, decision variables or coefficients of selection of the plants. $P_s(t)$ represents the instantaneous power produced by the solar plant. repre- sents the instantaneous power produced by the wind plant. $P_h(t)$ represents the instantaneous power produced by the hydropower plant.

The power generated by a photovoltaic system depends on two fundamental parameters, namely, the solar irradiation and the ambient temperature, as shown by (5) (Acakpovi and Hagan [10], Villalva et al. [11], Ramos-Paja et al. [12], and Tsai et al. [13]). Consider

$$P(t) = \eta A G(t), \qquad (5)$$

where η is the PV generation efficiency, A (m^2) is the PV generator area, and $G(t)$ is the solar irradiation in tilted module plane (W/m^2).

The efficiency η further relates to the temperature as follows:

$$\eta = \eta_r [1 - \beta (T_c - T_{cref})]. \qquad (6)$$

η_r is the reference module efficiency, β is the temperature coefficient, and T_{cref} is reference cell temperature in degree Celsius. For this study, $T_{cref} = 25°C$, $\beta = 0.05\%/°C$, and $\eta_r = 25\%$.

The output power of a wind turbine generator system is usually given by (7) as illustrated by Khajuria and Kaur [14], Abbas and Abdulsada [15], and Acakpovi and Hagan [16]:

$$P_m(t) = \frac{1}{2}\rho_a A C_p(\lambda, \alpha) V_w^3(t), \qquad (7)$$

where (i) C_p is the coefficient of performance also called power coefficient, (ii) A is the swept area by the turbine's blades (m^2), (iii) ρ_a is the air density (kg/m^3), (iv) V_w is the wind speed (m/s), (v) λ is the tip ratio, and (vi) α is the pitch angle.

The tip speed ratio λ is defined as the ratio of the angular rotor speed of the wind turbine to the linear wind speed at the tip of the blades [17] and can be expressed as follows:

$$\lambda = \frac{\omega_r R}{V_w}, \qquad (8)$$

where ω_r is the mechanical angular velocity of the turbine rotor in rad/s and V_w is the wind speed in m/s and R is the radius of the area swept by the turbine's blades [17]. The rotational speed n (r/min) and angular velocity ω_r are related by (9) which is given as follows:

$$\omega_r = \frac{2\pi n}{60}. \qquad (9)$$

Based on Khajuria and Kaur [14], for a VSWT, the coefficient C_p is calculated as follows:

$$C_p(\lambda, \alpha)$$
$$= 0.73 \left[\frac{151}{\lambda_i} - 0.58\alpha - 0.002\alpha^{2.14} - 13.2 \right] e^{-18.4/\lambda_i}, \qquad (10)$$

with λ_i being a coefficient given by (11) as follows:

$$\lambda_i = \frac{1}{1/(\lambda + 0.02\alpha) - 0.03/(\alpha^3 + 1)}. \qquad (11)$$

The objective is to design a wind turbine that can produce power between 3 kW and 5 kW. The following values have been assigned to parameters in the formula of C_p to satisfy the objective of generating power between 3 and 5 kW.

$$n = 100\,\text{rpm},$$
$$R = 2\,\text{m},$$
$$\alpha = 0°, \qquad (12)$$
$$\rho = 1.225\,\text{kg/m}^3.$$

The values of the power coefficient have therefore been computed on monthly basis for different wind speed recorded in the year 2013 for the area of Accra. The computed values are shown in Table 1.

The general formula for the determination of mechanical power produced by a hydraulic turbine is shown by Fuchs and Masoum [18] and Hernandez et al. [19] as follows:

$$P_h = \eta_t \rho_w g H Q(t), \qquad (13)$$

where η_t is the efficiency of the turbine (a value of 80% was assumed in this study), ρ_w is the density of water (1000 kg/m^3), g is the acceleration due to gravity (9.81 m/s^2), Q is the water flow passing through the turbine (m^3/s), and H is the effective pressure head of water across the turbine (m).

TABLE 1: Values of power coefficient for monthly wind speed recorded at Accra in 2013.

Month	Jan.	Feb.	Mar.	Apr.	May	Jun.	Jul.	Aug.	Sep.	Oct.	Nov.	Dec.
V_w (m/s)	2.60	2.60	2.60	2.60	2.10	2.10	4.60	5.10	5.10	2.60	4.60	2.10
$C_p(\lambda, \alpha)$	0.34	0.34	0.34	0.34	0.27	0.27	0.44	0.45	0.45	0.34	0.44	0.27

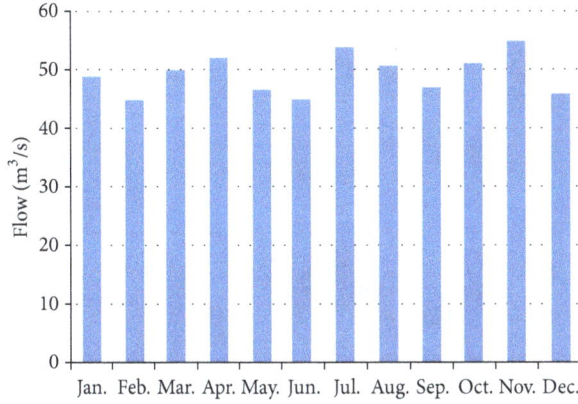

FIGURE 3: Average water flow per month.

Besides, an average water flow of 50 m³/s and an average head of 10 m were assumed with some random variability for different months. The average flow of 50 m³/s, which is relatively low, was selected with the aim of configuring pico hydro that will generate a maximum power of 4 kW. The randomly distributed profile of water flow is shown in Figure 3.

Replacing the various powers by their expressions, (14) is obtained as follows:

$$C_E = a_s C_{us} \int_0^T \eta AG(t) \, dt$$
$$+ a_w C_{uw} \int_0^T \frac{1}{2} \rho_a A C_p V_w^3(t) \, dt \qquad (14)$$
$$+ a_h C_{uh} \int_0^T \eta_t \rho_w g H Q(t) \, dt.$$

With the assumption that $G(t)$, $V_w(t)$, and $Q(t)$ are all constant over period T, the cost function in (14) becomes

$$C_E = a_s C_{us} \eta AGT + a_w C_{uw} \frac{1}{2} \rho A C_p V_w^3 T$$
$$+ a_h C_{uh} \eta_t \rho g H Q T. \qquad (15)$$

The next paragraph proposes an illustration to the optimization problem using a linear programming approach.

2.3. Case Study of a Solar, Wind, and Hydro Hybrid System Using a Linear Optimization Approach.
Following formulas (1) to (15), the optimization problem is posed as follows.

Minimize cost of electricity subjected to the following constraints:

(1) The energy generated by the hybrid system should meet the energy demand at any given time as expressed as follows:

$$a_s \cdot E_{GS}(t) + a_w \cdot E_{GW}(t) + a_h \cdot E_{GH}(t) \geq E_d(t). \qquad (16)$$

(2) The total energy generated should be within range of minimum and maximum energy that can be generated:

$$E_{min} \leq a_s \cdot E_{GS}(t) + a_w \cdot E_{GW}(t) + a_h \cdot E_{GH}(t) \qquad (17)$$
$$\leq E_{max}.$$

(3) Variables should also stay between bounds as follows:

$$0 \leq a_s \leq N_s,$$
$$0 \leq a_w \leq N_w,$$
$$0 \leq a_h \leq N_h,$$
$$0 \leq a_s, a_w, a_h, \qquad (18)$$
$$G_{min} \leq G \leq G_{max},$$
$$V_{wmin} \leq V_w \leq V_{wmax},$$
$$Q_{min} \leq Q \leq Q_{max}.$$

With the assumption that the irradiation G, the wind velocity V_w, and the water flow Q are all constant during period T, the problem can be considered as a linear optimization function subjected to linear inequalities constraints. E_{GS}, E_{GW}, and E_{GH} are the output energies, respectively, generated by the solar, wind, and hydropower generators.

The proposed artificial intelligence solves the optimization of the hybrid system using linear programming approach.

2.4. Data Collection for Accra.
The data collected for the study consists of the following:

(i) Wind speed and direction.

(ii) Solar radiation.

(iii) Temperature.

(iv) Location parameters including longitude, latitude, and elevation.

(v) Loads variation with time.

TABLE 2: Data collected from RETScreen software for the area of Accra (2013).

Month	Jan.	Feb.	Mar.	Apr.	May	Jun.	Jul.	Aug.	Sep.	Oct.	Nov.	Dec.
Accra												
R_s	4.10	4.59	5.21	5.08	5.02	3.97	3.70	3.84	4.59	5.19	4.79	3.86
S_w	2.60	2.60	2.60	2.60	2.10	2.10	4.60	5.10	5.10	2.60	4.60	2.10
T	27.4	27.8	28.0	28.1	27.9	26.6	25.0	24.6	25.1	26.2	27.2	27.3

(i) R_s: solar radiation in kWh/m^2/day.
(ii) S_w: wind speed in m/s.
(iii) T: temperature in °C.

TABLE 3: Data collected on electricity bills of a school at Accra (2013).

Month	Jan.	Feb.	Mar.	Apr.	May	Jun.	Jul.	Aug.	Sep.	Oct.	Nov.	Dec.
Energy (kWh)	4800	4100	4000	4400	5000	4500	4000	4100	4600	4000	4200	4250

Reliable data on wind speed, solar radiation, temperature, and location parameters are provided by the RETScreen Plus software that covers the period from 1997 to 2013. According to [20], the RETScreen Plus is a Windows-based energy management software tool that allows project owners to easily verify the ongoing energy performance of their facilities. It is developed by the Ministry of Energy in Canada in collaboration with NASA. References [21, 22] also describe RETScreen International as an innovative and unique renewable energy awareness, decision support, and capacity building tool developed by CEDRL with the contribution of more than 85 experts from industry, government, and academia. The data collected from RETScreen (2013) for the area of Accra is presented in Table 2.

Load. Data is collected from a tertiary school building located at Accra. The electricity bills issued by the Electricity Company of Ghana (ECG) were used to retrieve the amount of energy consumed every month for the year 2013. The collected data is shown in Table 3 and plotted in Figure 4.

Proposed Linear Programming Approach. Linear optimization problems are mostly solved by graphical means, where all the constraints are plotted separately and the optimal point is determined by means of rigorous observations. With the advent of advanced software such as Matlab, built-in functions have been created to handle linear optimization problem. The most recommended function to solve this problem is the "linprog" function of Matlab which is applied with the following expression:

$$[x_{\text{fval}}] = \text{linprog}\left(f, A, b, A_{\text{eq}}, b_{\text{eq}}, \text{lb}, \text{ub}\right), \quad (19)$$

where the optimized value is kept in the variable fval and the other variables are defined as follows: (i) f is the objective function, (ii) A is a k-by-n matrix, where k is the number of inequalities and n is the number of variables, (iii) b is a vector of length k, (iv) A_{eq} is the matrix summarizing all equality constraints, (v) b_{eq} is a vector of length m, (vi) ub is the matrix of upper bounds applied to the variables, and (vii) lb is the matrix of lower bounds applied to the variables.

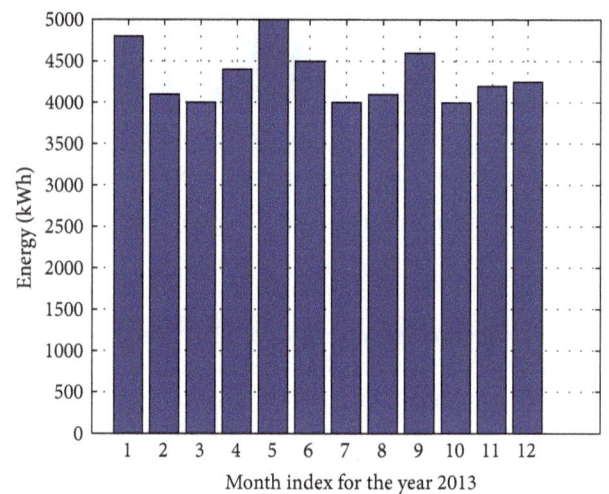

FIGURE 4: Energy consumed by a school complex at Accra for the year 2013.

The solution to our optimization problem is constructed around the "linprog" function of Matlab and can be described by the following algorithm:

(1) Initialize an index variable to N that will serve for iteration and set the period T to one month.

(2) Get the input load data, wind velocity, solar irradiation, and hydro data (water flow and total head) as well as necessary data to evaluate the unit cost of electricity per individual sources.

(3) Calculate the energy generated by individual sources of renewable energy generator using the models described above.

(4) Create decision variables for indexing.

(5) Define lower and upper bounds for all variables.

(6) Define linear equality and linear inequality constraints.

(7) Define the objective function.

(8) Solve the linear optimization problem with the function linprog of Matlab.

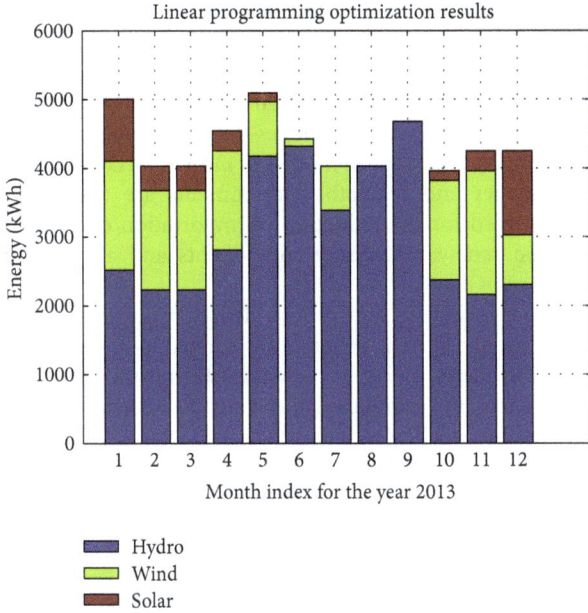

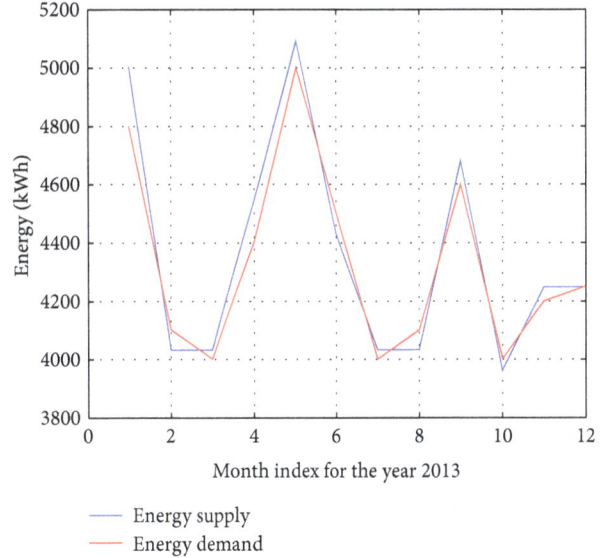

FIGURE 6: Energy supply versus energy demand (case of Accra).

FIGURE 5: Distribution of energy supply based on linear programming approach (case of Accra).

(9) Save result.

(10) Increase the index N by 1.

(11) If index N is less than or equal to 10, repeat processes from 2 to 10.

(12) Display result.

(13) Stop.

3. Results and Discussion

3.1. Presentation of Results. The data were fed into the developed Simulink model and simulation results were obtained in terms of graphs. Three basic graphs are plotted comprising the following:

(a) A bar chart shows the dynamic contribution of individual sources in meeting the load. This chart is necessary to prove the dynamism of the developed algorithm as a solution to the proposed optimization problem. Different input conditions must yield different optimization result and this must transpire in the first chart.

(b) The second graph represents the total supplied energy versus the load which also indicates the unmet loads.

(c) The third graph represents the dynamic estimation of the unit cost of electricity over time.

3.2. Interpretation of Results. Figure 5 shows a mix of individual sources to the total energy supplied to the load. It is apparent that the sources of lower cost are used to supply the load, gradually followed by the sources that are more costly. In this case, the hydro is always selected first, followed by the

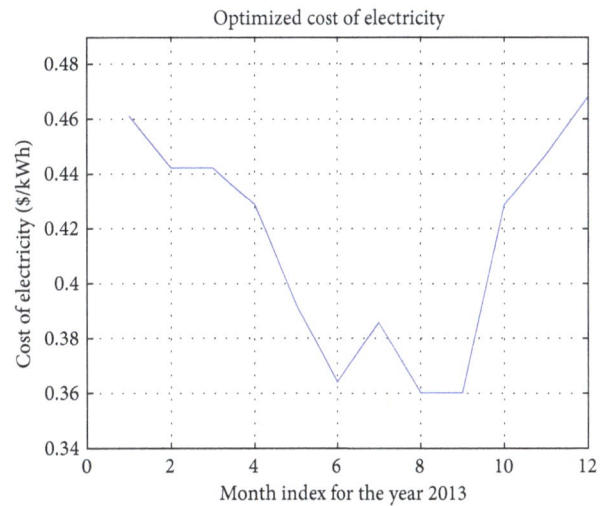

FIGURE 7: Cost of hybrid electricity (case of Accra).

wind and subsequently by the solar, since this is the highest in cost. On the other hand, the graph on cost of electricity in Figure 7 also shows the same trend. The cost appears to be high at the origin but diminishes after some time and remains low from the sixth month to the ninth month. This pattern is directly correlated to the previous graph on the contribution of individual sources. When the energy is solely supplied by the hydropower plant, the cost is low, but as far as the solar and wind are used the cost of electricity automatically goes higher.

Figure 6, on the other hand, shows the energy demand versus the energy supply. The energy supplied in blue colour is almost always above the demand which is in red with some few exceptions. This explains the reliability of the proposed solution. The few cases, months 2, 6, and 10 precisely, where the required loads were not met, can be attributed to the unavailability of enough primary resources needed for energy generation.

3.3. Discussion. A general framework for hybrid energy systems has been developed. This framework encompasses the direct model of individual sources of energy generation at the difference of some previous studies. It further adopts a linear optimization programming approach under Matlab software. The framework was tested with data collected from RETScreen International software for the location of Accra. Results showed interesting dynamics in selecting different sources to supply the load and at the same time obvious minimizing of hybrid energy cost. Also the required energy demands were consistently supplied at the exception of some few cases.

The findings of this study on the proposed framework for HES are congruent with the assertion of Du et al. [17] in the sense that both frameworks propose solution to constrained optimization of hybrid problem taking into consideration the multiple-input and multiple-output aspect. Moreover, the same consideration of capital cost and operation and maintenance cost together with the common use of the Matlab software consists of great factors that confirm the closeness or similarities with the study of Du et al. [17].

Additionally, the findings of [23] corroborate with the proposed framework in this paper which also combines multiple sources of energy with optimization based on cost. Histograms showing mix of energy which satisfies a desired load were equally seen as performance measures for both frameworks.

The proposed framework embeds the models of the hybrid energy components into the framework with the consideration of MIMO system and the use of advanced linear optimization solvers. The involvement of the model of components in the framework is novel and this leads to a significant theoretical contribution. Theoretically, previously developed framework without modelling the HES components should be improved with new models of components. This also encourages the adoption of green energy and improves HES in general.

4. Conclusion

A framework for hybrid system optimization has been proposed in this study. The proposed framework considers, in a first stage, input variables that are made of different renewable sources. These sources are fed into their individual energy generation models in stage 2. The calculated energies for individual sources are then fed to artificial intelligence box which estimates the cost of electricity per individual sources. The artificial intelligence box also monitors the load and decides on the best combination possible to feed the load at the lowest cost. To achieve this, different optimization techniques were proposed in the artificial intelligence box, which were further elaborated in a separate flowchart. The decisions from the artificial intelligence box are implemented through controlled switches that close the link between the selected sources and the load. Stage four of the proposed

framework handles the addition of the selected sources before feeding them to the load.

The framework is seen to be novel for the fact that it embeds the modelling of individual sources of energy generation and, therefore, this implies the development of more robust HES with complete models of individual components. New policies and regulatory instruments are encouraged in this regard for the reduction of importation/exportation duties on renewable energy components and also for net metering in Ghana.

Furthermore, it is recommended that optimization of hybrid energy systems should be carried out with more generation plants relying on metaheuristic optimization techniques such as Guaranteed Convergence Particle Swarm Optimization (GCPSO), Hybrid GAPSO (HGAPSO), Combined PSO, and Simulated Annealing Algorithms.

Competing Interests

The author declares that there are no competing interests.

References

[1] C. Ackwa, *Electricity Insecurity and Its Impact on the Economy of Ghana*, Institute of Statistical Social and Economic Research, University of Ghana, Accra, Ghana, 2015.

[2] S. C. Gupta, Y. Kumar, and G. Agnihotri, "Optimal sizing of solar-wind hybrid system," in *Proceedings of the IET-UK International Conference on Information and Communication Technology in Electrical Sciences (ICTES '07)*, pp. 282–287, Tamil Nadu, India, December 2007.

[3] R. Rawat and S. S. Chandel, "Simulation and optimization of solar photovoltaic-wind stand alone hybrid system in hilly terrain of India," *International Journal of Renewable Energy Research*, vol. 3, no. 3, pp. 595–604, 2013.

[4] P. G. Dalwadi and C. R. Mehta, "Feasibility study of solar-wind hybrid power system," *International Journal of Emerging Technology and Advanced Engineering*, vol. 2, no. 3, 2012.

[5] M. Muralikrishna and V. Lakshminarayana, "A 10 kW combined hybrid (wind and solar photovoltaic) energy systems for isolated generating system," *ARPN Journal of Engineering and Applied Sciences*, vol. 6, no. 6, pp. 99–104, 2011.

[6] B. Y. Ekren and O. Ekren, "Simulation based size optimization of a PV/wind hybrid energy conversion system with battery storage under various load and auxiliary energy conditions," *Applied Energy*, vol. 86, no. 9, pp. 1387–1394, 2009.

[7] D. K. Lal, B. B. Dash, and A. K. Akella, "Optimization of PV/wind/micro-hydro/diesel hybrid power system in HOMER for the study area," *International Journal on Electrical Engineering and Informatics*, vol. 3, no. 3, pp. 307–325, 2011.

[8] A. Acakpovi, E. B. Hagan, and M. B. Michael, "Cost benefit analysis of self-optimized hybrid solar-wind-hydro electrical energy supply as compared with HOMER optimization," *International Journal of Computer Applications*, vol. 114, no. 18, pp. 32–38, 2015.

[9] A. Acakpovi, E. B. Hagan, and M. B. Michael, "Cost benefit analysis of self-optimized hybrid solar-wind-hydro electrical energy supply as compared to HOMER optimization," *International Journal of Computer Applications*, vol. 114, no. 18, pp. 32–38, 2015.

[10] A. Acakpovi and E. B. Hagan, "Novel photovoltaic module modeling using Matlab/Simulink," *International Journal of Computer Applications*, vol. 83, no. 16, pp. 27–32, 2013.

[11] M. G. Villalva, J. R. Gazoli, and E. R. Filho, "Comprehensive approach to modeling and simulation of photovoltaic arrays," *IEEE Transactions on Power Electronics*, vol. 24, no. 5, pp. 1198–1208, 2009.

[12] C. A. Ramos-Paja, E. Perez, D. G. Montoya, C. E. Carrejo, A. Simon-Muela, and C. Alonso, *Modelling of Full Photovoltaic Systems Applied to Advanced Control Strategies*, Universidad Nacional de Columbia, Bogotá, Colombia, 2010.

[13] H. L. Tsai, C. S. Tu, and Y. J. Su, "Development of generalized phottovoltaic model using MATLAB/SIMULINK," in *Proceedings on the World Congress on Engineering and Computer Science (WCECS '08)*, p. 6, 2008.

[14] S. Khajuria and J. Kaur, "Implementation of pitch control of wind turbine using simulink (Matlab)," *International Journal of Advanced Research in Computer Engineering and Technology*, vol. 1, no. 4, pp. 196–200, 2012.

[15] F. A. R. Abbas and M. A. Abdulsada, "Simulation of wind-turbine speed control by MATLAB," *International Journal of Computer and Electrical Engineering*, vol. 2, no. 5, pp. 1793–8163, 2010.

[16] A. Acakpovi and E. B. Hagan, "A wind turbine system model using a Doubly-Fed Induction Generator (DFIG)," *International Journal of Computer Applications*, vol. 90, no. 15, pp. 6–11, 2014.

[17] W. Du, H. E. Garcia, and C. J. J. Paredis, "An optimization framework for dynamic hybrid energy systems," in *Proceedings of the 10th International Modelica Conference*, pp. 767–776, Lund, Sweden, March 2014.

[18] E. F. Fuchs and M. A. S. Masoum, *Power Conversion of Renewable Energy Systems*, Springer, New York, NY, USA, 2011.

[19] G. A. M. Hernandez, S. P. Mansoor, and D. L. Jones, *Modelling and Controlling Hydropower Plants*, Advances in Industrial Control, Springer, Berlin, Germany, 2013.

[20] R. D. Ganoe, P. W. Stackhouse, and R. J. DeYoung, *RETScreen Plus Software Tutorial*, NASA Technical Report Servers, 2014.

[21] G. J. Leng, *RETScreenTM International: A Standarized Tool for Assessing Potential Renewable Energy Projects*, Natural Resources Department of Canada, CANMET Energy Diversification Research Laboratory (CEDRL), 2014.

[22] E. Martinot and O. McDoom, "Promoting energy efficiency and renewable energy: GEF climate change projects and impacts," Pre-Publication Draft, Global Environment Facility, Washington, DC, USA, 1999.

[23] T. R. Ender and C. Haynes, "A hybrid energy systems modelling and simulation framework for regional cooperation through renewable and alternative energy," in *Proceedings of the International Conference on Integrating Central Asia into the World Economy*, Washington, DC, USA, 2007.

8

A Comfort-Aware Energy Efficient HVAC System Based on the Subspace Identification Method

O. Tsakiridis,[1] **D. Sklavounos,**[2] **E. Zervas,**[1] **and J. Stonham**[2]

[1]*Department of Electronics, TEI of Athens, Egaleo, 12210 Athens, Greece*
[2]*Department of Engineering and Design, Brunel University, Kingston Lane, Middlesex UB8 3PH, UK*

Correspondence should be addressed to O. Tsakiridis; odytsak@teiath.gr

Academic Editor: Aleksander Zidansek

A proactive heating method is presented aiming at reducing the energy consumption in a HVAC system while maintaining the thermal comfort of the occupants. The proposed technique fuses time predictions for the zones' temperatures, based on a deterministic subspace identification method, and zones' occupancy predictions, based on a mobility model, in a decision scheme that is capable of regulating the balance between the total energy consumed and the total discomfort cost. Simulation results for various occupation-mobility models demonstrate the efficiency of the proposed technique.

1. Introduction

As the control and limitation of the energy consumption remain a field with an exceptional technological and economical interest, areas that have been considered as high energy consumers comprise very challenging research issues. According to the US Energy Information Administration from 2013 through 2040 the electricity consumption in the commercial and residential sectors will be increasing by 0.5% and 0.8% per year [1]. It is well established and widely accepted through research studies that the main energy consumers in the commercial and residential buildings are the Heat Ventilation and Air Conditioning (HVAC) systems, as well as the lighting systems.

Buildings contributed a 41% (or 40 quadrillion btu) to the total US energy consumption in 2014. On an average, about 43% of the energy consumption in a commercial and residential building is due to HVAC systems [2]. Therefore, due to the high rate of energy consumption, the necessity of the demand-driven control in the HVAC systems has become inevitable. In modern buildings several sophisticated systems have been applied aiming at providing this type of control in the HVAC systems. The state-of-the-art technology of the HVAC control considers the occupancy of the zones a very

important parameter, playing a key role in methods aiming at reducing energy consumption. The detection and prediction of the zones' occupancy are a very challenging research field and several techniques have been proposed based on historic statistical data as well as on probabilistic models.

Another equally important factor taken into account in advanced HVAC control systems is the thermal comfort of the occupants. The objective of modern HVAC control systems is to reduce energy consumption without compromising the comfort of the occupants. Wireless sensor networks, equipped with temperature, humidity, and occupancy detection sensor nodes, are nowadays the basic platform to build automated HVAC control systems. A number of methods aiming to maintain the thermal comfort while saving energy have been proposed. Some of them are described in Section 2. Towards this direction, a novel technique is proposed in this paper, which aims at balancing the comfort and energy costs in a multizone system. The decisions on heating the zones or not may be taken either centrally or in a distributed manner by wireless sensor nodes scattered in the multizone system. In any case temperature and zone occupancy information must be exchanged between a node, responsible for a zone, and its neighboring nodes. The decision process itself relies on two kinds of predictions: (a) temperature-time predictions for the zones and (b) the zones' occupancy profile. The emphasis in this paper is on the zones' temperature predictions and to this end a deterministic subspace identification method is used for modelling the thermal dynamics of each zone. That is, each zone is modelled by a simple state-space model capable of producing accurate predictions based on the surrounding temperatures, the heating power of the zone, and the current state that summarizes the temperature history of the zone. For the zones' occupancy predictions we consider a semi-Markov model, where occupants (moving as a swarm) stay in a zone for a random period of time and then move to adjacent zones with given probabilities. What is needed by the decision process is the distribution of the first entrance time to unoccupied zones. Aiming at a proactive action, the proposed method periodically computes the risk of activating the heater or not and decides in favor of the action that produces the smaller risk. The computation of the risks relies on the relative weights of the energy and discomfort costs so that the balance between the total energy consumed and the total discomfort cost may be regulated.

The structure of the paper is as follows: In Section 2, prior and related work of the field of the demand-driven HVAC systems that utilize occupancy and prediction methods is presented. In Section 3, the proposed comfort-aware energy efficient mechanism is described, along with the subspace identification method, used for obtaining temperature-time predictions, and a discussion on the distribution of the first entrance time to unoccupied zones. Section 4 contains simulation results and assesses the effectiveness of the proposed algorithm. Finally, Section 5 summarizes the paper and proposes research directions for future work.

2. Prior Work

The necessity of demand-driven HVAC systems, for energy efficient solutions, has orientated the researchers towards occupancy-based activated systems. In multizone spaces the reactive and proactive activation of the heating/cooling of the zones contributes to significant energy savings and improves the thermal comfort for the occupants. Several research works with valuable results of energy savings have utilized the occupancy detection and prediction, in order to control the HVAC system appropriately. For the occupancy detection several types of sensors are used (CO_2, motion, etc.), while, for the prediction, combinational systems are usually applied, utilizing mathematical predictive models (e.g., Markov chains) alongside with detected real data.

The authors in [3] proposed an automatic thermostat control system which is based on an occupancy prediction scheme that predicts the destination and arrival time of the occupants in the air-conditioned areas, in order to provide a comfortable environment. For the occupants' mobility prediction the cell tower information of the mobile telephony system is utilized and the arrival time prediction is based on historical patterns and route classification. For the destination prediction of locations very close to each other (intracell), the time-aided order-Markov predictor is used. The authors in [4] proposed a closed-loop system for optimally controlling HVAC systems in buildings, based on actual

occupancy levels named the Power-Efficient Occupancy-Based Energy Management (POEM) System. In order to accurately detect occupants' transition, they deployed a wireless network comprising two parts: the occupancy estimation system (OPTNet) consisting of 22 camera nodes and a passive infrared (PIR) sensors system (BONet). By fusing the sensing data from the WSN (OPTNet and BONet) with the output of an occupancy transition model in a particle filter, more accurate estimation of the current occupancy in each room is achieved. Then, according to the current occupancy in each room and the predicted one from the transition model, a control schedule of the HVAC system takes over the preheat of the areas to the target temperature. In [5] the authors used real world data gathered from a wireless network of 16 smart cameras called Smart Camera Occupancy Position Estimation System (SCOPES) and in this way, they developed occupancy models. Three types of Markov chain (MC) occupancy models were tested and these are the single MC, the closest distance, and finally the blended (BMC). The authors concluded that BMC is the most efficient and thus they embodied it as the occupancy prediction method in their proposed "OBSERVE" algorithm, which is a temperature control strategy for HVAC systems.

The authors in [6] have developed an integrated system called SENTINEL that is a control system for HVAC systems utilizing occupancy information. For the occupancy detection and localisation, the system utilizes the existing Wi-Fi network and the clients' smart phones. The occupancy localization mechanism is based on the access point (AP) communication with a client's smart phone, so that if an occupant's phone sends packets to an AP then he is located within the range of the AP. By classifying the building areas into two main categories, namely, the personal and the shared spaces, the proposed mechanism activates the HVAC system when an owner of a personal space (e.g., office) has been detected within an area where his/her office is located or when occupants are detected within shared spaces. In [7] an integrated heating and cooling control system of a building is presented aiming to reduce energy consumption. The occupancy behavior prediction as well as the weather forecast, as inputs to a virtual (software based) building model, determines the control of the HVAC system. The occupancy detection technique utilizes Gaussian Mixture Models (GMM) for the categorization of selected features, yielding the highest information gain according to the different number of occupants. This categorization was used for observation to a Hidden Markov Model for the estimation of the number of occupants. A semi-Markov model was developed based on patterns comprised by sensory data of CO_2, acoustics, motion, and light changes, to estimate the duration of occupants in the space. The work in [8] proposes a model predictive control (MPC) technique aiming to reduce energy consumption in a HVAC system while maintaining comfortable environment for the occupants. The occupancy predictive model is based on the two-state Markov chain, with the states modelling the occupied and occupied condition of the areas. The authors in [9] propose a feedback control algorithm for a variable air volume (VAV) HVAC system for full actuated (zones consisting of one room) and underactuated (zones consisting of more than one room) zones. The proposed algorithm is called MOBSua (Measured Occupancy-Based Setback for underactuated zones) and it utilizes real-time occupancy data, through a WSN, for optimum energy efficiency and thermal comfort of the occupants. Moreover, the algorithm can be applied on conventional control systems with no need of occupancy information and it is scalable to arbitrary sized buildings.

3. Comfort-Aware Energy Efficient Mechanism

We consider a multizone HVAC system consisting of Z zones z_i, $i = 1, 2, \ldots, Z$. Each zone is equipped with a wireless sensor node capable of conveying the zone's temperature and occupancy information to its neighbors or to a central processing unit. The occupancy information may be very simple, that is, binary information indicating the presence or absence of individuals in the zone, or more advanced like the number of occupants in the zone. The objective is to minimize the total energy and discomfort cost defined as

$$
\begin{aligned}
\mathscr{E} &= \mathscr{E}_{\text{energy}} + \mathscr{E}_{\text{discomfort}} \\
&= \sum_{i=1}^{Z} \int_{-\infty}^{\infty} W_i I \left(H_i = \text{on} \right) dt \\
&\quad + \sum_{i=1}^{Z} \int_{-\infty}^{\infty} C_i I \left(T_i < T_c \right) dt.
\end{aligned}
\tag{1}
$$

Note that only the heating problem is addressed in this paper. Thus, the case of spending energy for cooling the zones and keeping the temperature T_i within a certain comfort zone is not treated for simplicity. In (1) $I(A)$ denotes the indicator function which takes the value 1 or 0 depending on whether condition A is true or false. W_i is the power of a heater that covers zone z_i and H_i is the state of the heater; that is, the heater is on or off. Thus, the first term of (1) is the total energy consumed by the multizone system. For the discomfort cost we define the cost per unit time C_i and the target comfort threshold T_c. As long as the zone's temperature T_i is higher than T_c the occupants do not feel discomfort. The parameter C_i may depend on the zone, the number of the occupants (e.g., the total system's discomfort cost is proportional to the number of people experiencing discomfort), and the difference of the zone's temperature T_i and the comfort threshold T_c. It is worth noting that there is a tradeoff between energy savings and discomfort cost. We may preheat all zones to the comfort level thus rendering the discomfort cost equal to zero. This policy is inefficient due to the large amount of energy consumed. In the other extreme we could act reactively by heating a zone only upon detecting occupants in it. In this case the energy cost would be as small as possible but the discomfort cost could increase dramatically. In the sequel we will describe a method that acts proactively by taking periodically decisions on whether to heat a zone or not. If the entrance to a zone is delayed then we postpone heating until the next decision epoch. If, on the contrary, it is anticipated that a zone will be occupied before the zone's temperature reaches the comfort level, then we preheat the zone.

The decision of whether or not the heater of an unoccupied zone z_i should be turned on depends on (a) the current state of the zone, (b) the relative value of the energy cost (W_i) and discomfort cost (C_i) per unit time, (c) the temperatures of the surrounding zones, and (d) an estimate of the time that the zone will become occupied. This decision may be taken either centrally from the central processing unit that gathers information by all zones or in a distributed manner by each zone's node after collecting the relative information. There is also the possibility of dividing the decision process functionality between the central unit and the wireless sensor network nodes. For example, the prediction of the first entrance times to unoccupied zones may be taken by the central unit that is aware of the location of people in the multizone system, and then the predicted values may be communicated back to the zones' nodes for the final decision. The specific implementation of the decision process is irrelevant to this paper and it will not be further analyzed.

We assume that time is discretized $t = mT_s$, T_s is the sampling period, and that each node takes its decisions periodically every D sampling periods. Note that there is no need for the nodes to be synchronized to each other. Let $Y_i(t)$ denote the random variable that models the remaining time from the current time instant t until entrance to the unoccupied zone z_i. It is obvious that the random variables $Y_k(t)$, for the various unoccupied zones z_k, do not follow the same distribution. We further assume that the node is capable of making predictions for the time it takes to exceed the comfort threshold T_c. To this end, let $S_i^{\text{on}}(t)$ denote the time period that is required to exceed T_c if the heater is turned on immediately. Similarly, we define $S_i^{\text{off}}(t)$ as the predicted time to reach the comfort threshold if the heater remains off for a period D and then, at the next decision epoch, it is turned on. The relation of the aforementioned parameters is shown in Figure 1.

Suppose now that at time t the node of the unoccupied zone z_i has to take a decision whether to turn the zone's heater on or remain in the off state. At the decision epoch t the risk to turn the heater on is

$$
\begin{aligned}
R_i^{\text{on}}&(t) \\
&= W_i D P \left\{ S_i^{\text{off}}(t) \leq Y_i(t) \right\} \\
&\quad + W_i \left(Y_i(t) - S_i^{\text{on}}(t) \right) P \left\{ S_i^{\text{on}}(t) < Y_i(t) \leq S_i^{\text{off}}(t) \right\}.
\end{aligned}
\tag{2}
$$

That is, if $S_i^{\text{off}}(t) \leq Y_i(t)$, there is plenty of time to reach the comfort threshold even if we start heating at the next decision epoch and therefore we consume unnecessarily $W_i D$ units of energy (this is the case depicted in Figure 1). Similarly, if $S_i^{\text{on}}(t) < Y_i(t) \leq S_i^{\text{off}}(t)$ we waste $W_i(Y_i(t) - S_i^{\text{on}}(t))$ units of energy. Note that there is no risk of turning on the heater if the entrance to the zone happens sooner than it is predicted. In a similar fashion we calculate the risk of keeping the heater off. In this case

$$
\begin{aligned}
R_i^{\text{off}}&(t) \\
&= C_i D P \left\{ Y_i(t) \leq S_i^{\text{on}}(t) \right\} \\
&\quad + C_i \left(S_i^{\text{off}}(t) - Y_i(t) \right) P \left\{ S_i^{\text{on}}(t) < Y_i(t) \leq S_i^{\text{off}}(t) \right\}.
\end{aligned}
\tag{3}
$$

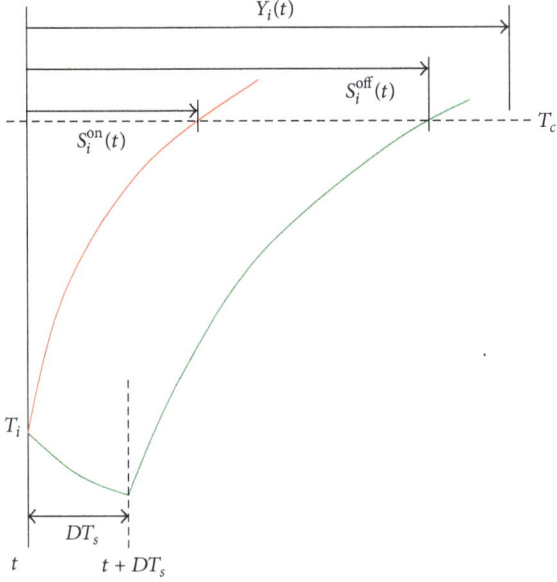

FIGURE 1: Predicted time intervals related to the decision process.

Using (2) and (3) the decision criterion takes the form

$$R_i^{\text{on}}(t) \underset{\text{off}}{\overset{\text{on}}{\lessgtr}} R_i^{\text{off}}(t). \tag{4}$$

Consider for the moment that there is no randomness on $Y_i(t)$, that is, strict time scheduled occupation of zones. Typical examples are the classrooms in a university campus. In this case, criterion (4) takes a very simple form since one of the involved probabilities is equal to one and the rest assume the value of zero. Thus, if $Y_i(t) \leq S_i^{\text{on}}(t)$, then the heater is turned on, if $S_i^{\text{off}}(t) < Y_i(t)$, the heater remains off, and if $S_i^{\text{on}}(t) < Y_i(t) \leq S_i^{\text{off}}(t)$ the heater will be turned on/off depending on the values of the relative energy and discomfort cost. The balance between these two costs is regulated by the weights W_i and C_i.

If $Y_i(t)$ is random, then criterion (4) needs to be slightly modified as the value of $Y_i(t)$ is unknown. In this case we replace $Y_i(t)$ with its conditional expected value given that $S_i^{\text{on}}(t) < Y_i(t) \leq S_i^{\text{off}}(t)$. Therefore, in criterion (4), we substitute $Y_i(t)$ by $E[Y_i(t) \mid S_i^{\text{on}}(t) < Y_i(t) \leq S_i^{\text{off}}(t)]$.

From the previous discussion, it becomes clear that the decision process needs the estimates $S_i^{\text{on}}(t)$, $S_i^{\text{off}}(t)$ and the distribution of the process $Y_i(t)$. The former is the subject of the next subsection, whereas a discussion on the distribution of $Y_i(t)$ is left for Section 3.2.

3.1. Temperature-Time Predictions Based on SID.
It is conceivable that in order to make predictions about a zone's temperature evolution we need a model able to capture the thermal dynamics of the tagged zone. To this end we resort to the deterministic subspace identification method. Each zone is treated separately and is represented by a discrete time, linear, time invariant, state-space model. That is,

$$x_{k+1} = Ax_k + Bu_k + w_k, \tag{5}$$

$$y_k = Cx_k + Du_k + v_k, \tag{6}$$

where y_k is the zone temperature at time k and is in general an ℓ-dimensional vector if ℓ temperature sensors scattered in the zone have been deployed. Assuming a uniform zone temperature we use only one sensor node and thus y_k is scalar. v_k represents the measurement noise due to imperfections of the sensor. u_k is a vector of input measurements and its dimensionality varies from zone to zone. This vector contains the measurable output of sources that influence zone's temperature. Such sources are the power of a heater located in the zone, the air temperature at various positions outside the zone, and so forth. The vector x_k is the state vector of the process at the discrete time k. The states have a conceptual relevance only and they are not assigned physical interpretation. Of course a similarity transform can convert the states to physical meaningful ones. The unmeasurable vector signal w_k represents noise due to the presence of humans in the zone, lamps' switching on and off, and so forth. For the rest of the paper we neglect the effects of state and output noise and thus we deal with a deterministic identification problem, which is the computation of the matrices A, B, C, and D from the given input-output data. The no noise assumption makes sense since we need temperature-time predictions for the empty zones only which are "free" from occupants and other disturbances.

In order to estimate the state-space matrices A, B, C, and D of each zone, we need a training phase during which all the nodes in the structure report their relative measurements to a central computing system. These measurements consist of the temperature of the zone and the current power of a heater, if the sensor node is in charge of a heated zone, or only the temperature if the sensor node monitors an area immaterial to the decision process but relevant to other zones, that is, the exterior of a building. The training phase should be performed only once but under controlled conditions, for example, no extra sources of heating and closed windows. Moreover, the length of this period should be quite large in order to exhibit several variations in the input signals to excite the modes of the system. In the simulation part we considered a training period of 24 h, that is, 86400 samples with sampling period 1 sec. The sampling period of 1 sec is too small for all practical reasons. However, as the dominant factor of performance degradation is the estimation of the residual arrival times we consider an ideal temperature prediction model (high accuracy sensors and temperature noiseless zones) and we access only the uncertainty of the occupation model.

After receiving all the measurements, the central system organizes them to input-output data for each zone. For example, if a zone is named **A**, neighbors zones are named **B**, **C**, and **D** and the exterior to the building zone is named **E**, then the temperature of zone **A** is the output of the system to be identified, whereas the temperatures of zones **B**, **C**, **D**, and **E** as well as the power profile of the heater of zone **A** are the input data for the system. A deterministic subspace identification process, described in the following paragraphs, is then run for each zone. The relevant matrices A, B, C, and D of the state-space model of each zone are communicated back to the sensor node responsible for the monitoring of the zone.

Following the notation and the derivation in [10] we define the block Hankel matrices

$$U_p = U_{0|i-1} = \begin{pmatrix} u_0 & u_1 & \cdots & u_{j-1} \\ u_1 & u_2 & \cdots & u_j \\ \vdots & \vdots & \vdots & \vdots \\ u_{i-1} & u_i & \cdots & u_{i+j-2} \end{pmatrix},$$

$$\qquad\qquad\qquad\qquad\qquad\qquad\qquad (7)$$

$$U_f = U_{i|2i-1} = \begin{pmatrix} u_i & u_{i+1} & \cdots & u_{i+j-1} \\ u_{i+1} & u_{i+2} & \cdots & u_{i+j} \\ \vdots & \vdots & \vdots & \vdots \\ u_{2i-1} & u_{2i} & \cdots & u_{2i+j-2} \end{pmatrix}$$

which represent the "past" and the "future" input with respect to the present time instant i. Stacking U_p on top of U_f we have the block Hankel matrix

$$U_{0|2i-1} = \begin{pmatrix} U_{0|i-1} \\ \hline U_{i|2i-1} \end{pmatrix} = \begin{pmatrix} U_p \\ \hline U_f \end{pmatrix} \qquad (8)$$

which can also be partitioned as

$$U_{0|2i-1} = \begin{pmatrix} U_{0|i} \\ \hline U_{i+1|2i-1} \end{pmatrix} = \begin{pmatrix} U_p^+ \\ \hline U_f^- \end{pmatrix} \qquad (9)$$

by moving the "present" time i one step ahead to time $i + 1$. Similarly, we define the output block Hankel matrices $Y_{0|2i-1}$, Y_p, Y_f, Y_p^+, Y_f^- and the input-output data matrices

$$W_p = W_{0|i-1} = \begin{pmatrix} U_{0|i-1} \\ \hline Y_{0|i-1} \end{pmatrix} = \begin{pmatrix} U_p \\ \hline Y_p \end{pmatrix},$$

$$\qquad\qquad\qquad\qquad\qquad\qquad\qquad (10)$$

$$W_p^+ = \begin{pmatrix} U_p^+ \\ \hline Y_p^+ \end{pmatrix}.$$

The state sequence X_m is defined as

$$X_m = (x_m \quad x_{m+1} \quad \cdots \quad x_{m+j-1}) \qquad (11)$$

and the "past" and "future" state sequences are $X_p = X_0$ and $X_f = X_i$, respectively. With the state-space model (13), (15) we associate the extended observability matrix Γ_i and the reversed extended controllability matrix Δ_i, where

$$\Gamma_i = \begin{pmatrix} C \\ CA \\ CA^2 \\ \vdots \\ CA^{i-1} \end{pmatrix},$$

$$\qquad\qquad\qquad\qquad\qquad\qquad\qquad (12)$$

$$\Delta_i = \begin{pmatrix} A^{i-1}B & A^{i-2}B & \cdots & AB & B \end{pmatrix}.$$

The subspace identification method relies on determining the state sequence X_f and the extended observability matrix Γ_i directly from the input-output data u_k, y_k and based on them, in a next step, to extract the matrices A, B, C, and D. To this end we define the oblique projection of the row space of Y_f along the row space of U_f on the row space of W_p:

$$\mathcal{O}_i = Y_f/_{U_f} W_p = \left[Y_f/U_f^\perp \right] . \left[W_p/U_f^\perp \right]^\dagger W_p, \qquad (13)$$

where $\dagger$ denotes the Moore-Penrose pseudoinverse of a matrix and $A/B^\perp$ is a shorthand for the projection of the row space of matrix A onto the orthogonal complement of the row space of matrix B; that is,

$$A/B^\perp = I - AB^T \left(BB^T \right)^\dagger B. \qquad (14)$$

As it is proved in [10] the matrix $\mathcal{O}_i$ is equal to the product of the extended observability matrix and the "future" state vector

$$\mathcal{O}_i = \Gamma_i X_f. \qquad (15)$$

Having in our disposal the matrix $\mathcal{O}_i$ we use its singular value decomposition

$$\mathcal{O}_i = (U_1 \quad U_2) \begin{pmatrix} S_1 & 0 \\ 0 & 0 \end{pmatrix} \begin{pmatrix} V_1^T \\ V_2^T \end{pmatrix} = U_1 S_1 V_1^T \qquad (16)$$

and we identify the extended observation matrix Γ_i and the state vector X_f as

$$\Gamma_i = U_1 S_1^{1/2} T,$$

$$\qquad\qquad\qquad\qquad\qquad\qquad\qquad (17)$$

$$X_f = T^{-1} S_1^{1/2} V_1^T = \Gamma_i^\dagger \mathcal{O}_i,$$

where T is an $n \times n$ arbitrary nonsingular matrix representing a similarity transformation.

One method [10] to extract the matrices A, B, C, and D uses in addition the oblique projection

$$\mathcal{O}_{i-1} = Y_f^-/_{U_f^-} W_p^\dagger = \Gamma_{i-1} X_{i+1}, \qquad (18)$$

where Γ_{i-1} is obtained from Γ_i by deleting the last ℓ rows ($\ell = 1$ in our case). Similar to the previous derivation we have

$$X_{i+1} = \Gamma_{i-1}^\dagger \mathcal{O}_{i-1} \qquad (19)$$

and the matrices A, B, C, and D are obtained by solving (in least squares fashion) the system

$$\begin{pmatrix} X_{i+1} \\ Y_{i|i} \end{pmatrix} = \begin{pmatrix} A & B \\ C & D \end{pmatrix} \begin{pmatrix} X_i \\ U_{i|i} \end{pmatrix}. \qquad (20)$$

After completion of the training phase the state-space model matrices A, B, C, and D for each zone are communicated back to the node responsible for the zone. Upon reception of the matrices the nodes enter a "convergence" phase; that is, starting from an initial state x_0, they update the state (5) for a predetermined period of time, using the measurements of the surrounding zones and the zone's heater status. Note that the training and the convergence phase need to be executed only once prior to the normal operation of the nodes. After the convergence phase the nodes enter the "decision" phase. During this phase the nodes update the state of the model (as in the convergence phase) and periodically they take decisions on whether to turn the zone's heater on or not. Suppose that at the decision epoch t the state of the tagged unoccupied node is x_t. The node bases its decision on the estimates $S_i^{on}(t)$ and $S_i^{off}(t)$. Recall that $S_i^{on}(t)$ is the time needed to reach the comfort level T_c if the heater is on. That is,

$$S_i^{on}(t) = \min \ell : y_{t+\ell} \geq T_c. \tag{21}$$

The node "freezes" the input-vector u_k to its current value u_t (more recent temperatures of the surrounding zones and the tagged zone's heater is on) and starting from state x_t repeats (5), (6) until y_{t+k} exceeds the value T_c. Note that

$$y_{t+k} = CA^k x_t + CA^{k-1} Bu_t + CA^{k-2} Bu_{t+1} + \cdots$$
$$+ CABu_{t+k-2} + CBu_{t+k-1} + Du_{t+k} \tag{22}$$

and for $u_t = u_{t+1} = \cdots = u_{t+k}$ we obtain

$$y_{t+k} = CA^k x_t + C(I + A + \cdots + A^{k-1}) Bu_t + Du_t$$
$$= CA^k x_t + C(I - A)^{-1}(I - A^k) Bu_t + Du_t. \tag{23}$$

If we set $y_{t+k} = T_c$ and

$$w = T_c - C(I - A)^{-1} Bu_t - Du_t, \tag{24}$$

then (23) takes the form

$$w = CA^k x_t - C(I - A)^{-1} A^k Bu_t$$
$$= CV\Lambda^k V^{-1} x_t - C(I - A)^{-1} V\Lambda^k V^{-1} Bu_t, \tag{25}$$

where we have considered the diagonalization of the matrix A as $A = V\Lambda V^{-1}$, with Λ being the diagonal matrix of the eigenvalues and V being the matrix of the corresponding eigenvectors. Numerical methods can be used to solve (25) for k. In a similar manner we compute $S_i^{off}(t)$. In this case, however, we first repeat (5) D times with input-vector u_k reflecting the fact that the heater is off, and then starting from the new state x_{t+D} we estimate the time needed to reach T_c if the heater is on.

3.2. On the Distribution of $Y_n(t)$.

The random variable $Y_n(t)$ expresses the remaining time from the current time instant t until entrance to zone z_n. That is, we assume that occupants move around as a swarm, visiting zones either in a predetermined order or in a random fashion and finally they end up in zone z_n. Each time the occupants visit a zone z_ℓ they

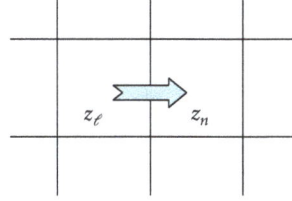

FIGURE 2: Movement of occupants from zone z_ℓ to the adjacent unoccupied zone z_n.

stay in it for a nonzero time period which is itself a random variable with distribution $F_\ell(x)$. We use the variable t to designate calendar time, the time elapsed since the process started, and τ to designate the time elapsed since entry to a zone. The zone occupation time distribution $F_\ell(x)$ may be given parametrically either a priori or inferred from examination of historic data by suitable fitting of the distribution's parameters. We may use the empirical distribution function (e.d.f.) instead which is the nonparametric estimate of $F_\ell(x)$. That is, if a sample of N occupation periods of zone z_ℓ, $\{x_1, x_2, \ldots, x_N\}$ is available, then

$$\widehat{F}_\ell(x) = \frac{1}{N} \sum_{m=1}^{N} I(x_m \in [0, x]), \tag{26}$$

where $I(\cdot)$ is the indicator function.

Let us consider first the simple case of two adjusting zones z_ℓ, z_n with zone z_ℓ being occupied and z_n unoccupied as it is shown in Figure 2. We assume that the occupants remain to zone z_ℓ for a random period of time X_ℓ which is distributed according to $F_\ell(x)$ and then upon departure they enter zone z_n. In this case $Y_n(t)$ expresses the remaining waiting time to zone z_ℓ and follows the residual life distribution which is defined by

$$\overline{R}_{Y_n(t)}(y) = 1 - R_{Y_n(t)}(y) = P\{X_\ell > \tau + y \mid X_\ell > \tau\}$$
$$= \frac{\overline{F}_\ell(\tau + y)}{\overline{F}_\ell(\tau)}. \tag{27}$$

Next we consider the case that occupants are currently in a zone (call it z_0) and upon departure they visit in cascade zones $z_1, z_2, \ldots, z_n$ as it is shown in Figure 3. Then

$$Y_n(t) = Y_0 + X_1 + X_2 + \cdots + X_{n-1}, \tag{28}$$

where Y_0 is the remaining time in zone z_0 and X_i is the occupation period of zone z_i. It is well known that the distribution of $Y_n(t)$ is the convolution of the distributions of the random variables $Y_0, X_1, \ldots, X_{n-1}$. That is

$$F_{Y_n(t)}(y) = P\{Y_n(t) \leq y\}$$
$$= R_0(y) \star F_1(y) \star F_2(y) \star \cdots \star F_{n-1}(y) \tag{29}$$

with $R_0(y) = 1 - \overline{F}_0(\tau + y)/\overline{F}_0(\tau)$. A more compact form is obtained by considering a transform of the distributions, that

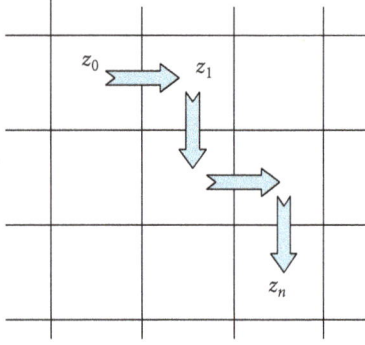

FIGURE 3: Movement of occupants in cascade from zone z_0 to zone z_n.

is, the Laplace transform, the moment generating function, or the characteristic function. Using the moment generating function (mgf) which for a distribution $F_X(x)$ of a nonnegative random variable X is defined by

$$F_X(s) = \int_0^\infty e^{sx} dF_X(x), \qquad (30)$$

(29) takes the form

$$F_{Y_n(t)}(s) = R_0(s) \cdot F_1(s) \cdot F_2(s) \cdot \ldots \cdot F_{n-1}(s). \qquad (31)$$

In some cases (unfortunately few) (26) can be inversed analytically in order to calculate the distribution $F_{Y_n(t)}(y)$. For example, if the occupation periods X_i, $i = 0, \ldots, n-1$, are exponentially distributed with corresponding parameters λ_i, then Y_0 is also exponentially distributed with parameter λ_0 due to the memoryless property of the exponential distribution, and the probability density function of $Y_n(t)$ is (assuming that the parameters λ_i are all distinct)

$$f_{Y_n(t)}(x) = \sum_{i=0}^{n-1} \frac{\lambda_0 \cdots \lambda_{n-1}}{\prod_{j=0, j\neq i}^{n-1}(\lambda_j - \lambda_i)} \exp(-x\lambda_i). \qquad (32)$$

(The expression for the general case, of nondistinct parameters, is slightly more involved but still exists.) If a closed form for the density $f_{Y_n(t)}(x)$ does not exist or it is difficult to be obtained, we resort to the saddlepoint approximation [11] to invert $F_{Y_n(t)}(s)$. To this end, for a random variable X with cumulative distribution function (cdf) $F_X(x)$, we define the cumulant generating function (cgf) $K_X(s)$ as the logarithm of the corresponding mfg $F_X(s)$; that is,

$$K_X(s) = \log F_X(s). \qquad (33)$$

Then, the saddlepoint approximation of the probability density function (pdf) $f_X(x)$ is

$$f_X(x)$$
$$\approx \left(\frac{1}{2\pi K_x''(\widehat{s}(x))}\right)^{1/2} \exp\{K_X(\widehat{s}(x)) - \widehat{s}(x)x\}, \qquad (34)$$

where $\widehat{s}(x)$ is the solution to the saddlepoint equation

$$K_X'(s) = x. \qquad (35)$$

Estimation (34) relies on numerical computations based on cgf $K_X(S)$, which in turn depends on $F_X(s)$. Note that we may use the empirical mgf $\widehat{F}_X(s)$ instead, so that a nonparametric estimation of the pdf $f_X(x)$ is possible.

The more general case is to model the movement of the occupants in the multizone system as a semi-Markov process. The states of the process are the zones of the system and at the transition times they form an embedded Markov chain with transition probabilities p_{ij}. Given a transition from zone z_i to zone z_j, the occupation time in z_i has distribution function $F_{ij}(x)$. The matrix $\mathbf{Q}(x)$ with elements $Q_{ij}(x) = p_{ij}F_{ij}(x)$ is the so called semi-Markov kernel. The occupation time distribution in zone z_i independent of the state to which a transition is made is $F_i(x) = \sum_j Q_{ij}(x)$. Consider now the case that at time t occupants exist in zone z_i and the node of zone z_n has to make a decision on whether to start heating the zone or not. For the decision process the distribution of time until entrance to zone z_n, $Y_n(t)$, is needed. $Y_n(t)$ is the sum of two terms. The first term is the residual occupation time at zone z_i and the second term is the time to reach z_n upon departure from zone z_i. The distribution $R_i(x)$ of the first term is given by (27), whereas for the second term we have to consider the various paths taken upon departure from state z_i. Thus, if zone z_i neighbors N_i zones, namely, z_k, $k = 1, \ldots, N_i$, then the time to reach zone z_n upon departure is distributed according to

$$G_{in}(x) = \sum_{k=1}^{N_i} p_{ik} H_{kn}(x), \qquad (36)$$

where $H_{kn}(x)$ is the distribution of the first passage from zone z_k to zone z_n. Pyke [12] and Mason [13, 14] provided solutions for the first passage distribution in a semi-Markov process. For example, Pyke's formula states that $H(s)$, the elementwise transform of the matrix $H(x) = \{H_{ij}(x)\}$, is given by

$$H(s) = \mathscr{T}(s)[I - \mathscr{T}(s)]^{-1}\left\{\left([I - \mathscr{T}(s)]^{-1}\right)_d\right\}^{-1}, \qquad (37)$$

where $\mathscr{T}(s)$ is the transmittance matrix with the transforms $\mathscr{T}_{ij}(s)$ of the distributions $F_{ij}(x)$. Notation M_d denotes the matrix that is formed by the diagonal elements of M. A simplified version of (37) is obtained if we let $\Delta(s) = \det(I - \mathscr{T}(s))$ and $\Delta_{ij} = (-1)^{i+j}\det(I - \mathscr{T}(s))_{ij}$, the ijth cofactor of $I - \mathscr{T}(s)$. Using this notation, for a semi-Markov process of n states, we get

$$H_{1n}(s) = \frac{\Delta_{n1}(s)}{\Delta_{nn}(s)}. \qquad (38)$$

By relabeling the states $1, \ldots, n-1$ of the process we can find the transform of the first passage from any zone to zone z_n.

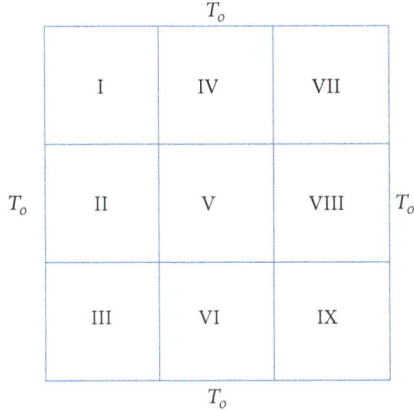

FIGURE 4: Multizone model.

TABLE 1: Training and testing parameters for outer temperature model and heat gain.

Phase	Walter's model			Heat gain
	T_{max}	T_{min}	T_m	$q(t)$
Training	15	2	7	600
Testing	16	1	6	700

TABLE 2: Training and testing zones' target temperatures.

Zone #	Target temperatures								
	I	II	III	IV	V	VI	VII	VIII	IX
Training	16	18	16	15	14	13	12	16	14
Testing	17	20	17	17	17	14	18	19	21

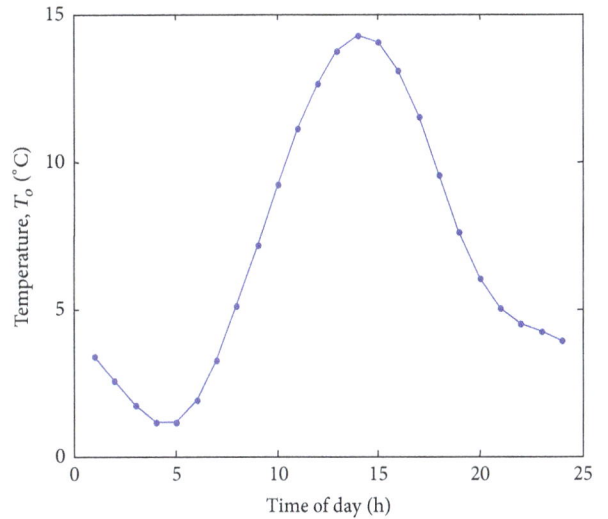

FIGURE 5: Empirical estimation of T_o.

In summary,

$$F_{Y_n(t)}(s) = R_i(s) \cdot G_{in}(s) = R_i(s) \cdot \left(\sum_{k=1}^{N_i} p_{ik} H_{kn}(s) \right), \quad (39)$$

with $H_{kn}(s)$ computed using (38). Next, the saddlepoint approximation technique can be used to calculate the density $f_{Y_n(t)}(x)$.

We have to mention that the process of estimating the first passage distributions needs to be performed only prior to the decision process. Thus, given the topology of the multizone system, we posit a parametric family for the occupation time distributions and we estimate the distributions' parameters from sample data. Then, the transition probabilities are estimated based on the data and the occupation mgf $F_{ij}(s)$ is obtained. Having in our disposal the transforms $F_{ij}(s)$ we can form the transmittance matrix $\mathcal{T}(s)$ and solve for the first passage transform $H_{ij}(s)$ using (38). Finally, numerical methods may be used to invert these transforms in order to obtain the necessary probability density functions.

4. Simulation Results

We simulated a multizone system consisting of a squared arrangement of rooms (zones) where each room (zone) is equipped with a wireless sensor node as shown in Figure 4. We assume that heat transfer takes place between room's air and the walls as well as between room's air and the roof. Furthermore, we neglect the effects of ground on the room temperature. North and south walls have the same effect on the room temperature and we assume the same for the east and west walls. According to these assumptions, there is a symmetry in the dynamics of the zones. For example, zones I, III, VII, and IX exhibit the same behavior. Note that, for the multizone system of Figure 4 several state variables, such as wall temperatures, are common to the individual zone systems. In the simulations we treat the multizone system as a single system with 42 states and nine outputs (the zones' temperatures). The parameter values for this lumped capacity zone model were taken from [15] (see also [16]).

In Figure 4, T_o denotes the outside temperature which is assumed to be uniform with no loss of generality. For simulation purposes daily outside temperature variations are obtained using Walters' model [17]. For this model, the maximum temperature of day T_{max}, the minimum temperature of day T_{min}, and the mean of the 24 hourly temperatures T_m need to be provided in order to estimate T_o. As an example, if $T_{max} = 15°C$, $T_{min} = 2°C$, and $T_m = 7°C$, the daily variations of temperature in Figure 5 are obtained, which shows that temperature is higher at 2:00–3:00 pm.

4.1. Training Phase. For the deterministic subsystem identification we used 86400 samples with sampling period $T_s = 1$ sec. For the outside temperature we used Walter's model with parameters given in the first row of Table 1 and a heater gain equal to 600 W. We set the zones' target temperatures T_{tg} to those of Table 2 (first row) with a margin equal to $T_{mg} = 1°C$. That is, for a specific zone the heater is on until temperature $T_{tg} + T_{mg}$ is reached. After this point the heater is turned off until temperature hits the lower threshold $T_{tg} - T_{mg}$ and the whole process is repeated.

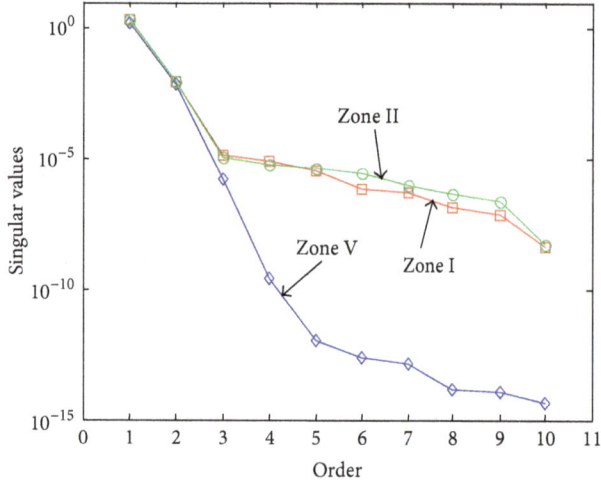

FIGURE 6: Singular values of zones I, II, and V.

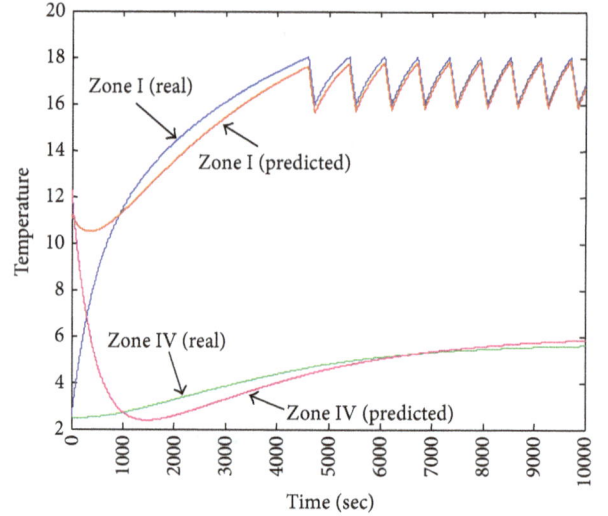

FIGURE 7: Real and predicted temperature of zones I and IV.

Having collected the data over the period of 24 hours, the measurements of the zones' temperature, the outer temperature, and the heat gains of each zone are organized into input-output data for the subsystem identification process. For example, for zone I, the outer temperature, the heating gain of zone I, and the temperatures of zones II and IV are the input data to the identification process whereas the temperature of zone I itself is the output data. Note that the number of input signals differs from zone to zone. Zones I, III, VII, and IX use 4 input signals, whereas the rest of the zones use 5 input signals. Based on the input-output data we identify the matrices A, B, C, and D for each zone as described in Section 3. During this process we have to decide on the order of the subsystems. This is achieved by looking at the singular values of the SVD decomposition in (20) and deciding on the number of dominant ones. Figure 6 depicts the singular values for zones I, II, and V. All subsystems exhibit similar behavior regarding the profile of their singular values and therefore we set the order of all subsystems equal to 2.

Next we run a test on the obtained state-space models. We set the outer temperature parameters as in the second row of Table 1 and the target temperatures as in the second row of Table 2. The initial state of the subsystem model was set to $[-10 \ 0]$, whereas the elements of the 42-state vector of the multizone system were set equal to 3 (the initial output temperature). Figure 7 shows the evolution of the actual and predicted temperatures for zones I and IV. For zone I the heater was on (heat gain equal to 700) whereas the heater for zone IV was off. As it is observed, after 5000 samples (appoximately 80 minutes period) the state of the subsystems has converged to one that produces almost the same output as the original system. After this point the WSN nodes can enter in the "decision" phase.

We should notice at this point that the "original" simulated system (42-state-space model) is also linear and thus the use of linear subspace identification may be questionable for more complex and possibly nonlinear systems. As the simulation results indicate, although the dynamics of each

zone are more complex (including the roof temperature,e.g.), a low dimensionality subsystem of order 2 can capture its behavior. For more complex systems we can choose the order of the identified subsystem to be high enough. For nonlinear systems the identification of a time varying system is possible by using a recursive update of the model.

Next we simulate the scenario of Figure 2. We assume that occupants enter zone I (at time instant 5000) and then after a random period of time they enter zone IV, where they remain for 3600 sec. Zone I is heated since the beginning of the process (heat gain 900 W) until the occupants leave the zone to enter zone IV. Zone IV uses a heater with gain 1200 W, which is turned on (or off) according to the decisions of its node. The target temperatures for zones I and IV are 17°C and 16°C, respectively. For the first set of experiments, the occupation period of zone I is Gamma distributed, that is,

$$f_X(x) = \frac{\beta^\alpha}{\Gamma(\alpha)} x^{\alpha-1} e^{-\beta x}, \qquad (40)$$

with the shape parameter $\alpha = 5$ and β such that the expected value of X, $E[X] = \alpha/\beta$, is 3600 and 7200. Decisions were taken every $D = 67$ seconds. Figure 8 shows the total comfort cost achieved for various values of the comfort weight C_i. The results are averages over 200 runs of the simulation. The case of $C_i = 0$ is the full reactive case where heating of zone IV is postponed until occupants enter the zone. On the other extreme, for high values of C_i. the decision process acts proactively by spending energy in order to reduce the discomfort level.

Figure 9 shows the total energy cost in KWh for the two cases of Figure 8. As it is observed, the higher the value of C_i the more the energy consumed since then heating of zone IV starts earlier. The difference between the two curves of Figure 9 is justified by the difference of the means of the occupation period, $E[X] = 3600$ and $E[X] = 7200$, for the two cases.

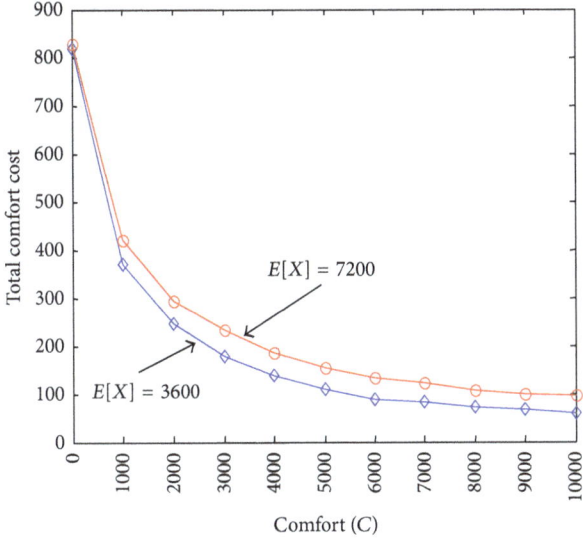

FIGURE 8: Total comfort cost versus C_i. The occupation time of zone I is Gamma distributed.

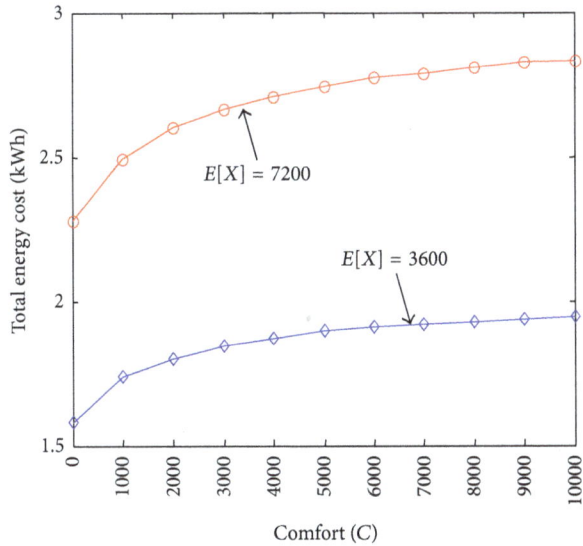

FIGURE 9: Total energy cost versus C_i. The occupation time of zone I is Gamma distributed.

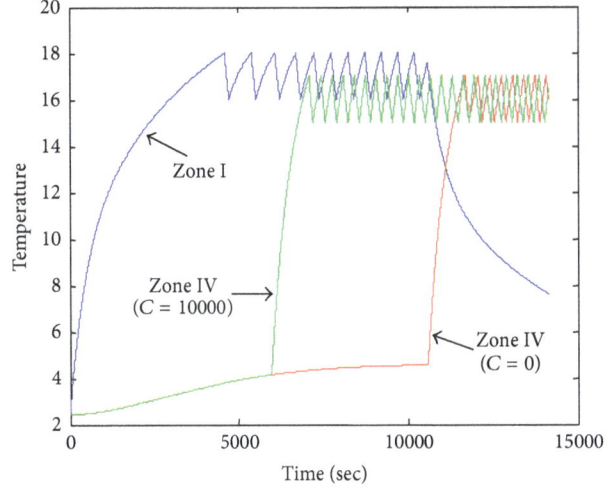

FIGURE 10: Temperature versus time for $C = 10000$ and $C = 0$.

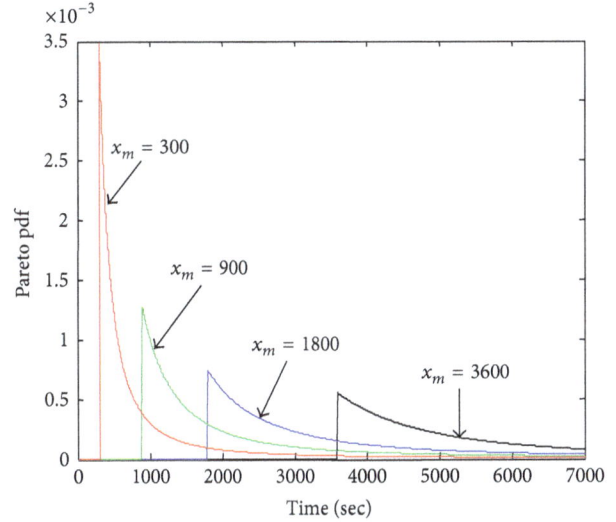

FIGURE 11: Pareto pdf for various values of x_m.

In Figure 10 we plot the temperature profiles for zones I, IV and for two different values of C. As it is observed the value of $C = 0$ corresponds to the reactive case; that is the heater of zone IV remains off until occupants are detected in the zone. On the other extreme a large value of C (10000) will force the heater of zone IV to be turned on as soon as possible in an effort to reduce the discomfort penalty.

The scenario of Figure 2 was also simulated for Pareto distributed occupation times of zone I. The pdf for Pareto distribution is

$$f_X(x) = \begin{cases} \dfrac{\alpha x_m^{\alpha}}{x^{\alpha+1}}, & x \geq x_m \\ 0, & x < x_m \end{cases} \tag{41}$$

with expected value

$$E[X] = \begin{cases} \dfrac{\alpha x_m}{\alpha - 1}, & \alpha > 1 \\ \infty, & \alpha \leq 1. \end{cases} \tag{42}$$

Figure 11 shows the Pareto pdf for four values of x_m ($x_m = 300, 900, 1800, 3600$) and α suitably chosen so that the mean value of the occupation period is 7200 in all cases.

Figures 12 and 13 show the total comfort cost and the total energy consumed, respectively, for various values of the comfort gain C_i. As it is observed, a 50% improvement on the comfort cost (compared to the reactive case) can be achieved for a value of C equal to 1000. However, there is a performance floor (this is evident for the case of $x_m = 300$), which is due to the nature of the Pareto distribution. Most of the samples for the occupation time period are concentrated close to x_m

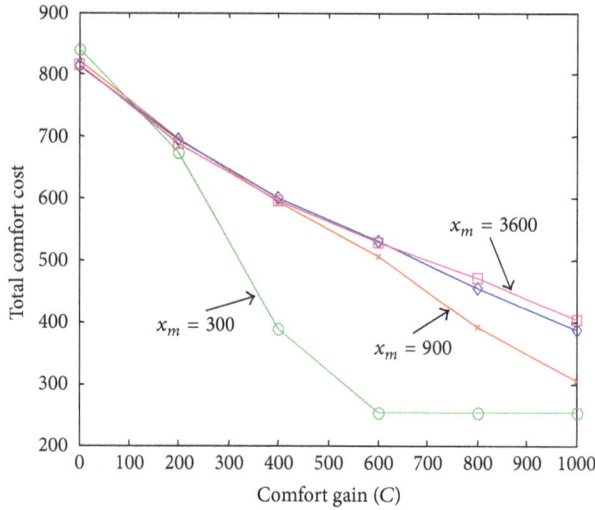

FIGURE 12: Total comfort cost for Pareto distributed occupation period.

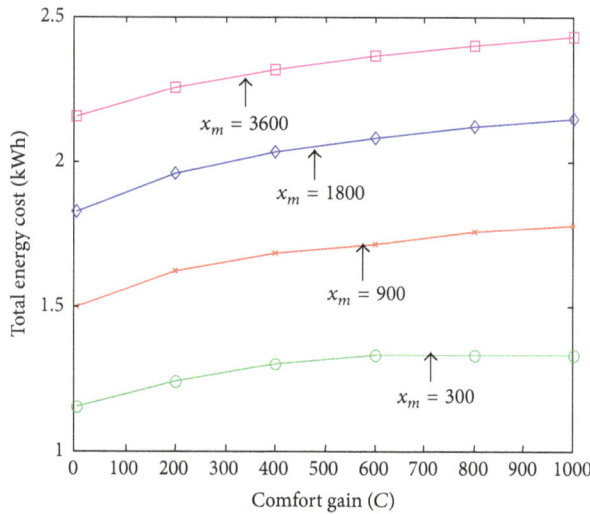

FIGURE 13: Total energy cost for Pareto distributed occupation period.

TABLE 3: Parameters used in cascade movement scenario.

Zone #	I	IV	V	VIII	IX	VII	III	II
Target °C	17	16	17	17	16	14	17	18
Power (KW)	0.9	1.2	0.9	0.9	0.9	1.2	0.9	1.2
Mean time $\times 10^2$	36	27	18	40	54	45	30	20

TABLE 4: Energy and comfort costs for constant occupancy times.

	Comfort weight C	500	1000	1500	2000
$D = 67$	Comfort cost	2615	72	67	67
	Energy cost	7.75	9.07	9.73	10.20
$D = 167$	Comfort cost	2856	72	71	64
	Energy cost	7.79	9.11	9.77	10.25

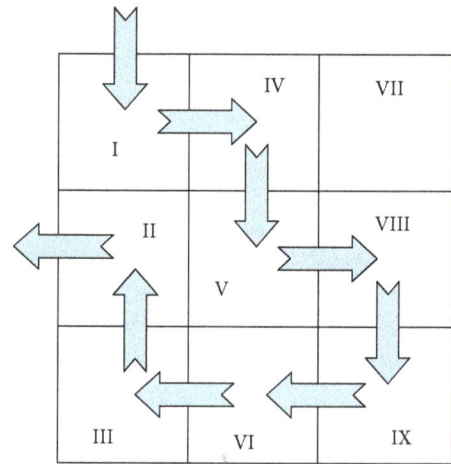

FIGURE 14: Movement of occupants in cascade (zones I, IV, V, VIII, IX, VI, III, and II).

and therefore there is not enough time to preheat the next visiting zone, which in turn implies high discomfort costs. Moreover, the heavy tail of the Pareto distribution causes some extremely high values of the occupation period which result in an increase of the total consumed energy, as it can be observed clearly from Figure 13.

Next, we simulate the scenario of Figure 3. The occupants move in cascade from zone to zone as it is depicted in Figure 14. The target temperatures for the zones are given in the first row of Table 3, whereas the heat gain of each zone is given in the second row of Table 3. For the occupancy time of each zone we considered an exponential random variable with mean provided in the last row of Table 3. Note that this assumption is the least favorable for adjacent zones due to the memoryless property of the exponential distribution.

The total comfort and energy costs for this scenario are plotted in Figure 15 for various values of the comfort weight C. The results are averages over 100 runs. The two curves for each cost correspond to $D = 67$ and $D = 167$, respectively. As it is observed, considerable comfort gains are achieved for a moderate increase of the consumed energy. In Figure 15 we also show the costs for a "fixed" proactive scenario in which the heating of a zone starts as soon as occupants are detected to its neighbor zone. For example, when occupants enter zone IV, the heater of zone V is turned on, and so on.

Table 4 shows the comfort and energy costs obtained when there is no randomness on the occupancy times for various values of the comfort weight C and for two different values of D ($D = 67$ and 167). The occupancy times of the zones were set to the mean values provided in the last row of Table 3. As it is observed, extremely low comfort costs can be achieved in this case which implies that the right modelling of the occupancy times is of paramount importance to the process.

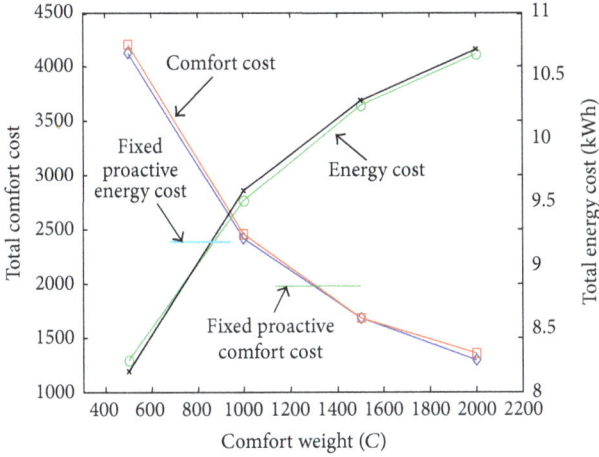

FIGURE 15: Total energy and comfort costs (cascade movement scenario).

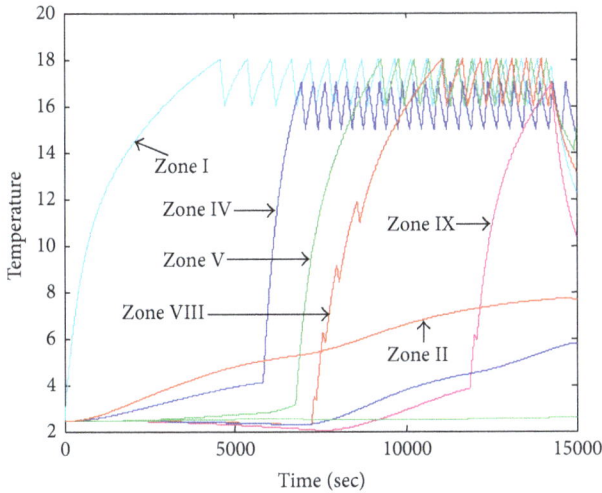

FIGURE 16: Temperature profiles of zones (cascade movement scenario).

Finally, Figure 16 shows a sample of the zones' temperature profile. Note that the temperature of zone II, although zone II is the last zone visited, starts increasing (at a slower rate) much earlier. This is because zone II neighbors zones I and V which are heated in earlier stages of the process.

5. Conclusions and Future Work

A method that fuses temperature and occupancy predictions in an effort to balance the thermal comfort condition and the consumed energy in a multizone HVAC system was presented. Temperature predictions are based on a subspace identification technique used to model the thermal dynamics of each zone independently. This technique is quite robust and accurate predictions are possible after a convergence period which may last 1-2 hours. With regard to the occupancy predictions the decision process makes use of the distribution of the first passage time from an occupied zone

to an unoccupied one. Simulation results for different first passage distributions demonstrated the effectiveness of the proposed technique.

Several extensions and modifications of the proposed method are possible. The comfort parameter C_i may depend on the zone, the number of occupants, and the current status of the zones. Moreover, this parameter may be time variable, that is, different values may be used for day and night hours. Regarding the decision process itself, there is no need that it be executed at regular time epochs. If the energy consumed by the wireless sensor nodes is an issue, we may take decisions at irregular time epochs, depending on the occupancy predictions of the zones. The computation of the risks, used by the decision process, may take into consideration additional parameters that affect energy consumption and/or the thermal comfort of the occupants. For example, open windows situation may be easily detected and incorporated suitably in the decision process.

Conflict of Interests

The authors declare that there is no conflict of interests regarding the publication of this paper.

References

[1] US Energy Information Administration, "Annual Energy Outlook 2015," DOE/EIA-0383, April 2015, http://www.eia.gov/.

[2] U.S. Energy Information Administration, June 2015, http://www.eia.gov.

[3] S. Lee, Y. Chon, Y. Kim, R. Ha, and H. Cha, "Occupancy prediction algorithms for thermostat control systems using mobile devices," *IEEE Transactions on Smart Grid*, vol. 4, no. 3, pp. 1332–1340, 2013.

[4] V. L. Erickson, S. Achleitner, and A. E. Cerpa, "POEM: power-efficient occupancy-based energy management system," in *Proceedings of the 12th International Conference on Information Processing in Sensor Networks (IPSN '13)*, pp. 203–216, ACM, Philadelphia, Pa, USA, April 2013.

[5] V. L. Erickson, M. Á. Carreira-Perpiñán, and A. E. Cerpa, "OBSERVE: occupancy-based system for efficient reduction of HVAC energy," in *Proceedings of the 10th ACM/IEEE International Conference on Information Processing in Sensor Networks (IPSN '11)*, pp. 258–269, Chicago, Ill, USA, April 2011.

[6] B. Balaji, J. Xuy, A. Nwokafory, R. Guptay, and Y. Agarwal, "Sentinel: occupancy based HVAC actuation using existing WiFi infrastructure within commercial buildings," in *Proceedings of the 11th ACM Conference on Embedded Networked Sensor Systems (SenSys '13)*, Rome, Italy, November 2013.

[7] B. Dong and K. P. Lam, "A real-time model predictive control for building heating and cooling systems based on the occupancy behavior pattern detection and local weather forecasting," *Building Simulation*, vol. 7, no. 1, pp. 89–106, 2014.

[8] J. R. Dobbs and B. M. Hencey, "Predictive HVAC control using a Markov occupancy model," in *Proceedings of the American Control Conference (ACC '14)*, pp. 1057–1062, Portland, Ore, USA, June 2014.

[9] J. Brooks, S. Kumar, S. Goyal, R. Subramany, and P. Barooah, "Energy-efficient control of under-actuated HVAC zones in commercial buildings," *Energy and Buildings*, vol. 93, pp. 160–168, 2015.

[10] P. Van Overschee and B. De Moor, *Subspace Identification for Linear Systems, Theory-Implementation-Applications,* Kluwer Academic, New York, NY, USA, 1996.

[11] H. E. Daniels, "Saddlepoint approximations in statistics," *Annals of Mathematical Statistics,* vol. 25, pp. 631–650, 1954.

[12] R. Pyke, "Markov renewal processes with finitely many states," *Annals of Mathematical Statistics,* vol. 32, pp. 1243–1259, 1961.

[13] S. J. Mason, "Feedback theory-some properties of signal flow graphs," *Proceedings of Institute of Radio Engineers,* vol. 41, no. 9, pp. 1144–1156, 1953.

[14] S. J. Mason, "Feedback theory—further properties of signal flow graphs," *Proceedings of the IRE,* vol. 44, no. 7, pp. 920–926, 1956.

[15] B. Tashtoush, M. Molhim, and M. Al-Rousan, "Dynamic model of an HVAC system for control analysis," *Energy,* vol. 30, no. 10, pp. 1729–1745, 2005.

[16] D. Sklavounos, E. Zervas, O. Tsakiridis, and J. Stonham, "A subspace identification method for detecting abnormal behavior in HVAC systems," *Journal of Energy,* vol. 2015, Article ID 693749, 12 pages, 2015.

[17] A. Walter, "Notes on the utilization of records from third order climatological stations for agricultural purposes," *Agricultural Meteorology,* vol. 4, no. 2, pp. 137–143, 1967.

9

Overcapacity as a Barrier to Renewable Energy Deployment: The Spanish Case

Pablo del Río[1] and Luis Janeiro[2]

[1]*Consejo Superior de Investigaciones Científicas, C/Albasanz 26-28, 28037 Madrid, Spain*
[2]*Ecofys, Kanaalweg 15-G, 3526 KL Utrecht, Netherlands*

Correspondence should be addressed to Pablo del Río; pablo.delrio@csic.es

Academic Editor: Umberto Desideri

Renewable energy sources (RES) play a critical role in the low-carbon energy transition. Although there is quite an abundant literature on the barriers to RES, the analysis of the electricity generation overcapacity as a barrier to further RES penetration has received scant attention. This paper tries to cover this gap. Its aim is to analyse the causes and consequences of overcapacity, with a special focus on the impact on RES deployment, using Spain as a case study. It also analyses the policies which may mitigate this problem in both the short and the longer terms.

1. Introduction

Climate change mitigation is arguably one of the greatest challenges facing humankind. Drastic emissions reductions will be required in order to put the world economy on emissions concentration path which minimizes the risk of collapse of the climate system. Generally, a 1.5- to 2-degree temperature increase above preindustrial levels is deemed compatible with this goal. In turn, this involves emissions concentration level of between 400 and 450 parts per million of CO_2 equivalent. Stabilizing concentrations at these concentration levels will require that global net emissions peak in the very short term, decline rapidly after that time, and reach zero soon after 2050. There is a wide consensus that the emissions reductions being required are substantial and that attaining such drastic reductions will require a mix of technologies. For example, the International Energy Agency [1] shows that, in order to put the world economy on an emissions path which is compatible with 2 degrees, emissions would have to be reduced by 40 $GtCO_2$ in 2050 with respect to a business-as-usual scenario. This reduction would be achieved in the least-cost manner through a combination of technologies, including renewables (34% of the required reduction), energy efficiency (33%), and carbon capture and storage (14%).

Renewable energy sources (RES) play a critical role in the low-carbon energy transition or energy revolution. There is quite an abundant literature on the relevance of RES in this energy transition as well as on the barriers to RES (see [2] for a review). However, to the best of our knowledge, the existence of electricity generation overcapacity as a barrier to further RES penetration has not been addressed in the literature so far. This paper tries to cover this gap. Its aim is to analyse the causes and consequences of overcapacity, with a special focus on the impact on RES deployment, using Spain as a case study. It also analyses the policies which may mitigate this problem in both the short and the longer terms. The focus on Spain is justified because this country has had both a substantial RES penetration in the past and a problem of overcapacity in its electricity generation sector. However, it is not the only country with this problem in the EU (see Section 2).

We rely on several sources of information. First, in order to identify the extent of the problem, official data on the electricity generation capacity and electricity demand have been used. Second, a deep literature review on the causes and consequences of this problem, relying on the views of a wide array of stakeholders, has been performed. Finally, a draft text was produced with the above-mentioned sources. It was sent to several stakeholders and their feedback and views were incorporated in the text.

Accordingly, this paper is structured as follows. The next section discusses the degree of the problem, that is, how much overcapacity exists in the Spanish electricity system. Section 3 is dedicated to the analysis of the causes, whereas Section 4 discusses the main consequences of overcapacity. Section 5 considers several policy options to address this problem. Section 6 provides the results of our interviews on the views of stakeholders. Section 7 concludes.

2. How Much Overcapacity Is There in the Electricity System?

Starting in the mid-90s, Spain was one of the first EU Member States to adopt effective support schemes for renewable energy in the power sector. By the time the current EU renewable energy directive (2009/28/EC) was adopted, Spain had already deployed more than 16 GW of wind power and more than 3 GW of solar photovoltaics, being one of the EU countries with a higher degree of penetration of renewables in its power system.

In 2010, Spain started a period of several profound energy policy changes in the direction of reducing support for RES. In January 2012, the Spanish Government decided to stop registration of new RES installations (Real Decreto-ley 1/2012, January 27th, 2012). In 2013, the Energy Reform was announced, introducing a completely new retribution scheme for RES, which included retroactive cuts in economic support for already existing installations (Real Decreto 413/2014, June 6th, 2014). These policy changes have resulted in a de facto halt of RES deployment in the country.

The main argument used to justify these decisions has been the need to contain the increasing costs of support for RES in order to correct the "tariff deficit" affecting the Spanish power system. (The tariff deficit refers to revenues (electricity prices) not covering regulated costs over the period. This tariff deficit reached 30000 million € in 2013, about 3% of Spain's GDP.) However, an often less mentioned contributing factor for this change in policy may be the fact that the Spanish power system suffers from a serious excess of generation capacity.

The electricity generation capacity installed in the last 15 years has increased significantly and virtually doubled between 2001 and 2014, from 55 GW in 2001 to 102 GW in 2014. Most new generation capacity came mainly from two technologies: wind power and natural gas-fuelled combined cycle gas turbines (CCGTs). Between 2003 and 2010, Spain installed 25.3 GW of natural gas-fuelled generation plants. From 2010 to 2014, the CCGT capacity has remained roughly constant. Over the same period 2003–2014, installed wind power increased from 6.2 GW to 22.8 GW. Figure 1 shows the evolution of the installed capacity per technology.

However, what seems more relevant is comparing the evolution of the installed capacity with the maximum electricity demand, which provides an indicator of the security of supply which the system has to safeguard.

Figure 2 shows the evolution of the installed generation capacity compared to the evolution of the hourly peak demand in the country (in the peninsular system). As it can

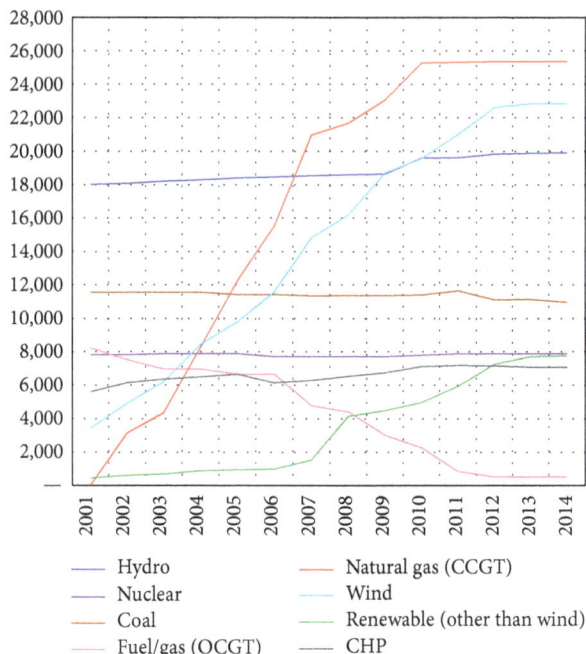

Figure 1: Evolution of the installed capacity (in MW) per technology. Source: our own elaboration from REE data.

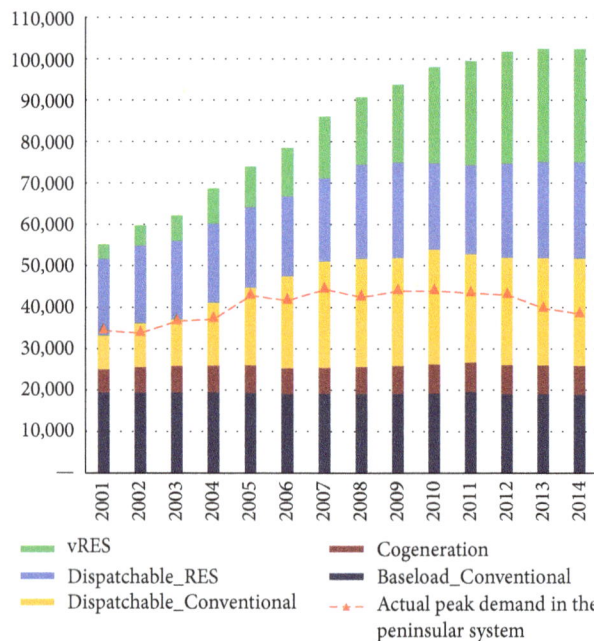

Figure 2: Peak power demand versus available capacity (MW). Source: our own elaboration. Data source: REE.

be seen from the graph, and in contrast to what happened in the last half of the 90s, the gap between the two has been steadily growing in the last 10 years. More importantly, the excess dispatchable capacity has also increased significantly over the same period.

Back in 2001, peak demand was roughly the same as the sum of all conventional installed capacities (including

TABLE 1: Installed power capacity and maximum peak power demand coverage.

	Installed power capacity as of 31 December 2013 (102 GW)	Coverage of maximum peak power demand 38.7 GW in 2014
Combined cycle	24.8	7%
Coal	10.7	2.5%
Nuclear	7.7	18.9%
Fuel/gas	0.5	0%
Cogeneration and others	7	10.8%
Hydro*	19.5	24%
Wind	22.3	34.5%
Solar PV	4.3	0.1%
Solar thermoelectric	2.2	0.1%
Renewable thermal	1	1.5%
Importer balance regarding international exchanges	—	0.6%

Source: Red Eléctrica de España [3].
*Pumped storage not included.

nuclear, coal, gas, and cogeneration plants). The 18 GW of dispatchable RES capacity available (mostly hydro) provided a "safety buffer" to guarantee security of supply. Additionally, there was an incipient, but already growing, amount of variable RES (vRES) in the system.

In 2014, the situation was radically different. Conventional capacity alone was already 13 GW larger than peak demand, which remained roughly at 2001 levels. Additionally, there are 23 GW of dispatchable RES capacity and 27 GW of variable RES capacity in the system. The installed capacity of RES in 2014 was even much higher than the historical peak demand, which occurred in 2007.

The most recent data show that installed power capacity more than doubles maximum peak demand in the day with the maximum electricity demand in the Spanish system (4 February 2014, between 8 p.m. and 9 p.m.); for example, compare columns 2 and 3 in Table 1.

Notwithstanding, the Spanish overcapacity problem has to be put in European context. Overcapacity problems are present or have been present in other EU countries. According to the "Leistungsbilanz" report [4], the day with the maximum load in Germany was December 5th 2013. On that day, the "secured capacity" was 116.3 GW and the maximum load was 79.1 GW. Therefore, there was a security margin 37.3 GW (47%).

However, the Spanish system has a larger excess capacity than other EU countries. Following the methodology of the European Network of Transmission System Operators for Electricity (ENTSO-E), the Spanish Ministry of Industry [5] determined that the "demand coverage ratio" of the Spanish system (1.73) was much higher than Italy (1.28), Germany (1.21), the UK (1.15), or France (1.09) (this study analysed values for the year 2011).

3. What Are the Factors Behind Overcapacity?

At a very general level, the causes of overcapacity are related to the factors affecting both sides of the problem, the supply, and the demand side, that is, factors influencing the substantial increase in capacity and other factors behind the stagnant evolution of electricity demand. We have identified a number of contributing causes, some of them interrelated to some extent, which we describe below.

3.1. Risks of Blackouts in the Late 1990s. A major risk in a context of highly increasing electricity demand (see below) is that this demand outpaces the installed generation capacity, leading to blackouts.

In the second half of the 90s, the gap between the "firm" generation capacity available (baseload and dispatchable plants) in the Spanish system and the peak hourly demand was closing at a very fast pace. Installed power in the "ordinary regime" (baseload plus dispatchable) was about 25000 MW in 1996, increasing to about 34000 MW in 2000. Peak hourly demand remained more or less flat in the same period at between 43000 and 45000 MW [6]. This resulted in some situations of high risks of blackouts in some regions of the country.

Indeed, in the early 2000s, several cuts in electricity supply occurred in Spain, which led governments to "urgently" approve a considerable number of CCGT projects [7]. The risk of a blackout was especially critical in regions located in the Mediterranean coast (Cataluña, Comunidad Valenciana, and Murcia) [8].

3.2. Planning: Wrong Projections of Electricity Demand. The aforementioned risk of blackouts led to a government response to increase the generation capacity (see below). Infrastructure planning is done years in advance and at that point demand was growing very quickly. Estimations of required new generation capacity were based on wrong projections of electricity demand, possibly influenced by the economic prosperity at the time, but apparently ignoring that economic ups and downs, in short cycles, are a main feature of virtually all economies. Indeed, the increasing and decreasing sides of the economic cycle are usually both more pronounced in Spain compared to the rest of EU countries.

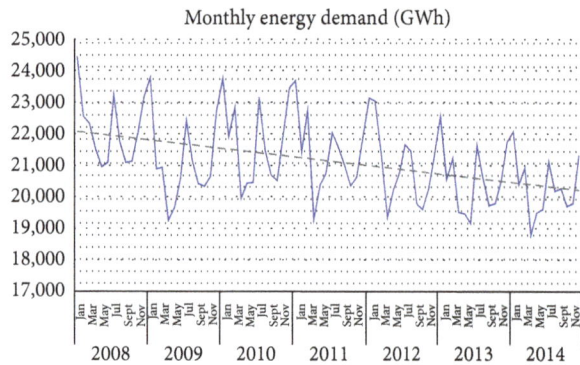

FIGURE 3: Monthly electricity demand 2008–2014. Source: REE [9] and REE [10].

TABLE 2: Objectives for RES-E in the two renewable energy plans (1999 and 2005).

	Objectives for 2010	
	PFER (1999)	PER (2005)
Small hydro (<10 MW)	2230	13521
Hydro (between 10 MW and 50 MW)	3151	3257
Large hydro (>50 MW)	13420	2199
Wind	8974	20155
Biomass	1897	1317
Cocombustion (biomass)		722
Biogas	78	235
Solar PV	144	400
Solar thermoelectric	200	500
Urban solid wastes	262	189
Total electricity	30355	42494

In contrast to the developments in installed capacity, the power demand forecasts done in the early 2000s turned out to be too optimistic. The 2005 revision of the Spanish Plan for Power and Gas Infrastructure 2002–2011 estimated that peak power demand would grow 24% in 6 years (from 43 GW in 2005 to 53 GW by 2011) (Plan Infraestructuras Eléctricas y Gasistas 2002–2011 (review of 2005)). These estimations, however, did not materialise in practice. The Spanish economy started to decelerate in 2007 and, after the global financial crisis in 2008, Spain entered a period of economic stagnation/recession. Actual peak power demand in 2011 was 44 GW (17.5% below estimations). Since 2011, the peak power demand has further decreased to 39 GW in 2014, a number comparable to the maximums reached a decade earlier.

According to the Ministry of Industry [5] itself, forecasts on the growth of GDP and electricity deviated considerably from the actual trends. GDP and electricity demand were forecasted to increase by an accumulated 25% and 24%, respectively, in 2013 compared to 2005 levels. In contrast, GDP grew by only 2% and electricity demand was reduced by 1% in such period. Figure 3 shows the evolution of the monthly electricity demand from 2008 to 2014.

3.3. Government Investment Signals.
Planning errors were not only restricted to the estimations of power demand. There was also a lack of appropriate coordination/monitoring of long-term power system requirements between renewable deployment objectives and developments in the conventional generation sector.

Through regulations or the approval of planning documents, the government undeniably influenced the expectations of investors in particular technologies. In particular, this was the case with the two technological categories undergoing the greatest growth rates in the 2000s. The general idea was to force an energy transition towards less polluting fuels and more efficient plants and away from coal and fuel-oil plants, that is, CCGTs and RES.

3.3.1. Investment in CCGTs.
The incentives driving investment in these two technologies (RES and CCGTs) were, however, very different. Investment in new conventional generation was mostly driven by expectations of substantial future increase in power demand in the country, as economic growth of the Spanish economy seemed robust in the period 2000–2007. Moreover, investment in gas-fuelled plants was perceived as a low risk investment due to their relatively low investment costs, shorter construction periods, and lower carbon risk as compared to coal plants. CCGTs were the main option of utilities to respond to the increasing electricity demand, in a context of cheap financing. These plants were less risky, easy, fast, and relatively cheap to be built compared to other alternatives [11]. The Plan for Electricity and Gas Infrastructures 2002–2011 gave an undeniable backing for the investment in CCGTs. This plan envisaged the construction of 14.8 GW of CCGTs, which was considered an "adequate" amount by the electricity sector association [11, 12]. This target was increased to 30 GW in 2005. It was an indicative target, that is, not binding neither for the state nor for the firms.

3.3.2. Renewable Energy Technologies.
Two main planning documents for RES were the Renewable Energy Plans in 1999 [13] and 2005 [14]. Table 2 includes the objectives set in both documents for electricity from renewable energy sources (RES-E).

3.4. Easy Granting of Administrative Authorisations.
The regions (or Autonomous Communities, AACC), which had competencies on the siting of electricity generation plants, were eager to grant permits to these plants, either because this allowed them to comply with the mandate that they should ensure their security of supply [7] or because this was regarded as a source of jobs for the region.

Regarding RES, there was no centralised authorisation procedure for RES plants in Spain. Administrative permits were easily and quickly granted by the AACC, given the high level of perceived local benefits and the fact that the costs were shared by all electricity consumers nationally [15].

3.5. Macroeconomic Factors.
Spain went under an economic boom in the late 1990s until 2007 and a deep crisis thereafter. Obviously, this economic cycle explains first the quick

increase in electricity demand (triggering the aforementioned concern of the government about the possibility of blackouts) and a stagnant, eventually depressed, electricity demand since 2008. Both industrial and residential demand increased significantly, and the latter also triggered by the housing boom.

Another important, and associated, key factor contributing to the overinvestment in the Spanish power sector was the availability of large volumes of cheap debt. Until 2008, banks were eager to lend money at relatively low interest rates, which favoured (over)investments across economic sectors, markedly the construction sector, but also in new power generation capacity.

4. What Are the Effects of Overcapacity?

The large overcapacity present in the Spanish power system has tangible economic consequences at system level (societal perspective) and for power market actors (both conventional and renewable generators).

At the system level, overcapacity results in the underutilisation of assets, which in turn results in a suboptimal allocation of economic resources.

Market actors also suffer the consequences of overinvestment. As a result of the substantial increase in generation capacity available, and the stagnation of (peak) power demand levels, natural gas-fuelled plants have seen a radical reduction in their utilisation factor over the last few years. In addition to the increased competition from other conventional generators, the substantial growth in renewable generation, which is granted priority dispatch in EU power markets, has further reduced the number of hours in which gas plants operate. Consequently, during the last years, the capacity factors for most CCGT plants have been well below those expected when investment decisions were made. Electricity generation from CCGTs has been reduced from 50 TWh in 2011 to 23 TWh in 2013 (see [12]). The reduction in nuclear, fuel oil, and coal has not been so dramatic. After the Royal Decree 134/2010, which supported national coal consumption, the situation was made even worse for CCGTs plants (Real Decreto 134/2010, de 12 de febrero, por el que se establece el procedimiento de resolución de restricciones por garantía de suministro y se modifica el Real Decreto 2019/1997, de 26 de diciembre, por el que se organiza y regula el mercado de producción de energía eléctrica). Given this situation, there is no expectation of investments in new capacity in the short term, except for renewable technologies and for fuel oil power plants substitution in the islands [16].

Besides the reduced number of hours of operation, overcapacity also puts downward pressure on wholesale market prices, further decreasing the business case for gas plants. There is a large amount of evidence showing this merit order effect in Spain (see, e.g., [17–19]).

The main consequence for conventional electric utilities has been stranded assets and a considerable reduction in their cash flows. Furthermore, investments on new generation capacity were highly leveraged. As a result, the Spanish firms (whether conventional or renewable) are among the most indebted in the EU, with debt as high as four times their EBITDA (Earnings Before Interest, Taxes, Depreciation, and Amortization) [7].

Similarly, for renewable generators, overcapacity results in two main negative effects: firstly, the downward pressure on wholesale power prices, while positive for electricity consumers, results in an increased competiveness gap (difference between the RES generation costs and the revenues obtained from the market), increasing the need for support per unit of energy generated; secondly, overcapacity hinders further penetration of renewables in the system since, under such conditions, it is difficult to justify politically the provision of economic support for new RES plants, which are not needed from a strict "security of supply" point of view.

5. What Are the Policy Options to Address the Overcapacity Issue and Open More Space for Further Renewable Deployment?

Given the dimensions of the problem described above, it seems very likely that its effects will last long and that the adjustments required to stabilise the situation will be deep.

Spain is equipped with a relatively young power plant fleet. During the last ten years, massive investments have been made in new natural gas-fired power plants. Almost no new coal power plants and also no new nuclear power plants have been brought online in Spain for at least twenty years.

According to Fichtner et al. [20], the retirement rate in the 2010–2020 period is about 24%. The installed capacity in 2010 amounted to 98000 MW, whereas the expected retirement in the period would be 23834 MW.

While it can be expected that the recovery of the Spanish economy will result in increasing levels of demand which will contribute to the correction of this issue in the long term, short-term (RES) energy policy decisions need to carefully take into account the current situation.

5.1. Measures Adopted So Far. Energy policy decisions in Spain seem indeed to have been influenced by this issue in the last few years.

The abrupt cancellation of economic support for all new RES plants in early 2012, aimed at containing increasing RES support costs, resulted also in an effective halt in new RES deployments in the country. Although a new RES support scheme is in place since 2014, this has not triggered a recovery in investments in the renewable sector so far.

In the Energy Reform of 2013, the Spanish Government tabled a regulatory framework for the "hibernation" (temporary closing) of combined cycle gas plants. Although the regulation has not been approved so far, the Spanish Government is considering the temporary closing of 6 GW of CCGTs in its infrastructure plan for the period 2015–2020 [21].

The Spanish Government has also been pushing actively in recent years for an expansion of the interconnection capacity with France, which could possibly result in increased electricity exports to northern EU countries, partially mitigating the effects of the Spanish overcapacity problem.

While these measures may help mitigate the effects of overcapacity in the short term, a long-term sustainable solution to the issue, enabling further penetration of renewable generation in line with 2030 RES objectives, would require a broader list of policy measures.

5.2. Alternative Policy Measures. Overcapacity is a serious short-term problem; however, the possible solutions to this problem may also open opportunities to accelerate a broader long-term transition to a low-carbon energy system. The excess generation capacity can be corrected by adopting measures both on the demand and on the supply side.

On the demand side, setting up policy incentives to increase domestic demand of electricity seems at first incompatible with EU energy efficiency and climate strategy and commitments. However, these policies could be justified if they offset primary energy consumption in other sectors. In this sense, the overcapacity in the electricity system may be an opportunity to accelerate the electrification of the heat and transport sectors, which is arguably a main trend for the decarbonisation of the whole economy [1]. Policy measures to move in this direction in the transport sector could include incentives for electric vehicles or "electricity-to-fuel" solutions. Similarly, in the heat sector, incentives to increase the penetration of efficient heat pumps could result in increased demand in the power sector offset by larger consumption reductions in the heat sector.

The supply side can also contribute to the correction. The excess generation capacity built in Spain has a positive side which is that the most polluting plants remaining in the system are no longer critical to guarantee security of supply. The Spanish Government could take this opportunity to require a gradual coal phase-out in the medium term, for example, deincentivise the continuation of the most polluting plants, for example, by establishing stricter environmental standards for combustion plants and removing existing subsidies for the consumption of domestic coal.

The Spanish Government could also reconsider the policy effectiveness and economic efficiency of existing capacity payments for dispatchable generation plants. The rationale for these payments is to incentivise investment in new plants when there are concerns about generation adequacy. Spain has currently the opposite problem. Capacity payments are hardly justified in these conditions except for strategic plants, for example, those placed in specific network hubs with an additional need for dispatchable capacity.

The goal set up in the Energy Union Package of a minimum interconnection target set for electricity at 10% of installed electricity production capacity of the Member States by 2020 and 15% by 2030 could certainly provide an opportunity for Spain to deliver electricity to other countries, such as Germany or Italy. In fact, interviews carried out for this project show that the improvements in the interconnection capacity with other countries are regarded as a main option to mitigate the overcapacity problem (see below).

5.3. Measures to Avoid New Overcapacities in the Future. Given the relatively low capacity factor of RES and the need

for backup capacity from conventional electricity sources, it can be argued that electricity systems with a large penetration of RES are inherently more capacity-intensive than others in order to produce a given amount of electricity and ensure the security of supply at all times during the day (baseload and peak-load).

Even if systems with high RES penetration may require somewhat larger shares of standby capacity, it is important to incorporate the lessons learned about the past conditions that created the current overcapacity problem into future long-term energy policy decisions.

In order to avoid building new unhealthy overcapacities in the future, developments on the conventional generation side need to be carefully coordinated with national renewable energy deployment objectives. Actual levels of deployment, both on the conventional and on the renewable side, need to be frequently monitored. It is critical that national authorities, in charge of long-term energy planning, and local and regional authorities, usually in charge of permitting procedures, coordinate very closely.

6. The Results of the Interviews on Stakeholders

6.1. Aim and Methodology. A survey among stakeholders was launched on October-November 2015. Its aim was to know the opinion of these stakeholders on the causes and consequences of overcapacity of electricity generation as well as the policies to address this problem. An open list of possible items in each category was provided to the interviewees, who were also free to mention other items. The Interview Guide (List of Items) is provided.

Causes

(1) Risk of black-outs at the end of the 90s.

(2) Problems in planning (wrong projections of electricity demand).

(3) Cheap and easy financing.

(4) The economic boom led to a considerable growth in demand.

(5) Positive signals by the government for investments in renewable energy.

(6) Positive signals by the government for investments in CCGTs.

(7) Easy granting of administrative permits by the regions (AACC).

(8) Overcapacity is an inherent problem for electricity systems with a high RES penetration.

Consequences

(1) Overcapacity leads to an suboptimal allocation of economic resources.

(2) It is rather a distributive issue: the losers are the investors in CCGTs.

TABLE 3: Responses of the interviewees on the causes of overcapacity.

Interviewee	Most relevant		Least relevant	
	First	Second	Second least-relevant	Least relevant
Utility A	4	2	7	1
Regulator A	2	5	1	7
Utility B	2	6	5	8
Regulator B	2	4	1	8
Regulator C	3	4	2	8
Regulator D	5	4	6	8
Expert A	2	6	—	—
Expert B	2	1	7	8
Expert C	—	—	—	—
Renewable energy association	2	6	8	7
Utility C	2	4	1	7

Note: (1) risk of blackouts at the end of the 90s, (2) problems in planning (wrong projections of electricity demand), (3) cheap and easy financing, (4) the economic boom led to a considerable growth in demand, (5) positive signals by the government for investments in renewable energy, (6) positive signals by the government for investments in CCGTs, (7) easy granting of administrative permits by the regions (AACC), and (8) overcapacity is an inherent problem for electricity systems with a high RES penetration.

(3) It is rather a distributive issue: the losers are the investors in renewable energy technologies.

(4) It is rather a long-term problem in the context of a transition towards a new energy model: overcapacity is a very important obstacle for future investments in renewable energy.

Policies

Already Adopted. Which of those measures not yet adopted are more effective to mitigate the problem in the short-term?

(1) The moratorium on renewable electricity generation investments adopted in 2012.

(2) The hibernation of CCGTs.

(3) The improvement in the interconnection capacity with other countries.

Not Yet Adopted. Which of those measures already adopted are more effective to mitigate the problem in the long-term?

(1) Incentives for the electrification of the transport sector and/or heat in buildings (efficient heat pumps).

(2) Opportunity to discourage the operation of the most polluting electricity generation plants.

(3) Termination of capacity payments for dispatchable generation plants.

(4) A better long-term energy planning.

Interviewees were asked to order those items in descending order according to their relevance, that is, from more important (1) to least important.

An e-mail interview with structured responses was sent to different types of stakeholders (see List of Stakeholders). These included energy experts, energy regulators, and the renewable energy associations. 15 people were contacted,

with no aim of representativeness. 11 accepted to respond to the interview. The following list provides a list of the interviewees. Three interviews belong or have belonged to an electric utility, four have been or are energy regulators, three are energy experts, and one comes from a renewable energy association. Their anonymity was guaranteed and, thus, their names cannot be provided.

List of Stakeholders includes

electric utility A,

former regional government official in charge of industry and energy matters,

electric utility B,

energy regulator (currently),

energy regulator (currently),

energy regulator (formerly),

energy expert,

energy expert,

energy expert,

formerly at electric utility C,

renewable energy association.

6.2. Main Results. In this section, the main results of the empirical analysis are provided, distinguishing between the causes and consequences of overcapacity as well the policies to mitigate this problem.

6.2.1. Causes. Table 3 provides a summary of the responses of the interviewees. The list of interviewees is included in the rows. The answers are grouped in two categories depending on the relevance attached to the different items (the two most and least relevant items). The numbers in the cell refer to the specific item (i.e., see "Interview Guide (List of Items)"). One expert did not answer the part of the interview on the

TABLE 4: Responses of the interviewees on the consequences of overcapacity.

Interviewee	Most relevant		Least relevant	
	First	Second	Second least relevant	Least relevant
Utility A	1	2	3	4
Regulator A	1	4	3	2
Utility B	1	4	3	2
Regulator B	1	2	4	3
Regulator C	1	2	3	4
Regulator D	1	4	2	3
Expert A	1	2	3	4
Expert B	1	4	3	2
Expert C	1	4	2	3
Renewable energy association	1	4	3	2
Utility C	1	4	2	3

Note: (1) it leads to a suboptimal allocation of economic resources, (2) it is rather a distributive issue: the losers are the investors in CCGTs, (3) it is rather a distributive issue: the losers are the investors in renewable energy technologies, and (4) it is rather a long-term problem in the context of a transition towards a new energy model: overcapacity is a very important obstacle for future investments in renewable energy.

causes and another did not provide his/her opinion on the least relevant causes of overcapacity.

Regarding the causes of overcapacity in Spain, an overwhelming majority of interviewees considered that lack of planning could be blamed for this problem. After that, the economic boom leading to a considerable increase in demand was pointed out as a main factor behind this problem. In contrast, the easy granting of administrative permits by the regional governments and overcapacity being an inherent problem in systems with a high RES penetration are deemed the least relevant causes of overcapacity. No noticeable differences across different types of actors can be discerned.

6.2.2. Consequences. Regarding the consequences (Table 4), the eleven interviewees unanimously indicated that overcapacity would lead to a suboptimal allocation of economic resources. This response was followed in importance by the answer "it was rather a problem for the future in the context of the transition towards a new energy model. Overcapacity is a very important obstacle for future investments in renewables." Therefore, overcapacity was clearly identified as a highly relevant barrier for further investments in renewable energy technologies. The interviewees were much less concerned about the distributive consequences of overcapacity in terms of losses for investors in CCGTs and, even more so, for those who invested in RES. Again, differences in the responses across actor types cannot be identified.

6.2.3. Policies. Finally, regarding the policies to address this problem, two types of policies were differentiated, depending on their short-term (effectiveness of already implemented measures) or long-term scope (what to do in order to mitigate the problem in the long-term).

Regarding the short-term policies (see Table 5), most interviews (6) considered that the improvement in the interconnection capacity with other countries could mitigate the problem. This was followed by "hibernation of CCGT plants" could mitigate the problem (3 interviewees) and

TABLE 5: Responses of the interviewees on the short-term policies to mitigate overcapacity.

Interviewee	Most relevant	Least relevant
Utility A	1	3
Regulator A	3	1
Utility B	2	1
Regulator B	3	1
Regulator C	1	3
Regulator D	3	1
Expert A	3	—
Expert B	2	1
Expert C	2	—
Renewable energy association	3	1
Utility C	3	1

Note: (1) the moratorium to renewable electricity generation investments adopted in 2012, (2) the hibernation of CCGTs, and (3) the improvement in the interconnection capacity with other countries.

"the moratorium to RES electricity generation could mitigate this problem" (2 interviewees). This last question was also considered to be the least relevant by an overwhelming majority of interviewees (7).

Regarding what to do to solve the problem in the long-term (Table 6), five interviewees argue that this is an opportunity to discourage the operation of more polluting power plants. Four claim that a better long-term energy planning could have improved this problem in the first place. Providing incentives for electrification of the transport and heat sectors and removing capacity payments for dispatchable power plants are not regarded as a solution to the problem. Differences in the responses across actor types for both short-term and long-term policies cannot be identified.

7. Conclusions

An excessive electricity generation capacity has been accumulated in Spain in the last decade. This overcapacity

TABLE 6: Responses of the interviewees on the long-term policies to mitigate overcapacity.

Interviewee	Most relevant		Least relevant	
	First	Second	Before last	Last
Utility A	4	1	2	3
Regulator A	2	1	4	3
Utility B	2	1	4	3
Regulator B	4	2	1	3
Regulator C	1	2	4	3
Regulator D	2	1	4	3
Expert A	4	—	—	—
Expert B	2	3	4	1
Expert C	4	1	2	3
Renewable energy association	3	2	1	4
Utility C	2	3	1	4

Note: (1) incentives for the electrification of the transport sector and/or heat in buildings (efficient heat pumps), (2) opportunity to discourage the operation of the most polluting electricity generation plants, (3) termination of capacity payments for dispatchable generation plants, and (4) a better energy long-term planning.

problem, caused by several factors, has several detrimental consequences and, most importantly, it can be considered a barrier for the future deployment of RES. Several policy alternatives have been proposed to deal with this problem, either in a short-term or in a long-term horizon. This paper has analysed those causes and consequences, using official data as well as interviews to relevant stakeholders. It has also tried to identify the opinions of different stakeholders with respect to policies which may mitigate the problem in the short or the long terms.

Overcapacity is clearly an economic problem in terms of the efficiency of the allocated resources (i.e., short term). But it is also a problem in a long-term horizon in so far as it could represent a barrier to an energy transition based on a large amount of RES.

A main lesson of this study for other countries is that energy planning should play a critical role to mitigate the possibility that a problem of overcapacity emerges. In particular, such planning should appropriately adjust to changes in the economic cycle. Several options to mitigate the overcapacity problem have been put forward in Section 6. Although stakeholders have their own preferences, the different alternatives should not be regarded as mutually exclusive. Indeed, in such a complex problem, with also difficult solutions, a combination of different options can be recommended.

Competing Interests

The authors declare that there are no competing interests regarding the publication of this paper.

Acknowledgments

This paper builds on the analysis conducted in the Intelligent Energy Europe (IEE) project "Dialogue on a RES policy framework for 2030 (Towards2030-dialogue)." The TOWARDS2030-dialogue project is an initiative that could be established thanks to the financial and intellectual support offered by the Intelligent Energy Europe (IEE) Programme of the European Commission, operated by the Executive Agency for Small and Medium Enterprises. For more details on the project, see http://towards2030.eu/.

References

[1] International Energy Agency, *Energy Technology Perspectives*, International Energy Agency, Paris, France, 2014.

[2] P. Del Río, "Analysing future trends of renewable electricity in the EU in a low-carbon context," *Renewable and Sustainable Energy Reviews*, vol. 15, no. 5, pp. 2520–2533, 2011.

[3] REE, *Informe del Sistema Eléctrico Espanol 2014*, 2014, http://www.ree.es/es/publicaciones/sistema-electrico-espanol/informe-anual/informe-del-sistema-electrico-espanol-2013.

[4] 50Hertz, Amprion, Tennet, and TransnetBW, Bericht der deutschen ÜÜbertragungsnetzbetreiber zur Leistungsbilanz 2014 nach EnWG § 12 Abs. 4 und 5, 2014, http://www.bmwi.de/BMWi/Redaktion/PDF/J-L/leistungsbilanzbericht-2014,property=pdf,bereich=bmwi2012,sprache=de,rwb=true.pdf.

[5] Ministry of Industry (MINETUR), The reform of the Spanish power system: towards financial stability and regulatory certainty, 2013, http://www.thespanisheconomy.com/stfls/tse/ficheros/2013/agosto/Power_System_Reform.pdf.

[6] CNE, Informe marco sobre la demanda de energía eléctrica y gas natural, y su cobertura, 2001.

[7] D. Lacallle, Por qué sube la luz si España tiene un 40% de sobre capacidad de producción? Inversión & Finanzas, 2014, http://www.finanzas.com/noticias/economia/20141114/lacalle-sube-espana-tiene-2804357.html.

[8] J. González, España tiene 107.615 MW de potencia eléctrica y sólo necesita la mitad ABC, October 2013, http://www.abc.es/economia/20131021/abci-espana-tiene-potencia-electrica-201310211202.html.

[9] REE, Informe del Sistema Eléctrico Espanol 2012, 2013, http://www.ree.es/es/publicaciones/sistema-electrico-espanol/informe-anual/informe-del-sistema-electrico-espanol-2012.

[10] REE, *Informe del Sistema Eléctrico Espanol 2014*, 2015, http://www.ree.es/es/publicaciones/sistema-electrico-espanol/informe-anual/informe-del-sistema-electrico-espanol-2014.

[11] A. Bolaños, "Mucho gas para tan poca luz," El País, 2013, http://sociedad.elpais.com/sociedad/2013/12/26/actualidad/1388082770_720869.html.

[12] UNESA, *Informe Eléctrico Memoria de Actividades Memoria Estadística*, 2013.

[13] IDAE, *Plan for the Promotion of Renewable Energy (PFER)*, Instituto Para la Diversificación y el Ahorro de la Energía, Spanish Ministry of Science and Technology, 1999.

[14] IDAE, *Renewable Energy Plan 2005–2010*, Instituto para la Diversificación y el Ahorro de la Energía, Spanish Ministry of Industry, Tourism and Trade, 2005.

[15] P. Del Río, A. Calvo, and G. Iglesias, "Policies and design elements for the repowering of wind farms: a qualitative analysis of different options," *Energy Policy*, vol. 39, no. 4, pp. 1897–1908, 2011.

[16] Store Project, Facilitating energy storage to allow high penetration of intermittent renewable energy. D5.2—SPAIN. Overview

of current status and future development scenarios of the electricity system, and assessment of the energy storage needs, 2014, http://www.store-project.eu.

[17] G. Sáenz de Miera, P. del Río González, and I. Vizcaíno, "Analysing the impact of renewable electricity support schemes on power prices: the case of wind electricity in Spain," *Energy Policy*, vol. 36, no. 9, pp. 3345–3359, 2008.

[18] L. Gelabert, X. Labandeira, and P. Linares, "An ex-post analysis of the effect of renewables and cogeneration on Spanish electricity prices," *Energy Economics*, vol. 33, no. 1, pp. S59–S65, 2011.

[19] M. T. Costa-Campi and E. Trujillo-Baute, "Retail price effects of feed-in tariff regulation," *Energy Economics*, vol. 51, pp. 157–165, 2015.

[20] Fichtner, "Study on incentives to build power generation capacities outside the EU for electricity supply of the EU," Final Report, 2012.

[21] MINETUR, Planificación Energética: Plan de Desarrollo de la Red de Transporte de Energía Eléctrica, Primera Propuesta, 2014, http://www.minetur.gob.es.

10

Benchmarking of Electricity Distribution Licensees Operating in Sri Lanka

K. T. M. U. Hemapala and Lilantha Neelawala

Department of Electrical Engineering, University of Moratuwa, 10400 Moratuwa, Sri Lanka

Correspondence should be addressed to K. T. M. U. Hemapala; udayanga@elect.mrt.ac.lk

Academic Editor: Jin-Li Hu

Electricity sector regulators are practicing benchmarking of distribution companies to regulate the allowed revenue. Mainly this is carried out based on the relative efficiency scores produced by frontier benchmarking techniques. Some of these techniques, for example, Corrected Ordinary Least Squares method and Stochastic Frontier Analysis, use econometric approach to estimate efficiency scores, while a method like Data Envelopment Analysis uses linear programming. Those relative efficiency scores are later used to calculate the efficiency factor (X-factor) which is a component of the revenue control formula. In electricity distribution industry in Sri Lanka, the allowed revenue for a particular distribution licensee is calculated according to the allowed revenue control formula as specified in the tariff methodology of Public Utilities Commission of Sri Lanka. This control formula contains the X-factor as well, but its effect has not been considered yet; it just kept it zero, since there were no relative benchmarking studies carried out by the utility regulators to decide the actual value of X-factor. This paper focuses on producing a suitable benchmarking methodology by studying prominent benchmarking techniques used in international regulatory regime and by analyzing the applicability of them to Sri Lankan context, where only five Distribution Licensees are operating at present.

1. Introduction

The regulators in distribution sector in the world expect to increase investments for increasing electrification with reductions of losses, reduction of the number of employees, and so forth. In Sri Lankan context the regulator is looking to reduce the tariff also as distribution sector is running with government funds. Therefore the main target of the Sri Lankan regulator is to provide the utility with incentives to improve their operating efficiency to ensure that customer will get quality electricity with low price. There are five electricity Distribution Licensees operating in Sri Lanka. In Sri Lanka the allowed revenue for a particular distribution licensee (DL) is calculated according to the allowed revenue control formula as specified in the tariff methodology of Public Utilities Commission of Sri Lanka (PUCSL). The Operational Expenditures (OPEX) component of the base allowed revenue needs to be adjusted at a rate defined by an efficiency factor per year. In successive tariff periods, the regulator (PUCSL) can revise the methodology for computing the efficient OPEX to be included in the distribution allowed revenue. A relative OPEX efficiency score obtained from a benchmarking study is an input to formulate this efficient factor (X-factor). PUCSL can decide on X-factor using the result of a benchmarking study. At present PUCSL does not take into account the X-factor when deciding the allowed revenue for each DL. The reason for not considering the X-factor is that there are no benchmarking studies that have been done on DLs to obtain relative OPEX efficiency scores. Without these relative efficiency scores (percentage values like 100% for one DL, 60% for another, etc.) X-factor cannot be obtained. Therefore PUCSL requires a suitable methodology to benchmark Distribution Licensees in Sri Lanka.

This paper describes a methodology to benchmark five Distribution Licensees in Sri Lanka, which facilitate PUCSL to regulate allowed revenue for each DL according to the relative OPEX efficiencies. The regulator can set differentiated price limits based on the companies' efficiency performance estimated from a benchmarking analysis [1]. And also it can decide which companies deserve closer examination, so that scarce investigative resources are allocated efficiently [2].

There are different benchmarking techniques used by international regulators [2–5]. Selecting the most appropriate benchmarking methodology is done after considering the principles discussed by CEPA's reports on benchmarking [3, 6]. The rest of the paper is organized as follows. A brief overview of different benchmarking techniques is presented in Section 2. Data Envelopment Analysis with methodologies is discussed in Section 3. Section 4 is presented results and discussion and finally the conclusions are given in Section 5.

2. Benchmarking Techniques

The following were identified as prominent techniques from the literature review for considerations against the principles discussed by CEPA's report [7]:

(i) Partial Performance Indicators (PPIs),

(ii) Ordinary Least Square (OLS),

(iii) Data Envelopment Analysis (DEA),

(iv) Stochastic Frontier Analysis (SFA).

2.1. Partial Performance Indicators (PPIs). These indicators are used to compare the ratios of single output to a single input of firms (e.g., energy sold per OPEX). They are often significantly affected by the capital substitution effects [7]. PPIs used in isolation are not possible to use the differences in the energy sector that directly impact on the market. For example, a utility may experience a relatively high or low unit cost simply because of the customer category. Therefore PPIs may not provide a meaningful comparison across different DLs as they are operating at different conditions [8].

2.2. Data Envelopment Analysis (DEA). Data Envelopment Analysis (DEA) is the prominent technique used by the researchers for benchmarking in the literature [9–13]. Thakur has used DEA and Malmquist Productivity Index to find the rate at which the efficiency frontier has moved over recent years after implementing the reforms process in India [14]. Cui et al. applied DEA and Malmquist Productivity Index to calculate the energy efficiencies of nine countries during 2008–2012 and explained the reasons for energy efficiency changing with respect to technical and management factors [15]. Javier Ramos-Real et al. have estimated the changes in the productivity of the Brazilian electricity distribution sector using Data Envelopment Analysis in terms of productivity change [16]. Chien et al. applied Malmquist Productivity Index comparing the performance of different thermal power plants in Taiwan [17].

DEA involves linear programming to determine the efficient firm(s) from a sample relative to the other firms in the sample [18–20] while the Malmquist Productivity Index evaluates the efficiency change over time [21]. In this nonparametric technique, the ratio of weighted outputs to the weighted input is maximized subjected to constraints (required to solve individual linear programming problems for each firm in the sample). The efficient firm is the one where no other firm or linear combination of other firms can produce more of all the outputs using less input [6].

It is important to select input output variables reflecting the use of resources and misspecification of variables can lead to wrong results [6]. DEA can also accommodate environmental variables that are beyond the control of the firms but can affect their performance (e.g., population density of a particular area of operation). This method is a multidimensional method and inefficient firms are compared to actual firms (or linear combinations of these) rather than to some statistical measure. This does not require specifying a cost or production function. Importantly DEA can be implemented on a small dataset, where regression analysis tends to require larger minimum sample size, but in case of small samples and high number of input or/and output variables there is a danger of overspecification of model and eventually "made-up" results for efficiency scores [22]. As more variables are included in the model, the number of firms on the efficient frontier increases.

2.3. Corrected Ordinary Least Squares (COLS). With this regression technique the most efficient firm or the frontier is estimated. This "corrected" form of ordinary least square has assumed that all deviations from the frontier are due to inefficiency [6]. This method requires the details of the cost or production function and assumptions about technological properties. COLS method is easy to implement and allow statistical inference about which parameters to include in the frontier estimation [6]. This method requires large data volume in order to create robust regression relationship and is sensitive to data quality.

2.4. Stochastic Frontier Analysis (SFA). Similar to COLS, SFA requires the specification of a production function based on input variables. But in this model the errors in parameters are incorporated into the model and do not assume that all errors are due to inefficiency [23]. A model of the form described under COLS is estimated with two error functions. The first of these will be assumed to have a one-sided distribution. The second error term has a symmetric distribution with mean zero. The Cobb-Douglas stochastic frontier model takes the form [24]

$$\ln q_i = \beta_0 + \beta_i \ln x_i + v_i - u_i, \quad (1)$$

where q_i is an output, x_i is an input, and v_i, u_i are error terms. SFA is theoretically the most appealing technique but the hardest to apply. Since it is difficult to implement in small samples, regulators traditionally have been reluctant to use SFA techniques in setting X-factors [6].

Further it is important to note that the reliable panel data of OPEX was not available. Unavailability of this published/audited historical OPEX data was mainly due to the fact that major 4 DLs are from the same legal entity having no separate audited accounts till the year 2010 (OPEX for year 2011 and 2010 was the only available data). This results in avoiding techniques that rely on panel data for this study.

3. Data Envelopment Analysis

There are a number of variables that can be considered when implementing any benchmarking technique as described in

Section 2. In regulators' point of view factors such as quality of the data, availability, ease of collection, relevance to the business, international practices/reviews, use of statistical indicators (such as correlation), nonredundancy to minimize overlapping, high discriminating power, and reflection of the scale of operation and cost drivers have to be considered when selecting variables.

Therefore the regulator must take care to keep the number of variables to minimum while those variables are strong cost drivers (i.e., OPEX). Relevant data should be accurate and importantly be practical to collect from the DLs timely. In order to find quality and feasible data several reports were analyzed. These include published reports by PUCSL [25–28] and Licensees [29–40]. After studying the above reports the following set of variables were collected:

(i) energy sold,

(ii) total number of consumers—this is the number of consumer accounts or the number of consumer connection points,

(iii) number of new connections provided,

(iv) number of employees,

(v) total distribution of lines' length—this includes MV and LV network length,

(vi) number of substations,

(vii) authorized operation area—this is a constant for each licensee,

(viii) operational expenditure.

Note that, in international benchmarking practices, the use of supply/service quality as a variable is rare. Most of the countries reviewed separately run a quality-of-service reward/penalty regime [23]. In Sri Lanka, the supply/service quality is to be determined according to the drafted electricity distribution performance regulations, where penalties have been introduced for underperformance [41].

3.1. Justification of Selected Variables

3.1.1. Cost Drivers. Cost is clearly depending on scale of the operation. Accurate data on energy distributed, production of the distribution business, the number of consumer accounts, network length (MV and LV line lengths), and the number of distribution substations can be timely obtained from DLs in Sri Lanka. Since data on the above-mentioned variables can be timely obtained, regulator can timely perform benchmarking exercise to figure out allowed revenue for each year.

3.1.2. Dispersion of Consumers. Distribution line length per consumer can be taken as indication of what extent the consumer concentration is. It is also an indication of the extent of rural electrification efforts taken by the DLs. For each DL this value is different. For example, DL5 is having a lower value indicating higher concentration of consumers, whereas DL4 is having a larger value as indicated in Table 1.

TABLE 1: Dispersion of consumers in each DL.

DL	Distribution line length per consumer (m)	Area per consumer (m²)
DL1	30.8	**21,425**
DL2	23.4	10,614
DL3	28.8	13,085
DL4	**31.3**	7,940
DL5	**8.8**	**727**

3.1.3. Correlation. Applying too many explanatory variables to a sample of few observations (i.e., the number of Distribution Licensees) would result in 100% efficient DLs. Therefore, it is necessary to combine several parameters into one single parameter in order to preserve sufficient degrees of freedom. It is important not to consider highly correlated variables simultaneously, in a benchmarking method. Correlation coefficients were calculated using past data from year 2006 to 2011 for each DL. The results are given in Table 2.

For example, correlation coefficient of energy delivered and number of consumer accounts is 0.9683, which is the highest correlation coefficient, while that of energy delivered and number of employees is having the second highest correlation. For further verification Figures 1 and 2 were plotted.

The energy delivered and the number of employees indicated higher correlation. It can be concluded that, from the selected set of variables, energy delivered and the number of consumers are having the acceptable correlation. It is sufficient to account for one variable from the energy delivered and number of consumers. Since energy delivered (output) is highly correlated with number of employees (input) it is justifiable taking number of employees as another input variable.

3.1.4. Input, Output, and Environmental Variables. Operational Expenditure (OPEX) has been taken as the main input variable to assess the efficiency. Energy delivered was used as the main output produced. The number of new connections provided was taken as an output, while the number of employees was taken as input variable. The number of employees includes management and operational staff. Demand for new connections depends on the conditions of the authorized area of operation of DLs. This is not under the direct control of the management of the DL. To provide the demanded connection the DL has to input its resources. Table 3 depicts the variation between DLs [41]. This reflects the variation in demand for new connections that is varying according to the area of operation. DLs need to meet this demand. Therefore DLs need to input their resources accordingly.

As given in Table 3, DL1 is giving 40 new connections per day (on average) whereas DL5 is only providing 6 new connections per day (on average). Obviously DL1 needs to input more resources than DL5 to cope with the demand for new connections. The demand for connection is out of the control of the DL's management. In some areas,

TABLE 2: Correlation coefficients.

Correlation coefficients	Energy delivered	Number of consumer accounts	Number of new connections	Number of employees	Network length	LV distribution substations
Energy delivered	1.0000	**0.9683**	0.8755	**0.9552**	0.7498	0.8245
Number of consumer accounts		1.0000	0.8769	0.8635	0.6750	0.7069
Number of new connections			1.0000	0.8635	0.7313	0.6198
Number of employees				1.0000	0.6750	0.6758
Network length					1.0000	0.7069
LV distribution substations						1.0000

TABLE 3: Average number of new connections provided by each DL.

Licensee	Average number of new connections provided per day (for year 2012)
DL1	**40**
DL2	**31**
DL3	**33**
DL4	**14**
DL5	**6**

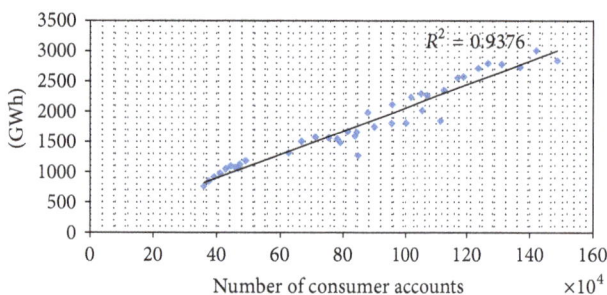

FIGURE 1: Energy delivered versus the number of consumers.

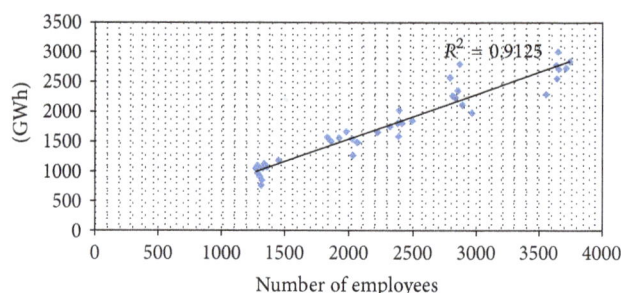

FIGURE 2: Energy delivered versus the number of employees.

a lot of infrastructure developments, resettlements, and rural developments are going on due to ending of the war with terrorists. This has caused high demand for new connections. Therefore, when evaluating the overall performance, the number of new service connections provided by respective DLs has to be considered.

Network length and substations can be considered as input or output either. Viewing the network length as an output runs the risk that a network that increases its length of lines is rewarded even if there is no impact on real world delivering of services to the customers [23]. In international regulatory practice network length has been considered as both input and output. Hence both scenarios were taken into consideration.

3.2. Selection of Benchmarking Techniques and Models. Results from application of benchmarking method will directly impact the allowed revenue of each DL. If the method itself is complicated and harder to understand, then there would be a doubt in the minds of DLs about the efficiency results. DEA, COLS, and PPIs fulfill the desirable characteristics such as easiness to compute and understand transparency and ability to implement in smaller samples.

If a benchmarking method requires higher number of data points, then it will be harder to implement with a smaller sample like five, as in the case of only five DLs in Sri Lanka. DEA can be easily implemented with five DLs, but care has to be taken to verify the results with other methods. International practice is that, for m number of inputs and n number of outputs, there has to be $n \times m$ number of DLs [42]. Otherwise all the DLs would get closer to 100% efficiency and discrimination could be difficult. In other words, with small sample and high number of input/output variables, there is a danger of receiving made-up results for efficiency scores [22]. When more variables are included in the model, the number of DLs on the efficient frontier increases. Feasibility of COLS has to be decided by practically implementing the COLS method with Cobb-Douglas cost function with the same set of variables, and also COLS implementation can be used to verify the results from DEA. To verify the results (efficiency scores) at least two different benchmarking methods must be used. Selected methods should have different characteristics so that the regulator can convince the DLs about the efficiency scores. In this case DEA and COLS are feasible to implement.

3.3. Implementation of Benchmarking Techniques Using DEA. After considering the factors discussed in Sections 3.1 and 3.2, energy sales, number of new connections, number of employees, OPEX, number of substations, area per consumer, and network line length per consumer were selected when implementing DEA. Note that if "*total network length*" is to be taken as an input, then "*number of substations*" has to be taken as input also. On the other hand if "*network length*" is to be taken as an output, then "*number of substations*" has to be taken as output also.

For each model (8 input/output variables to 3 input/output variables), the efficiency scores were obtained. Note that every possible input output configurations (models) were taken into consideration when obtaining results. Average efficiency scores of each DLs against different models are shown in Table 4.

It can be seen that the discrimination between each DL's efficiency score decreases with the number of variables considered. It is observed that DL2 is the lowest performer while DL5, DL1, DL3, and DL4 are ranked highest to lower according to the average efficiency scores. Even when considering 8 variables' models, it can be observed about 10% gap of efficiency with respect to all other DLs. Therefore it is possible to take the 8 variables' models as the base and take these efficiency values to calculate the X-factor. Note that the implementation is done using data corresponding to year 2011. The DL2 has high degree of freedom to improve its efficiency score since the model contains 8 variables.

If all DLs get closer to 100%, when implementing the DEA method with 8 variables models' with current values for respective variables (i.e., according to the year of implementation, values for the variables may get changed), then the reduced variables' models (starting from 7 variables to 3 variables) can be considered. This would allow higher discrimination between efficiency scores as it is observed in Table 4.

3.4. Implementation of Benchmarking Techniques Using COLS. Implementation of COLS method has been done according to the description given in Section 2.3. It is required to select suitable variables for "benchmark cost function." Variables should represent output produced by the business, input prices paid, and environmental conditions that affect the production cost.

In Sri Lanka, OPEX of DLs mainly consists of expenses for human resource. It is about 50% to 60% of their respective OPEX. Therefore cost per employee must be used as the main input price of the cost function. As energy sold (GWh) reflects the main output produced by the distribution business, it is included in the cost function. Five DLs have their designated area of operation. Accordingly the customer densities of DLs are different from each other. Table 5 illustrates the differences in customer densities as at year 2011.

Therefore the analysis must account for these differences in their business which is out of their (DLs) control. This variable is to capture the heterogeneity dimension of the distribution business [43, 44]. Further, the consumer density also can be accommodated in the model by using the consumers per unit network length, that is, number of consumers per kilometer of line length. Table 6 indicates the extent of heterogeneity. DL5 has a higher number since its area of operation is highly populated.

Efficiency scores with respect to the models are given in Table 7. The average results indicate more than 90% efficiencies for all DLs. Further, efficiency scores of all DLs lie in a band of 90.5% to 96.9%. Hence discrimination is lower.

TABLE 4: Average efficiency scores by model.

| Model | DEA average efficiencies of different models | | | | |
	DL1	DL2	DL3	DL4	DL5
8-Variable	100.0	89.7	100.0	100.0	100.0
7-Variable	100.0	89.9	100.0	98.2	100.0
6-Variable	100.0	86.3	99.7	96.3	100.0
5-Variable	100.0	84.4	99.1	93.0	100.0
4-Variable	100.0	82.2	97.4	88.8	100.0
3-Variable	99.7	77.1	94.6	82.9	100.0

TABLE 5: Differences in customer densities.

DL	Customer density (consumer accounts per km²)
1	47
2	94
3	76
4	126
5	1375

TABLE 6: Consumers per unit length of network.

Licensee	Customer per unit length of network (cons./km)
DL1	32.48
DL2	42.71
DL3	34.66
DL4	31.97
DL5	113.14

Therefore analysis carried out using three variables and the results are shown in Table 8.

It can be seen that the average efficiency scores are more dispersed than 4-variable models' average. Efficiency scores are stretched out in a band of 75.6% to 100%. Hence discrimination is higher. Note that, in each model in Table 8, DL2 is the lowest performer. Efficiency score of DL2 always ended up below 77%.

3.5. Implementation of Benchmarking Techniques Using PPIs. PPIs assume linear relationship between input and output. As explained in Section 2.1 they cannot measure the overall performance of the business. These partial indications can be misleading; therefore care should be taken to identify misleading information. PPIs were calculated for each DL by taking the OPEX and number of employees as inputs. Line lengths and number of substations were not taken into account, since those can be considered input or output either. On the other hand OPEX and number of employees can only be considered as inputs to the system while energy delivered to consumers and number of consumers can only be taken as outputs from the system. Table 9 depicts the results from PPIs.

Efficiencies obtained by PPIs are not used to directly conclude the relative efficiency score of a particular DL but to

TABLE 7: COLS with four variables.

Model	Energy sales GWh	Network length—total km	Consumer density—line Cons/km	Consumer density—area Cons/km²	Cost of employee LKR'000	Efficiency score				
						DL1	DL2	DL3	DL4	DL5
1	X			X	X	100.0	94.6	92.3	97.9	96.8
2	X		X		X	100.0	97.8	96.1	99.3	98.6
3	X	X		X		95.7	97.2	100.0	95.6	97.1
4	X	X	X			89.0	85.0	100.0	83.6	89.6
5	X	X			X	100.0	77.8	85.9	87.6	88.3
					Average	96.9	90.5	94.9	92.8	94.1
					Maximum	100.0	97.8	100.0	99.3	98.6
					Minimum	89.0	77.8	85.9	83.6	88.3

TABLE 8: COLS with three variables.

Model	Energy sales GWh	Network length—total km	Consumer density—line Cons/km	Consumer density—area Cons/km²	Cost per employee LKR'000	Efficiency score				
						DL1	DL2	DL3	DL4	DL5
1	X				X	100.0	76.6	94.6	85.4	88.8
2	X		X			100.0	74.8	93.5	81.6	90.3
3	X			X		100.0	74.7	93.4	79.7	91.9
4	X	X				100.0	76.4	96.1	84.0	89.8
					Average	100.0	75.6	94.4	82.7	90.2
					Maximum	100.0	76.6	96.1	85.4	91.9
					Minimum	100.0	74.7	93.4	79.7	88.8

TABLE 9: Efficiency scores from PPIs.

		DL1	DL2	DL3	DL4	DL5
Partial Performance Indicators						
Energy sales/OPEX	kWh/LKR	0.763	0.592	0.703	0.594	**0.773**
Number of consumers/OPEX	Nos/LKR Mil	345	310	**425**	397	321
Energy sales/employee	MWh	**976**	760	740	625	816
Number of consumers/employees	Nos	442	398	**447**	417	338
Corresponding relative efficiencies						
Energy sales/OPEX	%	98.8	76.6	91.0	76.9	100.0
Number of consumers/OPEX	%	81.2	72.9	100.0	93.3	75.4
Energy sales/employee	%	100.0	77.8	75.8	64.0	83.6
Number of consumers/employees	%	98.7	88.9	100.0	93.2	75.6
Average	%	94.7	79.1	91.7	81.8	83.6

qualitatively verify the results obtained from DEA and COLS. It can be seen that DL1, DL3, DL5, DL4, and DL2 are having the efficiencies from highest to lowest, respectively.

4. Results and Discussion

Let us consider the 4-variable model given under the DEA 3-variable method given in Table 10. In 4-variable model "energy sales" and "total network length" are taken as outputs of the electricity distribution business while OPEX is taken as the input. It is going to look at how efficiently (relatively) a DL has used its OPEX to provide electrical energy to its consumers and also to maintain the total network length owned by that DL.

In that case all DLs except DL2 have obtained relative efficiency score of 100%. DL2 has obtained a score of 77.4%. This means, only relative to each other, that DL2 is efficient by only about 77.4%. This does not imply that the other DLs with 100% efficiency score are strictly efficient. It is possible that DLs with 100% score could be operated more efficiently. DEA compares each DL with all other DLs and identifies those DLs that are operating inefficiently compared with other DLs' actual operating results. It achieved this by locating the best practice or relatively efficient DLs. This can be graphically

TABLE 10: Relative efficiency scores of each model under the DEA 3-variable method.

Model	Energy sales	New connections given	Area per consumer	Network line length per consumer	Total network length	Number of substations	Number of employees	OPEX	Relative efficiency score (%)				
									DL1	DL2	DL3	DL4	DL5
1	O	O						I	100	77.3	100	77.7	100
2	O		O					I	100	77	92.1	77.7	100
3	O			O				I	100	76.8	100	100	100
4	O				O			I	100	77.4	100	100	100
5	O					O		I	98.8	76.6	91	77.3	100
6	O				I			I	98.8	76.6	91	76.9	100
7	O					I		I	100	77.5	91.5	76.9	100
8	O						I	I	100	77.8	91	76.9	100

TABLE 11: Energy sales per OPEX and total network length per OPEX.

	Unit of measurement	DL1	DL2	DL3	DL4	DL5
Energy sales per OPEX	GWh/LKR million	0.76	0.59	0.70	0.59	0.77
Total network length per OPEX	km/LKR million	10.63	7.26	12.27	12.41	2.83

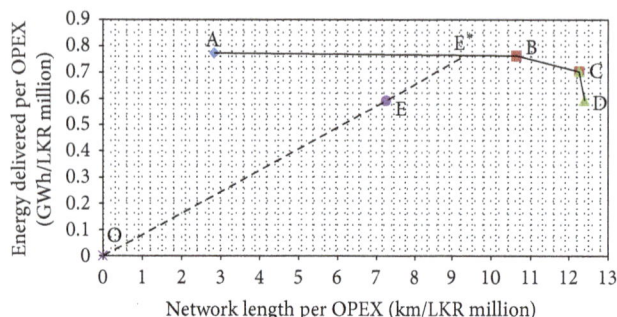

FIGURE 3: Graphical representation of DEA implementation.

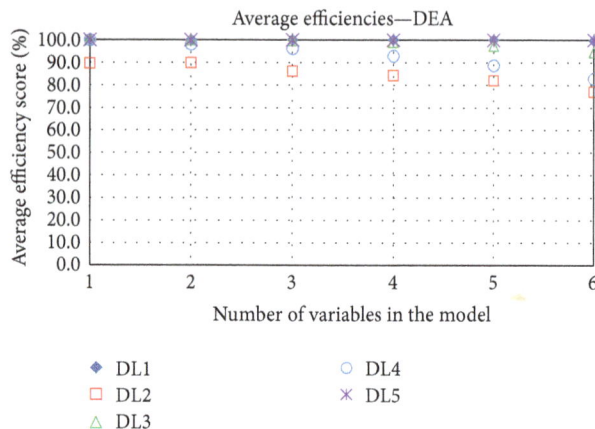

FIGURE 4: Increase of discrimination with reduction of variables in DEA.

illustrated (Figure 3) in following manner according to the ratios given in Table 11.

In Figure 3, points A, B, C, D, and E represent DL5, DL1, DL3, L4, and DL2, respectively. The 100% efficient boundary is demarcated by the line connecting ABCD. The target "efficient reference point" for DL2 (i.e., point E) is given by the point E* which is the intercept of line AB and extended line OE.

In other words this efficient reference point is point E*, against which DL2 was found to be most directly inefficient. DL2 (point E) was found to have inefficiencies in direct comparison to DL1 (point B) and DL5 (point A). The efficiency of DL2 can be obtained by the ratio of OE/OE* which is equal to 77.4%. DL2 (point E) can approach the point E* to become 100% relatively efficient, by increasing the respective output/input ratios. In this case, DL2 can reduce its OPEX by 22.6% while keeping the actual outputs in the same level, to be 100% relatively efficient.

According to the average efficiency scores obtained by DEA 3-variable models, DL5 is the efficient performer with 100% relative efficiency. DL5 is 100% efficient which means that it is relatively efficient only and not strictly efficient. That is, no other unit is clearly operating more efficiently than this DL5, but it is possible that all DLs, including DL5, can be operated more efficiently. Therefore, the efficient DL (DL5 in 3-variable models) represents the best existing (but not necessarily the best possible) practice with respect to efficiency. It can be pointed out that considering the small sample size (5 DLs) DEA is theoretically more appealing than COLS technique because COLS requires to estimate the number of coefficients leading to unsatisfactory results purely because of low sample size.

When more variables are included in the model, the number of DLs on the efficient frontier increases. To avoid lower discrimination of efficiency scores the 3-variable models are the most suitable in this context as shown in Figure 4. The models selected must be robust to changes in techniques

implemented. In particular, the ranking of firms, especially with respect to the "best" and "worst" performers, and the results must show reasonable stability and the different approaches should have comparable results. COLS and DEA are the main two different techniques used to measure the overall efficiency. Therefore robustness of the results obtained using those two techniques has to be analyzed.

Authors selected COLS 3-variable models over 4-variable models because 4-variable models results indicated average efficiency scores of all DLs lying in a band of 90.5% to 96.9% (i.e., low discrimination). COLS 3-variable technique indicated higher discrimination and the efficiency scores for all DLs lying in a band of 75.6% to 100%. Since authors have incorporated more variables (from 8 variables to 3 variables) in DEA, direct comparison with COLS results is not possible. The COLS method adopted used 3 variables including OPEX. Results from COLS method with 3 variables including OPEX can be compared with 3-variable model in DEA. This is because both methods used 3 variables as input and output; hence the degree of freedom is the same.

It can be seen that the results produced by DEA and COLS are robust for DL1, DL2, DL3, and DL4 as the differences are very low. For DL5 there is a considerable difference, but the efficiency score for DL5 is beyond 90% for both techniques. It is important to note that operation conditions of DL5 are extensively different than the remaining four DLs with respect to consumer density, authorized area of operation, and energy demand per consumer. According to the results given in Table 12 it can be concluded that average efficiency score given by DEA 3-variable models are robust and reliable.

4.1. Ranking of DLs according to Overall Efficiency. Since Sri Lanka is in the initial stage of electricity regulation

TABLE 12: Average efficiency scores.

	Average efficiency score				
	DL1	DL2	DL3	DL4	DL5
DEA (3-variable)	99.7	77.1	94.6	82.9	100.0
COLS (3-variable)	100.0	75.6	94.4	82.7	90.2
Difference	−0.3	1.5	0.2	0.3	9.8

TABLE 13: Ranking of DLs.

Rank	Ranking		
	DEA	COLS	PPI
1	DL5	DL1	DL1
2	DL1	DL3	DL3
3	DL3	DL5	DL5
4	DL4	DL4	DL4
5	DL2	DL2	DL2

(Electricity Act came into force in 2009), it is more important to peruse underperforming DLs to obtain at least the next level of efficiencies performing by peer DLs. Further, according to the efficiency scores, the regulator can decide which companies deserve closer examination, so that scarce investigative resources are allocated efficiently [2]. Table 13 depicts the ranking of each DL according to each technique used and also verification by using PPIs.

DL2 is the lowest performer while DL5, DL1, DL3, and DL4 are ranked highest to lower according to the average efficiency scores. It can be recommended that DL2 deserves closer supervision while DL4 also requires close supervision of the electricity regulator (i.e., PUCSL) as they are underperforming relative to other three DLs.

4.2. Influence on X-Factor. Regulator can officially obtain data for relevant variables and perform DEA analysis and use the average results from the DEA method which uses 3-variable models to obtain efficiency scores. Then it verifies those DEA results with efficiency sores obtained by COLS method using 3-variable method and verifies the rankings with PPIs. Then the average efficiency scores given by DEA 3-variable models can be used to decide on X-factor to persuade most underperforming DLs.

The regulator can decide on how to determine the X-factor (the translation of efficiency scores into X-factors), and the method of determining the X-factor may vary among the regulators [10, 13]. For example, X-factor can be calculated as shown in [11]. It is important to note that the relative efficiency scores resulting from this benchmarking exercise give an indication to the regulator (PUCSL) on how these DLs are operating relative to each other.

5. Conclusions

The relative efficiencies of five Distribution Licensees operating in Sri Lanka were analyzed using prominent benchmarking techniques. International practices in electricity distribution regulatory regime were considered when performing this benchmarking study. Techniques like Data Envelopment Analysis (DEA), Corrected Ordinary Least Squares (COLS) method, and Partial Performance Indicators (PPIs) method were utilized with several input output models in order to assess the efficiency in several angles. Care was taken to address the heterogeneity of the operating conditions such as consumer density and authorized area of operation of each DL which is out of the management control.

The efficiency scores obtained with respect to various possible models were scrutinized and came up with a suitable methodology to obtain efficiency scores considering the

data availability and low number of Distribution Licensees. The proposed methodology uses DEA with 3 input/output variables and gets the average efficiency scores as the final score, that is, having higher discrimination in the efficiency scores.

In parallel these efficiency scores were verified by the average results obtained by COLS method (3 variables including OPEX). Further, the ranking of Distribution Licensees is also verified with respect to DEA, COLS, and PPIs. It was revealed that for each method DL2 is the lowest ranked and DL4 is the next lowest ranked. DL1, DL3, and DL5 showed up more than 90% average efficiency for DEA and COLS.

Considering the fact that Sri Lanka is in its early stage in regulatory implementations, it is recommended to persuade underperforming DL. These efficiency scores would make a strong platform to the regulator when making the decision on X-factor in order to control the allowed revenue of Distribution Licensees. The methodology produced by this research have identified the inherent constraints prevailing in the context of Sri Lanka, such as low number of samples (i.e., 5 Distribution Licensees) and the unavailability of published/audited historical OPEX data (four DLs are coming under one legal entity. Therefore these four DLs do not have separate audited accounts till the year 2010). Hence, the electricity regulator can use the proposed methodology to start the evaluation of the efficiencies in order to begin incorporating efficiencies of Distribution Licensees in the electricity distribution revenue control formula. This would definitely encourage Distribution Licensees to minimize their inefficiencies in operations and maintenance. Further, the possible reduction in allowed revenue eventually would pass down to the consumers.

Conflict of Interests

The authors declare that there is no conflict of interests regarding the publication of this paper.

References

[1] M. Farsi, A. Fetz, and M. Filippini, "Benchmarking and regulation in the electricity distribution sector," CEPE Working Paper 54, 2007.

[2] G. Shuttleworth, "Benchmarking of electricity networks: practical problems with its use for regulation," *Utilities Policy*, vol. 13, no. 4, pp. 310–317, 2005.

[3] T. Jamasb and M. Pollitt, "Benchmarking and regulation: international electricity experience," *Utilities Policy*, vol. 9, no. 3, pp. 107–130, 2000.

[4] T. Jamasb, D. Newbery, M. Pollitt, and T. Triebs, "International benchmarking and regulation of European gas transmission utilities," Tech. Rep., Council of European Energy Regulators (CEER), 2006.

[5] T. Jamasb and M. Pollitt, "International benchmarking and regulation: an application to European electricity distribution utilities," *Energy Policy*, vol. 31, no. 15, pp. 1609–1622, 2003.

[6] Cambridge Economic Policy Associates (CEPA), "Background to work on assessing efficiency for the 2005 price control review," Final Report, Cambridge Economic Policy Associates (CEPA), 2003.

[7] W. Chung, "Review of building energy-use performance benchmarking methodologies," *Applied Energy*, vol. 88, no. 5, pp. 1470–1479, 2011.

[8] T. Jamas, *Incentive Regulation and Benchmarking of Electricity Distribution Networks: From Britain to Switzerland*, Faculty of Economics, University of Cambridge, 2007.

[9] T. Sueyoshi, M. Goto, and T. Ueno, "Performance analysis of US coal-fired power plants by measuring three DEA efficiencies," *Energy Policy*, vol. 38, no. 4, pp. 1675–1688, 2010.

[10] M. Ertürk and S. Türüt-Aşık, "Efficiency analysis of Turkish natural gas distribution companies by using data envelopment analysis method," *Energy Policy*, vol. 39, no. 3, pp. 1426–1438, 2011.

[11] T. Thakur, S. G. Deshmukh, and S. C. Kaushik, "Efficiency evaluation of the state owned electric utilities in India," *Energy Policy*, vol. 34, no. 17, pp. 2788–2804, 2006.

[12] T. Kuosmanen, A. Saastamoinen, and T. Sipiläinen, "What is the best practice for benchmark regulation of electricity distribution? Comparison of DEA, SFA and StoNED methods," *Energy Policy*, vol. 61, pp. 740–750, 2013.

[13] W. H. K. Tsui, A. Gilbey, and H. O. Balli, "Estimating airport efficiency of New Zealand airports," *Journal of Air Transport Management*, vol. 35, pp. 78–86, 2014.

[14] T. Thakur, *Performance evaluation of electric supply utilities in India [PhD Dissertation]*, Indian Institute of Technology Delhi (IIT Delhi), New Delhi, India, 2007.

[15] Q. Cui, H.-B. Kuang, C.-Y. Wu, and Y. Li, "The changing trend and influencing factors of energy efficiency: the case of nine countries," *Energy*, vol. 64, pp. 1026–1034, 2014.

[16] F. Javier Ramos-Real, B. Tovar, M. Iootty, E. F. de Almeida, and H. Q. Pinto Jr., "The evolution and main determinants of productivity in Brazilian electricity distribution 1998–2005: an empirical analysis," *Energy Economics*, vol. 31, no. 2, pp. 298–305, 2009.

[17] C.-F. Chien, W.-C. Chen, F.-Y. Lo, and Y.-C. Lin, "A case study to evaluate the productivity changes of the thermal power plants of the Taiwan power company," *IEEE Transactions on Energy Conversion*, vol. 22, no. 3, pp. 680–688, 2007.

[18] W. W. Cooper, S. Li, L. M. Seiford, K. Tone, R. M. Thrall, and J. Zhu, "Sensitivity and stability analysis in DEA: some recent developments," *Journal of Productivity Analysis*, vol. 15, no. 3, pp. 217–246, 2001.

[19] T. Jamasb, P. Nillesen, and M. Pollitt, "Strategic behaviour under regulatory benchmarking," *Energy Economics*, vol. 26, no. 5, pp. 825–843, 2004.

[20] T. Coelli and D. A. Lawrence, *Performance Measurement and Regulation of Network Utilities*, Edward Elgar Publishing, 2006.

[21] W. W. Cooper, L. M. Seiford, and J. Zhu, *Handbook on Data Envelopment Analysis*, vol. 164, Springer Science & Business Media, 2011.

[22] ERRA Licensing and Competition Committee, "Benchmarking in ERRA (Energy Regulators Regional Association) countries," Issue Paper, 2002.

[23] ACCC/AER, "Benchmarking opex and capex in energy," Working Paper 6, ACCC/AER, 2012.

[24] T. J. Coelli, D. S. Prasada Rao, C. J. O'Donnell, and G. E. Battese, *An Introduction to Efficiency and Productivity Analysis*, Springer, New York, NY, USA, 2nd edition, 2005.

[25] Public Utility Commission Sri Lanka, *Decision on Electricity Tariffs*, 2011.

[26] Public Utility Commission Sri Lanka, *Performance Report of Distribution Licensees*, 2012.

[27] Public Utility Commission Sri Lanka, *Tariff Methodology, Distribution Allowed Revenue*, 2011.

[28] Public Utilities Commission of Sri Lanka, *Performance Report of Distribution Licensees*, 2011.

[29] Ceylon Electricity Board, *Statistical Digest*, Ceylon Electricity Board, 2006.

[30] Ceylon Electricity Board, *Statistical Digest*, 2007.

[31] Ceylon Electricity Board, *Statistical Digest*, 2008.

[32] Ceylon Electricity Board, *Annual Report and Accounts*, 2008.

[33] Ceylon Electricity Board, *Annual Report and Accounts*, 2009.

[34] Ceylon Electricity Board, *Statistical Digest*, Ceylon Electricity Board, 2009.

[35] Ceylon Electricity Board, *Annual Report and Accounts*, Ceylon Electricity Board, Colombo, Sri Lanka, 2010.

[36] Ceylon Electricity Board, *Statistical Digest*, 2010.

[37] Ceylon Electricity Board, *Statistical Digest*, 2011.

[38] Lanka Electricity Company Limited, *Annual Report*, Lanka Electricity Company Limited, 2009.

[39] Lanka Electricity Company Limited, *Annual Report*, Lanka Electricity Company Limited, Colombo, Sri Lanka, 2010.

[40] Lanka Electricity Company, *Annual Report*, 2011.

[41] Public Utility Commission Sri Lanka, *Electricity Distribution Performance Regulation—Final Draft*, Public Utility Commission Sri Lanka, 2012.

[42] S. Talluri, "Data envelopment analysis: models and extensions," *Decision Line*, vol. 31, no. 3, pp. 8–11, 2000.

[43] M. Filippini, N. Hrovatin, and J. Zorič, "Efficiency and regulation of the Slovenian electricity distribution companies," *Energy Policy*, vol. 32, no. 3, pp. 335–344, 2004.

[44] M. Farsi and M. Filippini, "A benchmarking analysis of electricity distribution utilities in Switzerland," Working Paper 43, Centre for Energy Policy and Economics, Swiss Federal Institute of Technology, Zurich, Switzerland, 2005.

11

Recovery of Exhaust Waste Heat for ICE Using the Beta Type Stirling Engine

Wail Aladayleh and Ali Alahmer

Department of Mechanical Engineering, Tafila Technical University, P.O. Box 179, Tafila 66110, Jordan

Correspondence should be addressed to Ali Alahmer; a.alahmer@ttu.edu.jo

Academic Editor: Guobing Zhou

This paper investigates the potential of utilizing the exhaust waste heat using an integrated mechanical device with internal combustion engine for the automobiles to increase the fuel economy, the useful power, and the environment safety. One of the ways of utilizing waste heat is to use a Stirling engine. A Stirling engine requires only an external heat source as wasted heat for its operation. Because the exhaust gas temperature may reach 200 to 700°C, Stirling engine will work effectively. The indication work, real shaft power and specific fuel consumption for Stirling engine, and the exhaust power losses for IC engine are calculated. The study shows the availability and possibility of recovery of the waste heat from internal combustion engine using Stirling engine.

1. Introduction

Today, the energy researches take a wide place in the world; the automobile is so significant that it consumes more than half of the total energy used by all types of transportation combined. Numerically, the energy consumption of automobiles accounts for 52% of all energy used by the entire transportation; less than 35% of the energy in a gallon of gasoline reaches the wheels of a typical car; the remaining heat is expelled to the environment through exhaust gases and engine cooling systems [1]. Figure 1 illustrates the energy losses of internal combustion engine (ICE). The figure shows the thermal losses take 60% approximately, and 33% of the power is expelled with exhaust gases; in another mean two-thirds of our fuels' money was spent in the environment. Since most of the energy consumed by an internal combustion engine is wasted, capturing much of that wasted energy can provide more power and efficiency. Many researchers examine how to utilize that lost energy and many methods were used such as thermoelectric generation, piezoelectric generation, thermionic generation, thermophotovoltaic, and mechanical turbo [2]. But these entire components were considered as electric or electronic methods and they cannot

give largest power at high temperature. Another method now used to recover the heat from exhaust gas is called organic Rankin cycle (ORC). Figure 2 is based on the steam generation in a secondary circuit using the exhaust gas thermal energy to produce additional power by means of a steam expander. The principle working of the organic Rankine cycle is the same as that of Rankine cycle: the working fluid is pumped to a boiler where it is evaporated, passed through an expansion device (turbine or other expander), and then passed through a condenser heat exchanger which it is finally recondensed [3, 4]. Another method to recover exhaust heat is by the use of the Stirling engine technique. This method has more activity and it is considered an external combustion engine to produce mechanical work. The recovery and utilization of waste heat not only conserve fuel, usually fossil fuel, but also reduce the amount of waste heat and greenhouse gases damped to environment [2]. The use of Stirling engine has many advantages which can be summarized as follows [5–7]: high potential efficiency up to 45%, reversible operation, cleaner emissions, quiet operation, low vibrations, low maintenance, smooth torque delivery, and ability to run at different fuels; finally Stirling engine does not have valves, carburetor, ignition system, or boilers.

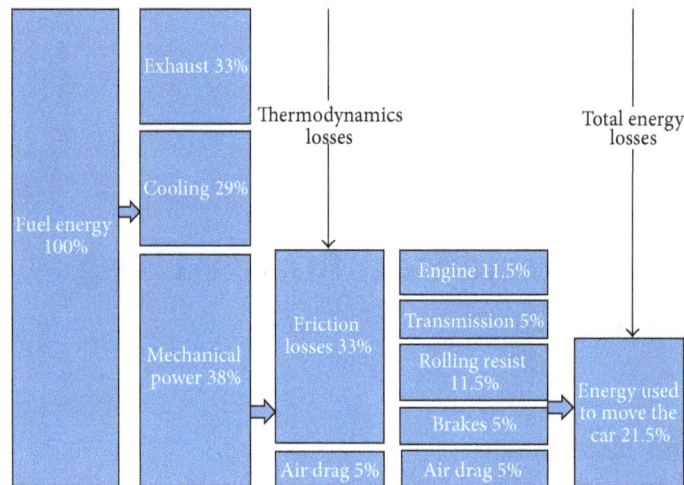

FIGURE 1: Energy losses of internal combustion engine.

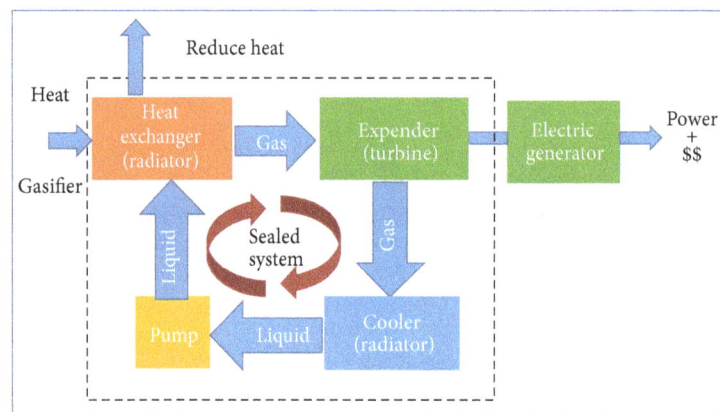

FIGURE 2: Organic Rankin cycle (ORC).

On the other side the main disadvantages and limitations can be concluded as long start-up time at cold starting, typically not self-starting, and finally being quite large and heavy.

The main objectives of this paper could be summarized in two points: using Stirling engine to recover the waste power through exhaust manifold to generate electrical power and also showing the effect of raising the entire operating pressure for Stirling engine to get more power in practical size for the automobile and internal combustion engine.

The body structure of this paper starts by highlighting objectives, advantages, limitations, and related research in Section one. The main factors that have an effect on the performance of Stirling engine were displayed in Section two. Thermodynamic model analysis in terms of Schmidt cycle, waste exhaust recovery, Stirling engine power, and exhaust temperature profile were covered in Section three. Our proposed exhaust's heat recovery system in Section four follows. Section five presented the experimental methodology and setup. The calculation and results were depicted in Section six. Finally, Section seven summarizes the entire paper and shows the main conclusion.

2. Stirling Engine Effective Factors

Usually the design point of a Stirling engine will be somewhere between the two limits of (1) maximum efficiency point and (2) maximum power point. There are many factors that may affect the out power and mechanical efficiency for the Stirling efficiency, which can be concise as the following. (i) Swept volume: the area under the P-V diagram indicates to the network that if the volume expands the power will be increased; (ii) regenerator efficiency: the regenerator has mesh wires to store the heat while the working gas transfers between the hot side and cold side; theoretically if the engine does not have a full regenerative, the major trouble will be in the stream flow losses through the regenerator; (iii) mean pressure: it is the average pressure inside the engine at the maximum and lower temperatures; the problem appeared when the inside pressure is more than the atmospheric pressure; unbalance on the piston will occur; (iv) working gas: the type of gas in the Stirling engine takes a major factor; to get more power, the filled gas must have high specific heat capacity so that the gas will gain and lose the heat rapidly;

then the piston is moving rapidly to produce a positive speed [5]. The hydrogen has the lowest molecular weight so it has great efficiency but low safety. Helium (He), N_2, and air were considered a working gas for Stirling engine; and finally (v) temperature difference: as any heat engine, the mechanical efficiency stands upon the hot temperature and cold temperature so more difference gives more efficiency.

3. Thermodynamic Analysis

3.1. Schmidt Cycle Analysis. The Schmidt cycle is defined as a Stirling cycle in which the displacer and the power piston or the two power pistons move sinusoidally and dead volumes are found. The assumptions upon which the Schmidt analysis was based are as follows [8]: (i) sinusoidal motion of parts; (ii) gas temperatures known and constant in all parts of the engine; (iii) absence of gas leakage; (iv) working fluid following perfect gas law; and finally (v) at each instant in the cycle the gas pressure being the same throughout the working gas.

In this paper the Schmidt cycle will be evaluated numerically. The performance of the engine can be calculated using P-V diagram. The volume in the engine is calculated using the internal geometry. When the volume, mass of the working gas, and the temperature are decided, then the pressure is calculated using an ideal gas method in PV = mRT equation.

Firstly, the volumes of the expansion and compression cylinder at a given crank angle will be determined. The instantaneous expansion volume V_E is

$$V_E = \frac{V_{SE}}{2} (1 - \cos \theta) + V_{DE}, \tag{1}$$

where V_{SE} is a swept volume of the expansion piston and V_{DE} is an expansion dead volume under the condition.

The instantaneous compression volume V_C is determined by

$$V_C = \frac{V_{SE}}{2} [1 - \cos \theta] + \frac{V_{SC}}{2} [1 - \cos (\theta - \varphi)] + V_{DC} - V_B, \tag{2}$$

where V_{SC} is a swept volume of the compression piston, V_{DC} is a compression dead volume, and φ is phase angle.

The total instantaneous volume is calculated in

$$V = V_E + V_R + V_C. \tag{3}$$

In the Beta type Stirling engine, the displacer piston and the power piston are located in the same cylinder. When both pistons overlap, an effective working space is created. The overlap volume V_B is

$$V_B = \frac{V_{SE} + V_{SC}}{2} - \sqrt{\frac{V_{SE}^2 + V_{SC}^2}{4} - \frac{V_{SE}V_{SC}}{2} \cos \varphi}. \tag{4}$$

The engine pressure P based on the mean pressure P_{mean}, the minimum pressure P_{min}, and the maximum pressure P_{max} is described in [8]:

$$P = \frac{P_{mean} \sqrt{1 - c^2}}{1 - c \cdot \cos (\theta - A)} = \frac{P_{max} (1 - c)}{1 - c \cdot \cos (\theta - A)}$$
$$= \frac{P_{min} (1 + c)}{1 - c \cdot \cos (\theta - A)}, \tag{5}$$

where

$$A = \tan^{-1} \frac{v \sin \varphi}{t + \cos \varphi + 1},$$
$$S = t + 2t x_{DE} + \frac{4t x_R}{1 + t} + v + 2 x_{DC} + 1 - 2 x_B,$$
$$x_B = \frac{V_B}{V_{SE}},$$
$$B = \sqrt{t^2 + 2(t - 1) v \cos \varphi + v^2 - 2t + 1}, \tag{6}$$
$$c = \frac{B}{S}, \quad t = \frac{T_C}{T_E}, \quad v = \frac{V_{SC}}{V_{SE}}, \quad x_{DE} = \frac{V_{DE}}{V_{SE}},$$
$$x_{DC} = \frac{V_{DC}}{V_{SE}}, \quad x_R = \frac{V_R}{V_{SE}},$$

where t is temperature ratio, v is a swept volume ratio, and x is dead volume ratio.

The net indicated work per cycle [8] is described by

$$W_{net} = \frac{P_{mean} V_{SE} \pi c (1 - t) \sin A}{1 + \sqrt{1 - c^2}}$$
$$= \frac{P_{min} V_{SE} \pi c (1 - t) \sin A}{1 + \sqrt{1 - c^2}} \cdot \frac{\sqrt{1 + c}}{\sqrt{1 - c}} \tag{7}$$
$$= \frac{P_{max} V_{SE} \pi c (1 - t) \sin A}{1 + \sqrt{1 - c^2}} \cdot \frac{\sqrt{1 - c}}{\sqrt{1 + c}}.$$

3.2. Waste Heat Energy Calculation. The quantity of waste heat contained in an exhaust gas is a function of both the temperature and the mass flow rate of the exhaust gas:

$$\dot{Q} = \dot{m} * c_p * \Delta T, \tag{8}$$

where $\dot{Q}$ is the heat loss (kJ/s); $\dot{m}$ is the exhaust gas mass flow rate (kg/s); c_p is the specific heat of exhaust gas (kJ/kg·K); and ΔT is temperature gradient in K.

The mass flow rate of exhaust gas $\dot{m}_E$

$$\dot{m}_E = \dot{m}_f + \dot{m}_a. \tag{9}$$

Mass flow rate of air ($\dot{m}_a$) can be evaluated according to

$$\dot{m}_a = \mu_v * \rho_a * v_s * N * 2. \tag{10}$$

Mass flow rate of fuel ($\dot{m}_f$)

$$\dot{m}_f = \frac{\dot{m}_a}{(A/F)_{ratio}}. \tag{11}$$

The volumetric efficiency (μ_v) has a range 0.8 to 0.9.

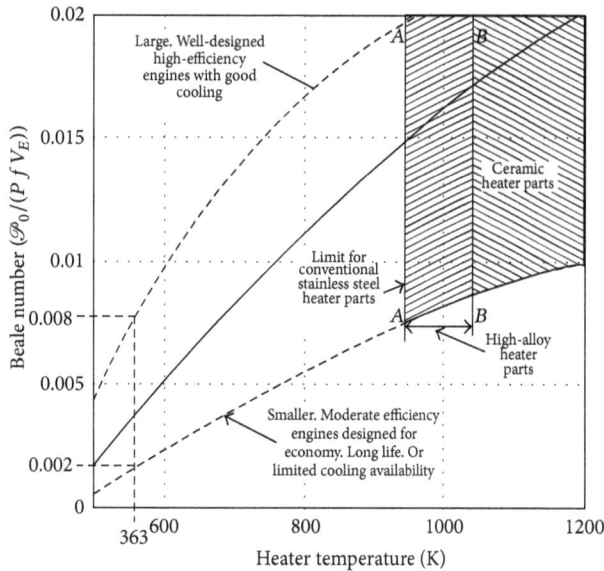

FIGURE 3: Graph of Beale number vs. heater temperature for a range of Stirling engines [5].

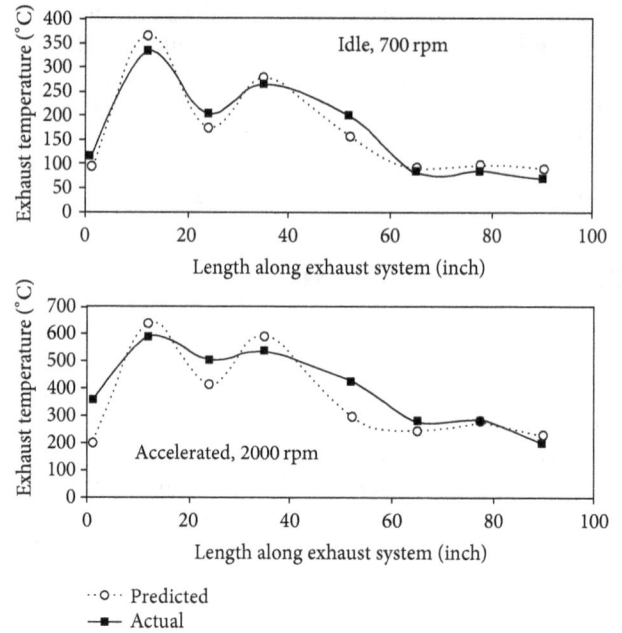

FIGURE 4: Illustration of the exhaust gas temperature versus tailpipe length [9].

3.3. Engine Power Output. Power output can be estimated using a variety of methods which takes into consideration many things like temperature difference, operating speed and pressure, expansion and compression space volumes, and regenerator effectiveness. The Beale and West numbers were used to determine the Stirling engine power.

3.3.1. Beale Number. It is an empirical number that characterizes the performance of Stirling engine. It is used to estimate the power output of Stirling through relating an indicated power, $\mathscr{P}_o$ (W), to mean pressure P (bar), operating frequency f (Hz), and expansion space volume V_E (cm^3) with the Beale number B_n:

$$\mathscr{P}_o = B_n P f V_E. \tag{12}$$

Beale number can be estimated from Figure 3, showing a graph plotted by measuring data from many Stirling engines. The solid line in the middle is typical of most Stirling engines while the upper and lower lines denote unusually high or low performing engines [5].

3.3.2. West Number. It is similar to the Beale number except it takes direct account of the temperature difference. The formula is expressed as

$$\mathscr{P}_o = W_n P f V_E \left(\frac{T_h - T_c}{T_h + T_c} \right), \tag{13}$$

where W_n is the West number, which has an average value of 0.25–0.35, and a higher number represents a more efficient engine.

3.4. Temperature Profile in Automotive Exhaust Systems. The flow rate and temperature profile of the burnt gases passing through the exhaust system are required to determine the most effective location for Stirling engine placement. Figure 4 shows the comparison of the estimated and measured temperatures for almost identical runs along the exhaust line of the test vehicle for both idling and accelerated engine speeds [9]. The surface temperature measurements method was used in nearly stable conditions after initial warm-up. Exhaust temperatures for idling conditions were much lower compared to accelerated conditions. When moving away from the exhaust valve, the gases temperatures will be decreased. Exhaust temperature varies with engine load; more loads or speed means more exhaust temperature due to decreasing in expansion cooling.

4. Proposed Exhaust's Heat Recovery System

Figure 5 presents the unit of heat recovery for exhaust system (3D/section) views. Stirling engine components are (1) exhaust gases outlet; (2) exhaust gases inlet; (3) hot heat exchanger; (4) displacer piston cylinder; (5) coolant jacket; (6) crank case container (to keep high pressure behind power piston); (7) displacer; (8) power piston; and finally (9) crank shaft and flywheel.

Afterwards, the exhaust gases leave the combustion chamber, they will enter through hot exchanger's pipe 1, and then they leave from heat exchanger pipe 2 and then to catalytic converter, muffler, and tailpipe.

Hot heat exchanger must be near to exhaust valve or isolate the inlet pipe in Rockwool to prevent the exhaust gases' heat escaping before it entered the heat exchanger. The coolant jacket is used to get difference in engine temperature also, to improve good contraction for working fluid for Stirling engine in cold side of Stirling engine; coolant jacket

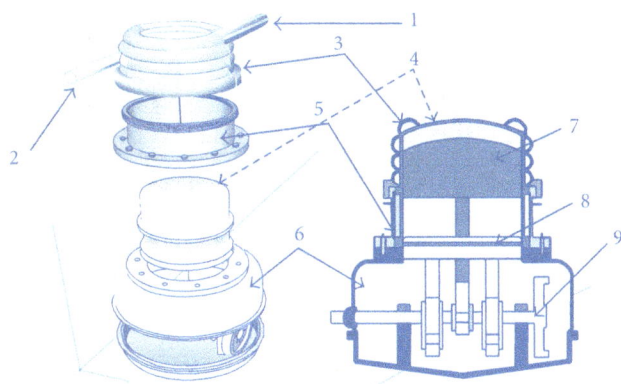

FIGURE 5: Exhaust's heat recovery unit.

FIGURE 6: SI Robin engine single cylinder, air cooled and with direct-injection.

must be connected directly with a separated radiator. The displacer owns a vertical hole with mesh material because of brief in engine size; the container also has compensated valve to modify entire pressure of the engine.

5. Experimental Setup

A number of experiments were carried out in Tafila Technical University in the automotive laboratories. A Robin engine single cylinder, air cooled and with direct-injection engine, was used in this work as shown in Figure 6. The engine specifications are listed in Table 1. To measure the gasoline engine torque, the engine was coupled to dynamometer. The reading of engine parameters was recorded after 135 sec of engine operation and depicted in Table 2. The Stirling engine was coupled to ICE. The specification of Stirling engine was listed in Table 3.

6. Results and Calculations

To evaluate the wasted exhaust power, the air and fuel flow through the combustion process should be estimated. From the recorded information during experiment in Table 2, the fuel consumption is 21.6 g in interval 135 seconds. So the

TABLE 1: Specifications and parameters of the internal combustion engine.

ICE specifications	
Brand	Robin engines single cylinder
Model	EX13D-4 stroke
Displacement	126 CC
Max. output	3.2 KW at 4000 rpm
Max. torque	8 N.m at 2500 rpm
Fuel	Unleaded gasoline
Spark plug	NGK B4 (recommended)
Cooling	Air cooling
Lubricant	API/SE or SAE 10W-30
ICE parameters	
Bore	58 mm
Stroke	48 mm
Air intake duct diameter	54.8 mm
Exhaust manifold diameter	64 mm

TABLE 2: Readings of ICE powered by Jordanian gasoline/octane rate 95 after 135 seconds at beginning of the engine operation.

ICE reading	
Exhaust gas temperature	200°C
Ambient temperature	24.5°C
Oil temperature	85.8 C
Fuel temperature	23.8 C
Engine speed	2585 rpm
Engine power	2.3 KW
Air flow speed	1.5 m/s

TABLE 3: Parameters and specifications of Stirling engine.

Stirling engine parameters	
Engine configuration	Stirling engine-Beta type
Hot swept volume	200 CC
Cold swept volume	150 CC
Hot dead volume	25 CC
Cold dead volume	20 CC
Regenerator volume	30 CC
Mean pressure	20 bars
Hot side temperature	200°C
Cold side temperature	35°C
Phase angle	90°
Engine speed	1500 rpm (25 Hz)
Working fluid	Air

fuel and air flow are 0.16 g/s and 4.32 g/s, respectively. The brake power and recovery exhaust power are 0.901 kW and the percentage difference for exhaust and brake power is 61%. The result of availability of heat recovery by Robin engine is

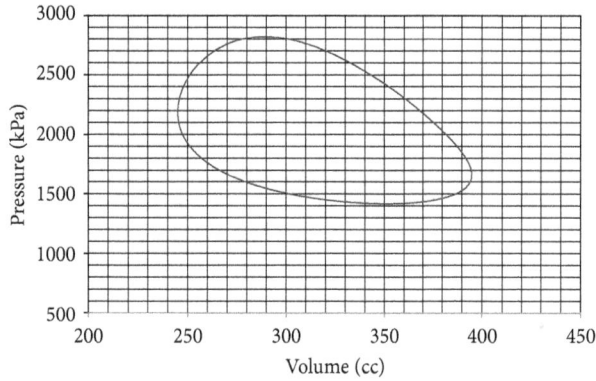

FIGURE 7: Relation between entire pressures and changing in volume for Stirling engine.

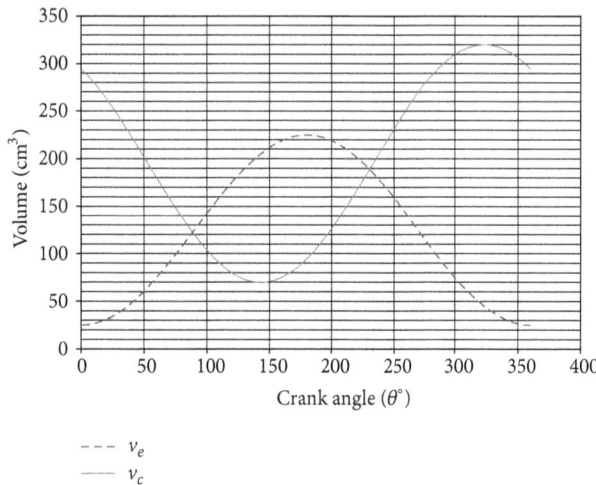

$--- v_e$

$--- v_c$

FIGURE 8: Relation for volume changing of expansion and contraction space versus crank angle.

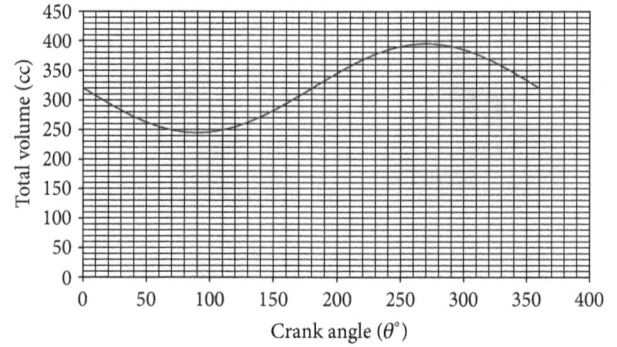

FIGURE 9: Changing of the total volume versus the crank angle for Stirling engine.

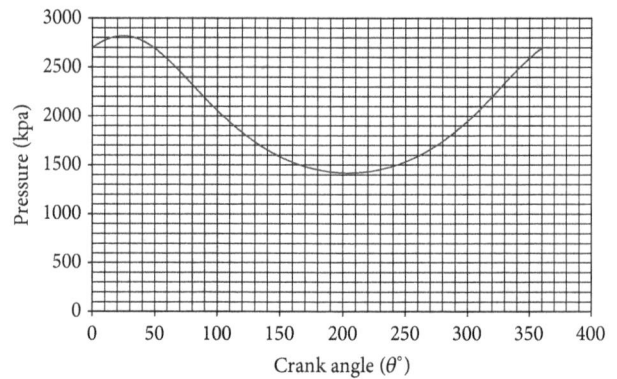

FIGURE 10: Changing of the pressure versus the crank angle for Stirling engine.

similar to 04 Prius 1NZ-FEX and BMW M135i which give (58% and 56%, resp.) percentage difference.

Figure 7 represents the relation between entire pressures and changing in volume for Stirling engine, the curve being formed as cam lobe due to dead volume affected and non-sinusoidal motion for the drive mechanism; Schmidt cycle gives ideal assumption, else dead volume and nonsinusoidal motion, too. The area enclosed in P-V curve was indicated work per one cycle. Figure 8 illustrates the relation for volume changing of expansion and contraction space with crank angle. From this diagram the location of the volumes of expansion can be determined and contraction space must be equal relatively to crank angle. Also it can be provided as an indication for critical point when the volume was converted or changed from increasing to decreasing or inverting versus crank angle. Figure 9 shows the relation between total volumes for Stirling engine versus crank angle. From this diagram the total volume changing relatively to crank angle can be determined; also the flow rate of working gas inside Stirling engine can be estimated at any instant. After that the flow losses through the engine can be evaluated. The

total pressure versus crank angle is depicted in Figure 10. The relation between pressure and total volume is inverse here. The expansion and contraction intervals cannot be determined precisely from P-V diagram. But on the other side, the expansion and contraction can be easily evaluated from V-θ or P-θ diagram.

According to the Schmidt cycle, the ideal expansion work, pumping work, total network, and ideal net power are 88.5 J, −56.6 J, 31.9 J, and 0.772 kW at 1500 rpm, respectively. For estimation the real output (shaft power) for Stirling engine is 0.42 kW based on the Beale method. From Figure 3, the Beale number is nearly 0.0042 at heater temperature 200°C. So, mechanical efficiency for Stirling engine is 51.8%.

To determine the specific fuel consumption using Stirling engine, the following equation was used:

$$\text{s.f.c.}_{\text{with Stirling}} = \frac{\text{s.f.c.}_{\text{without Stirling}}}{1 + \hbar}, \tag{14}$$

where $\hbar$ is lowering percentage for brake power; in this case it is equal to 18%, so the s.f.c. with Stirling engine will lower to 15% of s.f.c. without Stirling engine. $\hbar$ can be calculated by dividing real brake power for Stirling engine on brake power for IC engine:

$$\hbar = \frac{\mathscr{P}_{\text{br.sti}}}{\mathscr{P}_{\text{br.IC}}}. \tag{15}$$

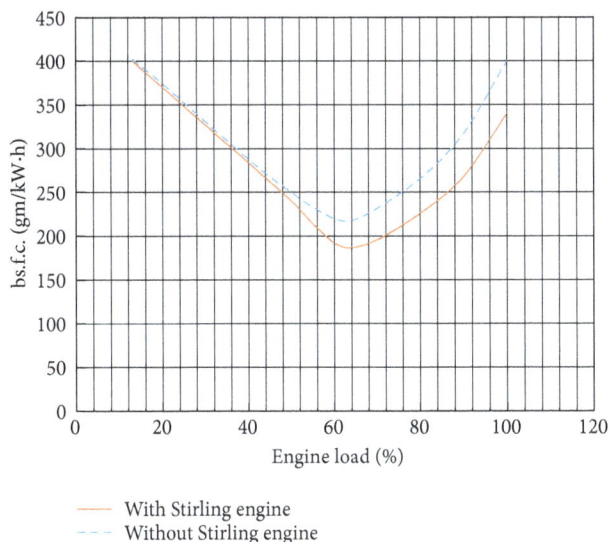

FIGURE 11: Showing brake specific fuel consumption without Stirling engine and with Stirling engine.

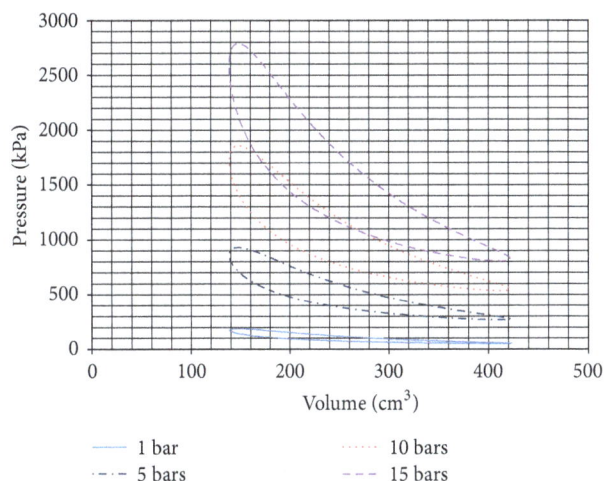

FIGURE 12: P-V diagram at different operating pressures.

The brake specific fuel consumption without Stirling engine and with Stirling engine is displayed in Figure 11. At beginning of engine loading, bs.f.c. for internal combustion engine with and without Stirling engine are identical due to lowering in exhaust temperature (low load); subsequently when the engine load was increased, exhaust temperature will also increase, so Stirling engine will work effectively to reduce specific fuel consumption. So at those moments bs.f.c. curves are not identical.

Figure 12 displays the P-V diagram at different operating pressures. The indicated work is represented by the area of enclosed curve for each operating pressure. As shown in the figure as the operating pressure increases, the area enclosed will increase and it will increase the indicated work.

7. Conclusion

The utilization of the exhaust waste heat for ICE by the use of Stirling engine was investigated. The study results can be summarized into the following points.

(i) Waste heat of internal combustion engine is considered great problem; two-thirds of that energy which entered through the engine was lost to the environment.

(ii) Waste heat recovery takes great benefits as raising fuel mileage and reducing greenhouse gases and fuel consumption, so that the IC efficiency will be increased.

(iii) Around 15% can be improved in vehicle fuel economy through installing Stirling engine cross exhaust manifold to recover waste heat in internal combustion engines.

(iv) The recovered power through Stirling engine can be converted to charge vehicle's batteries or to operate the mechanical auxiliary such as oil pump, water pump, A/C compressor, and power steering pump.

(v) Applications' range for this project is not trapped on the automobile only, but it can be applicable on electricity generation planets, mining application cement planets, and factories.

(vi) Three obstacles to using Stirling engine are as follows: (1) adding some weights to the automobile which is going to decrease its fuel efficiency, so in order to be viable it must be light; (2) backpressure through the exhaust system; (3) additional pumping power losses.

Conflict of Interests

The authors declare that there is no conflict of interests regarding the publication of this paper.

References

[1] J. Halderman, *Automotive Technology*, Prentice Hall, 4th edition, 2011.

[2] J. Jadhao and D. Thombare, "Review on exhaust gas heat recovery for I.C. engine," *International Journal of Engineering and Innovative Technology*, vol. 2, no. 12, pp. 93–100, 2013.

[3] N. Galanis, E. Cayer, P. Roy, E. S. Denis, and M. Désilets, "Electricity generation from low temperature sources," *Journal of Applied Fluid Mechanics*, vol. 2, no. 2, pp. 55–67, 2009.

[4] K. K. Srinivasan, P. J. Mago, and S. R. Krishnan, "Analysis of exhaust waste heat recovery from a dual fuel low temperature combustion engine using an Organic Rankine Cycle," *Energy*, vol. 35, no. 6, pp. 2387–2399, 2010.

[5] C. Lloyd, *A low temperature differential stirling engine for power generation [M.S. thesis]*, Department of Electrical and Computer Engineering, University of Canterbury, 2009.

[6] U. Ramesh and T. Kalyani, "Improving the efficiency of marine power plant using stirling engine in waste heat recovery systems," *International Journal of Innovative Research and Development*, vol. 1, no. 10, pp. 449–466, 2012.

[7] J. Ruiz, *Waste heat recovery in automobile engines: potential solutions and benefits [Master dissertation]*, Department of Mechanical Engineering, Massachusetts Institute of Technology, 2007.

[8] K. Hirata, "Schmidt theory for Stirling engines," Tech. Rep., National Maritime Research Institute, 1997.

[9] M. Ehsan, M. Shah, H. Hasan, and S. Hasan, "Study of Temperature Profile in automotive exhaust systems for retrofitting catalytic converters," in *Proceedings of the International Conference on Mechanical Engineering (ICME '05)*, Dhaka, Bangladesh, 2005.

12

A Critical Review on Wind Turbine Power Curve Modelling Techniques and Their Applications in Wind Based Energy Systems

Vaishali Sohoni, S. C. Gupta, and R. K. Nema

Department of Electrical Engineering, Maulana Azad National Institute of Technology, Bhopal 462051, India

Correspondence should be addressed to Vaishali Sohoni; vaishalisohonibpl@gmail.com

Academic Editor: Kamal Aly

Power curve of a wind turbine depicts the relationship between output power and hub height wind speed and is an important characteristic of the turbine. Power curve aids in energy assessment, warranty formulations, and performance monitoring of the turbines. With the growth of wind industry, turbines are being installed in diverse climatic conditions, onshore and offshore, and in complex terrains causing significant departure of these curves from the warranted values. Accurate models of power curves can play an important role in improving the performance of wind energy based systems. This paper presents a detailed review of different approaches for modelling of the wind turbine power curve. The methodology of modelling depends upon the purpose of modelling, availability of data, and the desired accuracy. The objectives of modelling, various issues involved therein, and the standard procedure for power performance measurement with its limitations have therefore been discussed here. Modelling methods described here use data from manufacturers' specifications and actual data from the wind farms. Classification of modelling methods, various modelling techniques available in the literature, model evaluation criteria, and application of soft computing methods for modelling are then reviewed in detail. The drawbacks of the existing methods and future scope of research are also identified.

1. Introduction

Wind energy has emerged as a promising alternative source for overcoming the energy crisis in the world. Wind power based energy is one of the most rapidly growing areas among the renewable energy sources and will continue to do so because of the growing concern about sustainability and emission reduction requirements. The uncertain nature of wind and high penetration of wind energy in power systems are a big challenge to the reliability and stability of these systems. To make wind energy a reliable source, accurate models for predicting the power output and performance monitoring of wind turbines are needed. The theoretical power captured (P) by a wind turbine is given by [1]

$$P = \frac{1}{2}\rho A_w C_P\left(\lambda, \beta\right) v^3. \tag{1}$$

The power production of a wind turbine (WT) thus depends upon many parameters such as wind speed, wind direction, air density (a function of temperature, pressure, and humidity) and turbine parameters [2]. Much complexity is involved in considering the effects of all the influencing parameters properly. It is therefore difficult to evaluate the output power using the theoretical equation given above. Power curve of a wind turbine, which gives the output power of turbine at a specific wind speed, provides a convenient way to model the performance of wind turbines. A typical power curve for a pitch regulated wind turbine is shown in Figure 1. In the first region when the wind speed is less than a threshold minimum, known as the cut-in speed, the power output is zero. In the second region between the cut-in and the rated speed, there is a rapid growth of power produced. In the third region, a constant output (rated) is produced until the cut-off speed is attained. Beyond this speed (region 4) the turbine is taken out of operation to protect its components from high winds; hence it produces zero power in this region.

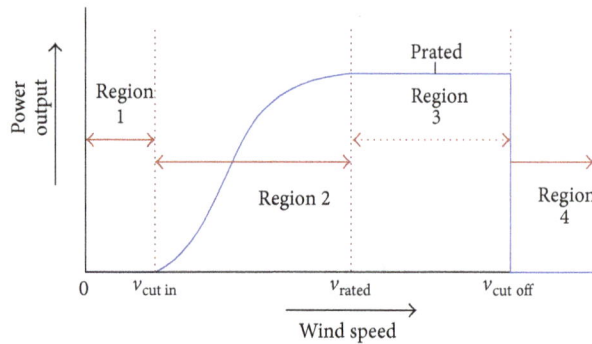

FIGURE 1: Typical power curve of a pitch regulated wind turbine.

The power curve of a WT indicates its performance. Accurate models of power curves are important tools for forecasting of power and online monitoring of the turbines. A number of methods have been proposed in various works to model the wind turbine power curve. These methods which use data from manufacturers' specifications and actual data from the wind farms have been utilized by many researchers in various wind power applications [3, 4]. The literature reviewed reveals that appropriate selection of power curve models can help in improved performance of wind energy based systems. This paper presents current status of research and future directions of different wind turbine power curve modelling approaches. The need of modelling, modelling methodology, classification of models, and methods of evaluation have been discussed. The impact of various parameters on these curves, the standard procedure for power performance measurement of wind turbines, and the need for developing site specific curves is also discussed. Various models proposed and used in various studies have been compared critically and finally inferences are drawn.

2. Need of Power Curve Modelling

The power curve reflects the power response of a WT to various wind speeds. Accurate models of the curves are useful in a number of wind power applications. The objectives of modelling the wind turbine power curve have been discussed here.

2.1. Wind Power Assessment and Forecasting. The WT power curve can be used for wind power assessment. Wind resource assessment of a region in terms of wind speed, wind power density, and wind energy potential is done to identify areas suitable for wind power development [3]. In this process, estimation of energy is done by using the available wind data and wind turbine power curve. Predicting the output power of the turbine at a candidate site is also required in sizing and cost optimization studies during the design stage of a wind energy based system. The accuracy in power prediction is important as an overestimation can result in poor reliability and an underestimation can lead to oversizing of the wind energy conversion system. Wind turbine operators who trade energy directly to the electricity market also need to forecast the power output of their turbines accurately, so that they will be able to deliver the traded amount of power [2].

Power curves are supplied by the manufacturers in a tabular or graphical form. However, a generic equation which represents this curve accurately is required in various problems of wind power systems. Derivation of an appropriate function to describe the actual shape of the curve is a very important task. However, the manufacturer's curves are created under standard conditions therefore they may not represent the realistic conditions of the site under consideration. The turbine performance at the wind farms is also not ideal due to wear and tear and aging of turbines. Another method to model the power curves is to derive them using the actual data of wind speed and power measured from the turbines [4]. The data of wind turbines collected by the SCADA (supervisory control and data acquisition) system can be utilized for this purpose. This method can incorporate the actual conditions at the wind farms, thus providing better accuracy in power prediction.

2.2. Capacity Factor Estimation. The capacity factor of a WT is defined as the ratio of the average power output to the rated output power of the generator and is an indicator of its efficiency [5]. It is used to estimate the average energy production of a WT required for the sizing and cost optimization studies, optimum turbine-site matching, and ranking of potential sites [5, 6]. The wind turbine power curve models are used to estimate the capacity factor of a WT. A comparative analysis of four power curve modelling methods in estimation of capacity factor of wind turbine generator is presented in [7].

2.3. Selection of Turbines. The power curve can be used to make generic comparison between models and can aid in the choice of turbine from the available options. The selection of the turbine characteristics which match with the wind regime of the site helps in optimizing the efficiency of wind energy system [8].

2.4. Online Monitoring of Power Curves. Power curves can be used for monitoring the performance of turbines. For this, a benchmark curve which represents the performance of a normally operating turbine is required. This reference curve can be extracted from measured power output and wind speed data of wind turbines. The actual curve of the turbine to be monitored can be compared with this benchmark curve. The deviations of the actual values from the expected output can indicate underperformance or faults [1]. The wind power output of a turbine can be affected by underperformance or various faults/anomalies of the turbine such as blade faults and yaw and pitch system faults [4, 9]. Different types of faults affect the turbine system differently and will cause the power curve to depart from the expected value in a different way. Tools which can characterize and quantify these departures can aid in early identification of faults. Statistical analysis of the outlier data can give indications of the specific reason of anomaly. Wind turbine condition monitoring by use of power curve copula modelling is suggested in [10, 11] and is a topic of further research. Early recognition of the emerging faults and timely repair and maintenance of the equipment can help in improving the performance of wind turbines.

3. Modelling Issues

A number of aspects require consideration while modelling the power curves of wind turbines. The selection of model and methodology adopted depends upon the purpose of modelling, available data, impact of various parameters on these curves, and other related issues.

The following important issues should be taken into consideration while modelling of the power curve.

3.1. Difference in Models. The power curves vary with different manufacturers and models. Therefore the model used to describe them should also be different [12]. Also, there is difference between pitch regulated and stall regulated turbines. Pitch regulated turbines maintain constant output from the rated to cut-off speed, whereas the stall regulated turbines have a decreased power output above the rated wind speeds (stall region).

3.2. Cut-In and Cut-Off Behaviour. The turbine behaviour near cut-in and cut-off wind speeds can be difficult to model [8]. These limits are different for different turbine models. When the power curve is derived using the measured data, some nonzero and negative values of power outputs below cut-in speed can be obtained. The cut-off hysteresis which occurs during the period between shut down and restart of turbine affects the productivity of turbine [13]. Hysteresis effects can be more significant with certain wind patterns and terrains such as unsteady and gusty winds requiring frequent starting and shutting down, resulting in considerable loss of energy production. Power curve correction which takes into account this behaviour of turbine at cut-off can reduce the power prediction errors.

3.3. Single versus Group of Turbines. The manufacturers' curves are suitable for predicting the power output of a single turbine of a specific type. In a big wind farm a number of turbines are spread over a wide area. Wind energy production involves uncertainties due to stochastic nature of the wind and variation of the power curve [14]. The speed and direction of wind encountered by the turbines of a wind farm may not be the same due to variation of wind. Hence, in a wind farm, power produced by turbines with identical specifications can also differ, even if the wind speed is the same. The shadowing effect of turbines causes this difference as the turbines which operate in wake of other turbines may get reduced wind speeds [15]. This difference can also happen due to factors such as wear and tear, aging, and dirt or ice deposition on blades. With the growth of wind energy projects it has become essential to develop methods to monitor the performance of not only a single turbine but also the wind farm as a whole. Therefore building appropriate models to obtain the relationship between wind speed and output power when a group of turbines are deployed on a wind farm is required.

3.4. Influencing Factors. A number of factors can cause the power curve to deviate from the theoretical value [1, 2]. The important influencing factors are given here and need due attention during modelling.

(i) Wind Conditions at the Site. Wind is highly stochastic in nature. The wind speed and direction change continuously. The wind at a particular site is affected by weather phenomena and topology of the site. The turbulence of wind at a given location affects the power production [16]. Obstacles like trees, buildings, and other high structures influence the wind.

(ii) Air Density. The pressure, temperature, and humidity of site affect the air density [17], hence affecting the power produced. Effect of varying air density has been considered for developing site specific curves [18]. It is shown in [2] that temperature has the highest influence on air density and considering its effect along with the wind direction resulted in improved performance of models.

(iii) Extrapolation of Wind Speed. The wind speed changes with height. This wind shear effect is affected by the roughness of terrain. The power curve uses the wind speed measured at the hub height of turbine, but this height varies with different models and manufacturers, and it is not always possible to measure the wind speed at this height. A number of methods have been used in the literature to express the variation of wind speed with height [12]. Also the wind speed measured at the masts is different from the speed at the turbine location and sometimes when the wind speed values at the particular site are not available the wind speed measurements from a nearby location are used to determine the wind profile of this site. The accuracy of conversion of the measured wind speed to wind speed at hub height and at the turbine location depends on factors such as the vertical wind profile at the site, position of masts relative to the turbine, and the method used for extrapolation.

(iv) Turbine Condition. The power curve is affected by the condition of turbine and associated equipment. Aging and wear and tear of turbine, anomalies and faults, blade condition, yaw and pitch misalignments, controller settings, and so forth cause the power curve to depart from actual values [1, 11].

3.5. IEC 61400-12-1 Standard [19]. IEC 61400-12-1, the commonly adopted international standard for power performance measurement, is found to be of relevance. The procedure for measuring the power performance characteristics of single wind turbines is specified in this standard. It is the most accepted standard for power curve measurement of single wind turbines. The standard describes the measurement methodology for the measured power curve which is determined by simultaneous measurement of wind speed and power output at the test site. A previous site calibration is required for certain terrain conditions. The annual energy production is calculated by applying the measured power curve to reference wind speed frequency distributions supplemented by sources of uncertainty and their effects. The standard prescribes derivation of power curve using the hub height wind speed measured with a cup anemometer in the suitable measurement sector, but if the wind speed has a large variation over the rotor swept area then there can be a significant difference between the hub height wind speed and wind speed averaged over the whole rotor swept area. The measurement methods and accuracy of measuring instruments can cause variance in measurements and can lead to large prediction errors. The impact of other measurement options such as consideration of rotor equivalent wind speed in which speed is measured at heights over the full rotor plane with the use of remote sensing technology (LIDAR and SONAR) and nacelle based anemometry is a topic of further research [20].

The IEC standard uses ten-minute averaged data grouped into wind speed intervals of 0.5 m/s (method of bins). This 10-minute averaging of data introduces systematic averaging errors and short wind fluctuations are killed off. Wind at a specific site can be affected by a number of factors such as topology of the site and obstacles and weather phenomena. Although the IEC power curve considers the wind condition of the current site it may not always be appropriate to apply to the wind conditions of other sites. Research efforts are therefore required to develop site specific power curves. These curves can incorporate the wind conditions of the particular site, thus giving better results [18, 19].

Appropriate selection of modelling method is an important requirement for during planning and operation stage of wind based system and helps in improving the performance of the system. The methods which consider only wind speed as input may not take into account the variance caused by various influencing parameters. Methods which consider the influence of these parameters on the power curve can result in more accurate models. Wind at a specific site can be affected by a number of factors such as topology of the site and obstacles and weather phenomena. It is shown in [18] that the use of developed site power curves, which used the knowledge of site and turbine parameters for modelling, resulted in more accurate energy assessment than the turbine power curve. The issues discussed above if addressed properly can result in efficient models of power curves.

4. Wind Speed Modelling

Wind power generated is highly correlated with the wind speed distribution across the region where the wind farm is situated and depends upon the type of WT deployed in the wind farm. The accuracy in prediction of wind energy can be achieved by modelling the wind speed and power simultaneously. The wind speed at a site varies randomly and its variation in a certain region over a period of time can be represented by different probability distribution functions (PDF). Selection of appropriate PDF to describe the actual wind speed distribution of the site is crucial for accuracy in power prediction.

The most commonly used and accepted distribution is the two-parameter Weibull distribution [5, 21]. It is a versatile PDF, is simple to use, and is found to be accurate for most of the wind regimes encountered in nature. However, Weibull distribution is not suitable for certain wind regimes, for example, those having high frequencies of null winds, and

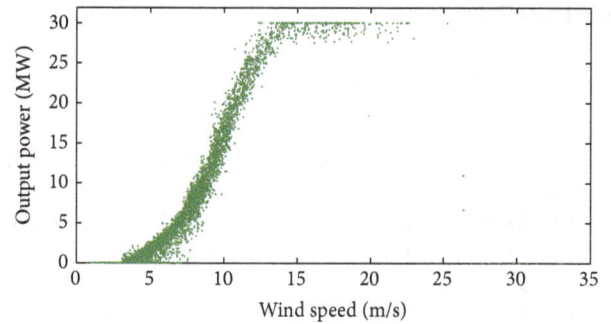

FIGURE 2: Power curve using actual data for a group of wind turbines at a wind farm (NREL) [66].

for short time horizons [1, 22]. The Weibull PDF is given by

$$f(v) = \frac{k}{c}\left(\frac{v}{c}\right)^{k-1} \exp\left(-\frac{v}{c}\right)^k. \tag{2}$$

Another widely used distribution is Rayleigh PDF [18] in which the shape parameter of (2) is taken as $k = 2$. It is a simple PDF and can describe the wind regime with sufficient accuracy when little detail is available about the wind characteristics of a site. The wind speed distribution has also been described in the literature using several other PDFs which include lognormal, beta, and gamma distributions [23]. A detailed review of different PDFs for wind speed modelling and techniques for estimation of their parameters is given in [22]. Appropriate PDF and parameter estimation technique should be selected for modelling the wind speed for a particular site.

5. Power Curve Models Classification

The power curve modelling methods can be classified into *discrete, deterministic/probabilistic, parametric/nonparametric,* and *stochastic* methods or they can be classified on the basis of data used for modelling.

5.1. Discrete Models. In this method as described in IEC 61400-12 all the wind speeds are discretized into 0.5 m/s bins [19]. The power output for each bin is then modelled. This is a simple method as it does not require mathematical functions for describing the curve. Also it takes into account the nonlinear wind speed-power output relation. However a large number of data are required in this method to develop a reliable model.

5.2. Deterministic and Probabilistic Models. A deterministic power curve model assumes a fixed relation between the output power and wind speed. But when a fleet of wind turbines are deployed on a wind farm, turbines of the same type may produce different amount of power even if the wind speed is the same (Figure 2). A probabilistic power curve model incorporates these power variations to characterize the relationship between wind speed and actual output powers. Most of the models available in the literature are of deterministic nature and are constructed by using the manufacturers' power curve data. A probabilistic model proposed in [14] characterizes the dynamics of output power by a normal distribution with varying mean and constant standard deviation. The method given in the paper accommodates the uncertainty of output power. The probabilistic nature of wind power output can also be modelled by deriving curves using actual data of power output and wind speed of turbines deployed in a wind farm. This method requires a large number of historical data but results in accurate models [4, 24].

5.3. Parametric and Nonparametric Models. A parametric model defines the relationship between input and output by a set of mathematical equations with a finite number of parameters. In a nonparametric model, no assumption is made about the functional form of the phenomenon under observation. Parametric models of WT power curve can be built by utilizing a set of mathematical expressions having a fixed number of parameters, which are usually collected together to form a single parameter vector $\theta = (\theta_1, \theta_2, \theta_3, \ldots, \theta_n)$. Nonparametric models are used when it is difficult to define the underlying theory upon which the parametric model can be constructed [24].

5.4. Models Based on Presumed Shape, Curve Fitting, and Actual Data. The models of power curves can be classified according to the data being used for modelling. Models of power curve based on presumed shape of curve utilize only the cut-in, cut-off, and rated speeds and the rated power of the selected turbine for calculating the parameters of expressions used in the model [12, 25, 26]. These ratings are available from the specifications of the turbines. When the manufacturer's power curve data is available, models can be developed by fitting one or more appropriate expressions to the actual curve. The parameters of the expression being fitted to the actual curve are generally calculated by using the least squares method [4]. The models derived from actual data of wind farm need the actual wind speed and power output data from an operational wind farm. If the effect of the influencing parameters is also included in the model, then the data of the included parameters is also required. This data can be obtained from the wind farm's SCADA system.

5.5. Stochastic Models. The stochastic method consists of characterizing the power performance of wind turbine by evaluating dynamic response against the fluctuating wind speed inputs [27, 28]. The dynamic power output is separated into a deterministic stochastic part in this model. In [29] the Markov chain theory is used to describe the power output of WT. The resulting model is independent of turbulence intensity; however, the effect of other influencing parameters is not taken into account in this method.

6. Power Curve Modelling Approaches

Various approaches have been used in the literature for modelling of WT power curve. These methods, their merits, limitations, and application areas are discussed here.

6.1. Parametric Models. The power delivered by a WT can be expressed as

$$P = \begin{cases} 0 & v < v_c, \ v > v_f \\ q(v) & v_c < v < v_r \\ P_r & v_r \leq v \leq v_f. \end{cases} \qquad (3)$$

The relationship between power output and wind speed of a WT between cut-in and rated speed is nonlinear (region 2 of Figure 1). The relation $q(v)$ can be approximated by various functions using polynomial and other than polynomial expressions. The pitch regulated turbines maintain a constant power output in region 3 of Figure 1, whereas the stall regulated turbines have decreased power output in this region; thus the power in this region for a stall regulated turbine should not be modelled as a constant. The governing equations for different approximations of power curve are given in Table 1.

6.1.1. Polynomial Function Approximation. The nonlinear wind speed-power relationship, $q(v)$, can be approximated by various polynomial expressions. Different models using linear, quadratic, cubic, and higher powers of speed or their combinations have been used in the literature.

  (i) The most simplified model based on a linear curve, which describes region 2 of power curve by a straight line, is used in many applications [12, 39–42].

  (ii) A quadratic model represents the nonlinear portion of the curve by an equation of degree 2. $q(v)$ has been approximated by a quadratic equation in [25] to describe the relation between output power and wind speed of a WT. A binomial expression discussed in [30] has been adopted by many researchers [43, 44] to determine output power of wind turbines. Regions 2 and 3 of Figure 1 for a stall regulated WT can be described by using two different binomial expressions as in [45].

  (iii) Model based on cubic law approximates region 2 of power curve by cubic law. A model which describes the nonlinear power-wind speed relationship by a cubic law is discussed in [46]. This model is used for power output calculations in [26, 31, 47]. A cubic expression for region 2 and a linear expression for region 3 are chosen to describe the power curve in [48].

The models given above use the WT specifications of rated power and cut-in, cut-off, and rated wind speed only to determine the equations for power curve.

  (i) A methodology based on Weibull's parameter is proposed in [6]. This model based on Weibull shape parameter is used by many researchers [49, 50] for calculating the power output of WT.

  (ii) A linearized segmented model discussed in [24, 51] carries out a piecewise linear approximation of the

TABLE 1: Expressions of parametric models.

Model	Expressions of P and q	Parameters
Linear [12]	$q(v) = P_r \dfrac{(v - v_c)}{(v_r - v_c)}$	—
Quadratic [25]	$q(v) = P_r \left(\dfrac{v - v_c}{v_r - v_c}\right)^2$	—
Binomial [30]	$q(v) = (a + bv + cv^2) P_r$	$a = \dfrac{1}{(v_c - v_r)^2}\left[v_c(v_c + v_r) - 4v_c v_r \dfrac{(v_c - v_r)^2}{2v_r}\right]$ $b = \dfrac{1}{(v_c - v_r)^2}\left[4(v_c + v_r)\dfrac{(v_c - v_r)^3}{2v_r} - 3(v_c + v_r)\right]$ $c = \dfrac{1}{(v_c - v_r)^2}\left[2 - 4\dfrac{(v_n + v_r)^3}{2v_r}\right]$
Cubic [31]	$q(v) = av^3 - bP_r$	$a = \dfrac{P_r}{(v_r^3 - v_c^3)}$ $b = \dfrac{v_r^3}{(v_r^3 - v_c^3)}$
Weibull based [6]	$q(v) = a + bv^k$	$a = \dfrac{P_r v_c^k}{(v_c^k - v_r^k)}$ $b = \dfrac{P_r}{(v_c^k - v_r^k)}$
Double exponential [32]	$P = \exp(-\tau_1 \exp(-v\tau_2))$	τ_1 and τ_2 to be estimated
4PL [1, 33]	$P = a\left(\dfrac{1 + ame^{v/\tau}}{1 + ame^{v/\tau}}\right)$	$\theta = (a, m, n, \tau)$ (1) From manufacturers' curve [33] $a = P_r$ $n = e^{2sv_{ip}/(P_r - P_{ip})}$ $m = n\left(\dfrac{2P_{ip}}{P_r} - 1\right)$ $\tau = \dfrac{(P_r - P_{ip})}{2s}$ (2) Parameters obtained by evolutionary techniques for SCADA data in [1]
4PL [34, 35]	$P = f(v, \theta) = D + \dfrac{(A - D)}{1 + (v/C)^B}$	$\theta = (A, B, C, D)$ A = minimum asymptote B = Hill slope C = inflection point (point of curve where the curvature changes direction) D = maximum asymptote (parameters obtained by evolutionary techniques)
5PL [24, 35, 36]	$P = f(v, \theta) = D + \dfrac{(A - D)}{\left(1 + (v/C)^B\right)^G}$	$\theta = (A, B, C, D, G)$ A = minimum asymptote B = Hill slope C = inflection point (point of curve where the curvature changes direction) D = maximum asymptote G = asymmetry factor (parameters obtained by evolutionary techniques)

curve by using the equation of a straight line. The resulting curve follows the actual curve more accurately.

(iii) The power curve of wind turbine has also been modelled by other than polynomial functions. The power curve that is modelled in [52] uses a exponential equation whereas a double exponential equation is used in [32].

(iv) A polynomial regression parametric model is developed from real data as a benchmark method in [53]. Three nonparametric methods are also proposed in this study.

6.1.2. Approximations Considering Inflection Point on the Curve. All the polynomial expressions given above do not take into account the inflection point on the power curves.

In most real power curves there is an inflection point on the curve at which its curvature changes sign. Models which consider this inflection point can describe the actual shape of the curve more accurately than the above models. A new formula for power curve interpolation which considers inflection point on the curve is proposed in [54]. A double exponential model is proposed in [32] to fit the data in two inflection zones using a single equation. Functions based on four- and five-parameter logistic approximations also consider this inflection point on the curve and are promising approaches for modelling of power curve.

(i) The shape of a wind power curve can be approximated by a four-parameter logistics (4PL) function [1, 34]. A procedure to obtain the parameters of the 4PL function model modelled from manufactures power curve data has been proposed in [33]. The four parameters of the function are obtained directly from the power curve instead of using an optimization process in this work. The paper also proposes manufacturer's power curve approximation using a three-parameter model. The power curve is derived from the SCADA data of wind farms using a 4PL approximation in [1, 4, 24]. It is shown in [1] that this model can be used for online monitoring of power curves. Another form of four PL expression is used for extracting power curve from actual wind speed and power curve data of a wind farm in [35]. The method is applied for wind energy estimation of the selected wind farm site. Literature reviewed reveals that the four-PL model produces less errors in representing the power curve than the methods based on polynomial approximation. However a 4PL curve is symmetric about the inflection point whereas the power curves are asymmetric. Models which can incorporate this asymmetry can therefore produce even better results. Cubic splines can be used for asymmetric data. A cubic spline is the smoothest curve that passes through the exact data points. As the power curves are quite smooth their asymmetry can be approximated by a cubic spline interpolation technique [55–57], but the disadvantage with a spline fit is that it does not represent the random variation of data.

(ii) A five-parameter logistic (5PL) approximation includes a fifth parameter (G in Table 1) to control the degree of asymmetry [58]. This method can model the asymmetry effectively and can be used for modelling the WT power curve [36]. A 5PL model however has the possibility of becoming ill-conditioned; thus evaluation of parameter vector becomes difficult. A 5PL model derived from the SCADA data of wind farm is applied for energy estimation of the farm in [35] and it is shown that it produces less error in the estimated energy compared to the 4PL model.

6.1.3. Model Based on Curve Fitting of Manufacturer's Curve. The power curve models obtained by means of curve fitting of the manufacturer's curve are used in several applications. The characteristic equation of wind generator is fitted with

three binomial expressions for $q(v)$ in [59] to get the accuracy in fitting. A ninth-order polynomial for power curve fitting has been used in [60] and it has been found that it gives accurate correlation with the real data, producing exclusively positive values for the generated power between cut-in to cut-off range. High-order polynomials may produce better fitting results for a particular set of data; however they may not represent the variance of data and should be used carefully. The power curve models by curve fitting of the manufacturer's curve are analyzed in [55–57].

6.2. Parameter Estimation. The parametric models of WT power curve express the shape of the curve by a set of mathematical equations. Determination of the coefficients of these equations requires fitting the data to the selected model. The techniques and algorithms used in various works for parameter estimation of power curve models are discussed here.

6.2.1. Techniques for Parameter Estimation

(i) Least Squares Method. The least squares method minimizes the summed square of residuals to obtain the parameters of the model and is the most commonly used and accepted method [4, 24].

(ii) Maximum Likelihood Method (MLM). Another approach used in the literature is to determine the parameters of power curve model by maximum likelihood method. In this method the parameters of a statistical model are estimated by maximizing the likelihood function. It was found in [1] that this method did not perform well in comparison to the least squares method.

6.2.2. Algorithms for Parameter Estimation. The parameters of the parametric models especially those using the 4PL and 5PL approximations are difficult to evaluate. The parameter estimation becomes more difficult when the models are derived from the actual data of wind turbines. Developing accurate models of wind turbines and optimization for huge data sets are a very complicated process. Modern nontraditional solution techniques for parameter estimation enhance the accuracy, reduce the computational time, and are easy to implement. Various evolutionary techniques have been applied for determining the parameter vector θ of logistic function based power curve models [4, 24].

6.3. Data Preprocessing. The power curve derived from actual wind speed and power output data of wind turbines uses SCADA data from the wind turbines. This data is prone to errors due to measurement, sensor, and communications system errors. The data is also affected by nonproduction of turbines when it is shut down by the control system for some reason other than anomalous operation. SCADA system can have null entries or erroneous data which can result in inaccurate models. Hence it is necessary to remove these misleading entries before using this data for further analyses. The most common method is to remove the data manually. These outliers can be identified by visual inspection [11] of wind speed power output plot and can be removed before proceeding for development of model. However the method can lead to inaccurate results as the data from SCADA system is voluminous and it is difficult to differentiate between correct and erroneous data. These outliers have been removed by different statistical methods in various works before development of models. In [4] the analysis of residuals together with control charts is used to filter potential outliers. The outliers can be detected by classical least mean square (LMS) method which minimizes the sum of the squares over all the measurements and if a measurement is found to be far away from the correct value it prevails in the resulting fitting; however, in this method a single outlier point can destroy the fitting. In [32] *least median of squares* method is used for data preprocessing in which instead of the sum as in LMS method sum of medians is minimized to identify the outliers. It is shown that this method is very robust; however it requires iterative solution. The wind data preprocessing is done in four steps in [63] which include validity check, data scaling, missing data processing, and lag removal. In [64] a probabilistic method developed around a copula-based joint probability model for power curve outlier rejection is proposed. A data mining approach to process the raw data has been proposed in [65]. Appropriate method for data preprocessing is important requirement for development of an efficient model.

6.4. Evaluation of Model. After developing the model from the data, it is important to determine whether this model appropriately represents the behaviour of the actual data for the power curve. The evaluation of the developed models in various applications is done on the basis of a number of performance metrics. The root mean square error (RMSE), goodness of fit statistics used by a wide number of researchers, estimates of the standard deviation of random component in the data, and a value closer to zero indicate a better fit. The R-squared statistic is the square of the correlation between the actual and the predicted values that measures how closely the fit explains variation in the data. An R-squared value closer to one indicates a good fit [38]. Other criteria used in the literature include the mean absolute error (MAE), mean absolute percentage error (MAPE), the sum of squares error (SSE), SD (standard deviation), and chi-square R [2, 4, 8, 11, 17]. The most commonly used criteria are given in Table 2.

Selection of appropriate model analyzed on the basis of suitable criteria is a very important task for improving the performance of wind power plants. Many models used in the literature have not been evaluated for goodness of fit with actual curve data or their suitability for specific applications. The polynomial models of [54, 57] are compared by observing the visual fit. Models in [2, 11] are compared by MAE, RMSE, MAPE, and SD performance metrics; however, suitability of these models for wind power applications is not evaluated. Also as these models will ultimately be used in wind energy applications it is not appropriate to judge their suitability on the basis of goodness of fit parameters alone, but it should also be examined how successfully they can be employed for the particular applications.

TABLE 2: Model evaluation indices.

	Expression
Absolute error (AE) [1]	$\mathrm{AE} = \left| y_m(i) - y_a(i) \right|$
Relative error (RE) [1]	$\mathrm{RE} = \dfrac{y_m - y_a}{y_a} \times 100$
Mean absolute error (MAE) [24]	$\mathrm{MAE} = \dfrac{1}{N} \sum_{i=1}^{N} \left| y_m(i) - y_a(i) \right|$
Mean absolute percentage error (MAPE) [37]	$\mathrm{MAPE} = \dfrac{1}{N} \sum_{i=1}^{N} \left| \dfrac{y_m(i) - y_a(i)}{y_a(i)} \right|$
Root mean square error (RMSE) [24]	$\mathrm{RMSE} = \left[\dfrac{1}{N} \sum_{i=1}^{N} \left(y_m(i) - y_a(i) \right)^2 \right]^{1/2}$
R-squared [38]	$R^2 = 1 - \dfrac{\sum_{i=1}^{N} \left(y_a(i) - y_m(i) \right)^2}{\sum_{i=1}^{N} \left(y_a(i) - y_{am}(i) \right)^2}$

N = number of data, $y_m(i)$ = ith modelled value, $y_a(i)$ = ith actual value, and $y_{am}(i)$ = mean of actual value.

6.5. Nonparametric Models. Various nonparametric methods can be used for modelling of WT power curves. The SCADA data collected from the wind farms is voluminous and usually contains errors. It may be difficult to obtain the relation between input and output using a functional form. Nonparametric methods can be suitable for deriving power curves from this data. After preprocessing of data, model extraction can be done using different methods. Nonparametric models can also incorporate the effect of parameters other than wind speed on the power curves more easily than the parametric models. The models can be trained by taking these other parameters as inputs to the models [17, 18]. Developments in soft computing techniques offer promising approaches for power curve modelling. Various advanced algorithms can be utilized to generate accurate nonparametric power curves for various applications. Table 3 summarises details of some nonparametric models from the literature.

6.5.1. Neural Networks. Artificial neural networks (ANN) inspired by biological nervous system emulate the natural intelligence of human brain [67] and can learn the nonlinear relationship between input and output data sets by use of activation function within the hidden neurons. Neural networks are used to estimate power generation of turbines at a wind farm in [17]. A separate multilayer perception (MLP) network for each turbine uses ten-minute averages of wind speed and direction from two meteorological towers as inputs and power generated by the turbine as the output. A comparative analysis of regression and ANN models for WT power curve estimation is done in [61] and it is shown that neural network models perform better than the regression models. However ANN has a black box approach and it is difficult to develop an insight about meaning associated with each neuron and weight [68]. ANN based multistage modelling has been used in [62] to model the wind turbine power curve. The wind speed and air density are used as inputs in the first stage and the normalized power output obtained from this stage, wind speed data turbulence intensity, is used to train the second ANN stage. It is claimed that this method has produced

TABLE 3: Details of some nonparametric models from various studies.

Ref.	Data set			Model		
	Interval	Data	Number of values	Model	Parameters	Structure/transfer function (TF)/training method
[4]	10 min	SCADA data 100 WTs	Total 4347 Training 3476 Testing 871	k-NN	$k = 100$	Euclidian distance metric
[17]	10 min	12 WTs (wind speed and direction from two meteorological towers)	Training 1500 patterns for each WT	ANN	Number of hidden layers = 1 Number of hidden layer neurons = 8	(i) Separate MLP network for each WT (ii) Training-pattern mode (iii) TF-hyperbolic (all layers)
[8]	—	Measured data 100 kW WT	—	Fuzzy clustering	Number of cluster centers = 8	(i) CFL (ii) Subtractive clustering
[24]	10 min	Data set 1 generated with method in [9]	Total 1008 Training 50% Testing 50%	ANN	Number of hidden layer neurons = 5	(i) Feed forward back propagation (ii)Training: Levenberg-Marquardt (iii) TF: hidden layer transig (iv) TF: output layer purlin
		Data set 2	Total 4388 Training 50% Testing 50%			
		Data sets 3, 4, and 5	Total 2208 Training 50% Testing 50%			
		Data sets 1–5 as above	As above	Fuzzy clustering	Number of cluster centers = 8	Fuzzy C-means
[2]	10 min	SCADA (three 2 MW WTs) (model type 1: wind speed Model type 2: wind speed and direction, temperature)	32796 Training 60% Validation 40%	Fuzzy clustering	Number of cluster centers Type 1 = 3 Type 2 = 6	CCFL
				MLP NN	Number of hidden layers = 2	(i) Training-gradient descent (ii) TF: hidden layer-sigmoid (iii) TF: output layer-linear
				k-NN	Type 1 $k = 150$ Type 2 $k = 3$	—
				ANFIS		(i) FIS structure Sugeno type (ii) Training-hybrid learning (iii) Membership functions Input space-generalized normal Output space-linear (iv) Number of MFs = 3

better results compared to the parametric, nonparametric, and discrete models.

6.5.2. Clustering Methods.

Clustering is grouping of similar data into classes or clusters. A wind farm having many wind turbine generators has variable power outputs due to variation of wind speed. Efficient power curve can be found by applying clustering methods. Power curve characterization by cluster centre, fuzzy C-means, and subtractive clustering methods is done in [69]. Fuzzy clustering applies the concept of fuzzy sets to cluster analysis and belongingness of each point of data set to a group is given by a membership function. The method has the advantage of adapting noisy data. *Fuzzy C-means clustering* uses fuzzy partitioning to partition a collection of vectors into c fuzzy groups and finds a cluster centre c_i in each group. The performance of FCM depends upon the initial cluster centres. The method proposed in [70] to decide the number of clusters and their initial values for initializing iterative optimization based clustering algorithms is used in [8] to set up cluster centre fuzzy logic (CCFL) model of power curve. This cluster estimation method is used as a basis for identifying fuzzy models and the effect of number of *cluster centres* and *cluster neighbourhood distance* on RMSE is calculated. On comparison of CCFL model with polynomial model fitted by least square method it is concluded that the RMSE obtained with the fuzzy logic method is much lower than that obtained with the least square polynomial model.

6.5.3. Data Mining.

Data mining refers to extracting or mining knowledge from large amounts of data [71]. Developments in data mining offer promising approaches for modelling power curves of wind turbines. Selection of appropriate data mining method and algorithm is important to get accurate, stable, and robust power curve. Data driven nonparametric models using multilayer perception (MLP), random forest, M5P, boosting tree, and k-nearest neighbour (k-NN) algorithms are developed in [1] along with the 4PL parametric models. Performance of these models for online monitoring of the power curves was analyzed and the least square 4PL parametric and k-NN models were found to be of high fidelity to be used as reference curves for monitoring of power curves. Control chart approach is used for detecting the outliers and indicating the abnormal conditions of the turbine. However in another study [2] the performance of k-NN was found to be poor. Different data mining algorithms, namely, MLP, REP tree, M5P tree, bagging tree, and k-nearest neighbour algorithms, are used to build models for power prediction and online monitoring in [4]. Principle component analysis and k-NN algorithm are used for data reduction and filtering of outliers is done by residual and control chart approach. In [2] four data mining techniques, namely, bagging, M5P algorithm, REP tree algorithm, and M5 rules, were used for constructing the nonparametric power curve models. Model trees are type of decision trees with linear regression functions at the leaves and are applied for modelling the power curves in a few applications [2, 24]. Much detail of these models is not given in these studies. More information on model trees can be found in [72, 73].

6.5.4. Adaptive Network-Based Fuzzy Inference System (ANFIS) Model.

Adaptive network-based fuzzy inference system (ANFIS) is a fuzzy inference system implemented in the framework of adaptive networks and thus integrates the best features of fuzzy systems and neural networks [68]. A fuzzy inference system using fuzzy *if-then rules* is based on human knowledge and reasoning processes. In ANFIS, tuning of nonlinear signal relations can be done by constructing a set of fuzzy rules with appropriate membership function parameters tuned in a training phase [67]. Application of ANFIS for wind turbine power curve monitoring is proposed in [2]. This method of modelling is compared with earlier best performing methods, namely,

ANN, CCFL, and k-NN methods, found in the literature. Effect of including direction of wind and ambient temperature on the prediction error is evaluated. Data for pitch regulated turbines is used for the modelling.

6.5.5. Joint Probability of Wind Speed and Power.

Another approach of modelling is to consider the joint probability distribution of power and wind speed instead of considering the implied function of the two variables. Considering joint probability of the two variables instead of their individual probabilities can incorporate measures of uncertainty into performance estimates [11]. SCADA wind speed and power measurement data from wind turbines are used to estimate bivariate probability distribution functions and construct power curve using copula modelling technique in [10]. The application of empirical copulas is proposed to approximate the complex form of dependency between active power and wind speed. Usefulness of copula analysis for turbine condition monitoring and early recognition of faults is suggested in this paper. It is shown that different fault modes produce different signatures in R, R^2, and chi^2 statistics and can be used to identify the type of turbine fault.

6.5.6. Wavelet Support Vector Machine.

Wavelet analysis is a technique to analyze the nature of signals and is a promising tool for nonstationary signals. In support vector machines the data is mapped into higher-dimensional feature space via nonlinear mapping. The power curve of WT is used for wind power prediction in [37]. A novel wavelet support vector machine based model for wind speed prediction is proposed in this paper and its performance is found to be effective for short time prediction.

A number of works in literature also include comparative analyses of various parametric and nonparametric methods [2, 8, 57, 74]. Seven different functions have been compared in [75] for modelling the power curves of six different turbines. In [53] four models are developed from available operational power output data. The polynomial regression based parametric model is used as a benchmark model. Three nonparametric methods in addition to the above methods, namely, locally weighted polynomial regression, cubic spline regression, and a penalized spline regression model, are proposed in this study.

7. Selection of Modelling Method

A number of models and modelling methodologies have been proposed in various works for modelling of WT power curve. The choice of appropriate model and methodology adopted for a specific application is important and is a difficult task. The model selection for a particular application is done on the basis of availability of data, complexity of model, desired accuracy, and type of turbine and its power curve. On the basis of reviewed literature the following points are identified for selection of modelling methodology.

  (i) Wind power curve models required for initial wind resource assessment need handy methods for estimation of energy. Wind power output calculation and energy estimation which are done during designing of wind based systems need a power curve model with fair degree of accuracy. When only specification values (cut-in, cut-off, and rated speeds and the rated power) for a wind turbine are available, the polynomial models based on presumed shape can be used. These models can also be used as a handy tool for calculation of wind turbine output during design stage of wind farms because of the simplicity of calculations. When the manufacturers' curve data is available it is preferable to fit a polynomial function to the data as it results in better accuracy. These models are thus suitable for modelling of single turbines for predicting power for small systems where fairly accurate accuracy is desired.

(ii) For power prediction and selection of turbines for designing of large wind based systems a very good accuracy is required as oversizing can result in loss of revenue and undersizing can hamper the reliability of the system. Accurate prediction is also required for wind farm operators for energy trading. The 4PL and 5PL functions may therefore be used for these applications to develop models from the manufacturers' curve data.

(iii) When the SCADA data from a nearby wind farm is available and it is desired to assess the power output of a prospective wind farm with good accuracy having a group of turbines or when the models is to be used for online monitoring of curves, it is appropriate to extract model from SCADA data of the wind farm with appropriate extrapolations. These models can be derived using the 4PL and 5PL parametric methods or one of the nonparametric methods such as ANN or ANFIS. The models can also incorporate the effect of other influencing parameters in the models. Choice of parameters depends upon terrain, wind conditions, obstacles, correlation of parameters, and so forth. The copula method of defining the power curve using a joint probability distribution function of wind speed and power can be used for condition monitoring applications.

8. Discussions and Prospects

Various methods of modelling of WT power curve have been reviewed. Several methods of modelling have been proposed and used in various studies (Figure 3). A summary of noteworthy contributions is given in Table 4. The salient features of these models are summarised in Table 5. The inferences drawn, the deficiencies, and the suggestions proposed are given below.

(i) Most of the parametric models used in the literature use polynomial approximations to model wind power curve models. These models are mostly used for predicting power output of turbines for sizing and cost optimization applications. These models do not consider the inflection point on the power curve accurately and can result in large prediction errors.

However they are simple to use and can be utilized for predicting power during initial resource assessment and designing of small systems if very good accuracy is not desired. 4PL and 5PL parametric models can follow the actual shape of the power curve more accurately. These novel methods can result in reduced power prediction errors and can be applied for power and energy assessment during design of large systems and forecasting of power for energy trading where good accuracy is a crucial requirement.

(ii) The deterministic methods which use manufacturer's data for modelling are suitable for single turbines and not appropriate for modelling a group of turbines. The probabilistic methods which consider the variation of both power and wind speed are suitable for modelling power curve for a fleet of turbines.

(iii) Power curve models extracted from actual data of wind farms can incorporate actual conditions at a site and are suitable for modelling a group of turbines. These curves can be derived from the available data by parametric and nonparametric methods.

(iv) The parametric models derived from actual wind farm data used in the literature include linear segmented and 4PL and 5PL models. Nonparametric methods used are neural networks, clustering methods, data mining, ANFIS, and copula models. Data mining techniques can offer good results as the data available from the wind farms is voluminous and frequent updating of data is easier. ANN and ANFIS models perform well for power prediction and online monitoring applications. These curves derived from actual data can help in minimizing power prediction errors.

(v) The 4PL and data mining technique based models extracted from actual data of wind farms have been analyzed in the literature for their application in online monitoring. It is indicated that the monitoring of the power curves can be used to detect anomalies and statistical analysis of the outlier data can give indications of the specific reason of anomaly. Application of 5PL model for online monitoring of curve is not yet researched.

(vi) The wind power output of a turbine can be affected by various faults/anomalies or underperformance of the turbine, such as blade faults and yaw and pitch system faults. Different types of faults affect the turbine system differently and cause the power curve to depart from the expected value in a different way. Tools which can characterize and quantify these departures can aid in early identification of faults. Likely link between copula statistics and WT faults/anomalies is indicated in the literature should be investigated. Further research should also focus on considering the joint probability distribution of these variables.

(vii) Applications of advanced algorithms for developing improved parametric and nonparametric methods need to be explored.

TABLE 4: Summary of noteworthy contributions.

Author	Models	Data	Evaluation	Features of applications (suggested/analysed/used)
Diaf et al. [12] Diaf et al. [25] Giorsetto and Utsurogi [30] Chedid et al. [46]	Linear Quadratic Binomial Cubic	Ratings of WT $(v_c, v_f, v_r, P_r,$ etc.)	— — — —	(i) All are polynomial models (ii) Do not follow curvature of power curve (iii) Accuracy is poor (iv) Mostly applied for power prediction for sizing of wind based systems
Powell [6]	Weibull based	Ratings of WT and k	—	(i) Requires shape parameter of site (ii) Accuracy depends on wind regime (iii) Applied for capacity factor evaluation
Ai et al. [59]	Manufacturer's curve fitting (3 binomial expressions)	Manufacturer's curve	—	(i) Needs many expressions to represent the shape of curve accurately (ii) Applied for sizing of hybrid wind PV system
Diaf et al. [55]	Manufacturer's curve interpolation by cubic splines	Manufacturer's curve	—	
Katsigiannis et al. [60]	Manufacturer's curve fitting 9th-order polynomial	Manufacturer's curve	—	(i) Accurate for the particular data set (ii) High-order polynomials may not represent the variance of data (iii) Applied for sizing of hybrid wind, PV, and biodiesel based system
Gottschalk and Dunn [58]	Linear, Weibull based, cubic splines, manufacturer's curve fitting	$v_c, v_f, v_r, P_r,$ and manufacturer's curve	Visual comparison, % error in energy production	(i) Polynomial models (ii) Fitting accuracy with goodness of fit indicators is not evaluated
Nand Kishore and Fernandez [26]	Linear segmented	Manufacturer's curve	—	(i) Better accuracy (ii) Needs many expressions
Liu [54]	New nonlinear formula	Ratings of WT	Visual comparison	(i) Considers inflection point on curve (ii) Follows shape of curve more accurately than polynomial models (iii) Simple method (iv) Applied for economic load dispatch, reliability analysis, and multiple turbines with and without correlation
Villanueva and Feijóo [33]	3PL and 4PL curve	Manufacturer's curve	MAE, MAPE, and RMSE	(i) Parameters are derived directly, no need of iterative procedure
Kusiak et al. [1]	4PL curve Parameter estimation: Technique: least squares and MLM Algorithm: ES Data mining: k-NN, MLP, random forest, M5P tree, and boosting tree	Actual data of wind farm	AE, RE	(i) Least square better than MLM method (ii) k-NN outperforms other nonparametric models (iii) MLP best accuracy (iv) Residual and control chart approach for online monitoring is presented (v) 4PL least squares and k-NN models proposed for detecting anomalies
Kusiak et al. [4]	4PL curve Parameter estimation: Technique: least squares Algorithm: ES Data mining: k-NN, MLP, REP tree, M5P tree, and bagging tree	Manufacturer's curve Actual data of wind farm	MAE, AE, RE, and mean relative error	(i) k-NN outperforms other nonparametric models (ii) Outlier filtering by residual approach and control chart (iii) Application for power prediction and online monitoring is suggested
Li et al. [17]	ANN (MLP)	Actual data of wind farm	MSE	(i) Better than the traditional model (ii) Applied for power prediction
Li et al. [61]	Regression ANN (MLP)	Actual data of wind farm	RMSE	(i) Comparison of both models (ii) NN model is more accurate (iii) Effect of wind direction is considered
Pelletier et al. [62]	ANN (multistage 2-layer MLPs)	Actual data of wind farm	Visual mean error, MAE	(i) Compared with parametric, nonparametric, and discrete models (ii) Six inputs variables considered
Üstüntaş and Şahin [8]	Least squares fitting (2nd-order polynomial) CCFL	Actual data of wind farm	RMSE	(i) CCFL model is more accurate than the least squares fitting based model

TABLE 4: Continued.

Author	Models	Data	Evaluation	Features of applications (suggested/analysed/used)
Lydia et al. [24]	Parametric model: linear segmented, 4PL and 5PL. Parameter estimation: Technique: least squares; Algorithms: GA, EP, PSO, and DE. Nonparametric model: ANN, clustering, data mining (model trees)	Actual data of wind farm	RMSE, MAE	(i) 5P logistic model with parameters estimated using DE is the most accurate among parametric models (i) ANN model is the most accurate among the nonparametric models
Lydia et al. [36]	5PL	Actual data of wind farm	RMSE, MAE	(i) Used for wind resource assessment
Sohoni et al. [35]	4PL and 5PL	Actual data of wind farm	% error in energy estimation	(i) Applied for wind energy estimation
Schlechtingen et al. [2]	CCFL, ANN, k-NN, ANFIS	Actual data of wind farm	MAE, RMSE MAPE, SD	(i) ANN and NN perform best (ii) k-NN worst performance (iii) Effect of wind direction, temperature consideration produces less errors (iv) Application in online monitoring
Stephen et al. [10]	Copula model	Actual data of wind farm	—	(i) Joint probability distribution of wind speed and power is considered (ii) Applications for condition monitoring are suggested
Gill et al. [11]	Copula model	Actual data of wind farm	R, R^2, and chi-square	(i) Joint probability distribution of wind speed and power is considered (ii) Applications for condition monitoring are suggested
Zeng and Qiao [37]	Wavelet SVM for wind speed prediction	Actual data of wind farm and manufacturer's power curve	MAE, MAPE, and SD	(i) Outperforms the persistence model for short-term prediction (ii) Applied for short time power prediction

TABLE 5: Comparison of modelling methods.

Models	Data required for modelling	Merits	Demerits	Applications
Polynomial models (linear, quadratic, binomial, cubic, and Weibull based)	v_c, v_f, v_r, and P_r of turbine	(i) Simplicity (ii) Limited data required (iii) Parameter calculation is easy	(i) Do not follow curvature of power curve (ii) Accuracy is poor (iii) Sometimes more than one expression are used to describe the shape of curve	Suitable for power prediction and energy estimation during initial resource assessment and designing of small systems
Manufacturer's curve fitting	Manufacturer's curve	(i) Need less data	(i) Requires manufacturer's power curve data (ii) Fairly accurate (iii) Many expressions may be required for accurate representation of curve	Suitable for power prediction and energy estimation during initial resource assessment and designing of small systems
Cubic splines	Manufacturer's curve	(i) Exact fit	(i) Variance of data is not taken into account	Power prediction
4PL model	Manufacturer's curve, actual data of wind farm	(i) Consider inflection point on curve; hence shape of curve is represented more accurately than the earlier models (ii) One expression is required	(i) Asymmetry of curve not modelled	Online monitoring; further research on power prediction during design and power forecasting applications is required
5PL model	Manufacturer's curve actual data of wind farm	(i) Consider inflection point on curve and asymmetry of curve is modelled more accurate than the earlier and 4PL models (ii) One expression is required	(i) Parameter estimation is difficult	Further research on power prediction during design and power forecasting and online monitoring applications is required
ANN	Actual data of wind farm	(i) Found to be accurate than other methods	(i) Black box approach	Wind power assessment for sizing and power forecasting, and online monitoring applications, suitable for group of turbines
Clustering	Actual data of wind farm	(i) More accurate than the regression method	(i) Accuracy depends on the number of cluster centres	Wind power assessment for sizing and power forecasting, and online monitoring applications, suitable for group of turbines
k-NN	Actual data of wind farm	(i) Performance variable in different studies	(i) Accuracy depends on value of k (ii) Less training time as instance based scheme	Wind power assessment for sizing and power forecasting, prediction, and online monitoring applications, suitable for group of turbines
Model trees(REP, M5P, and bagging tree)	Actual data of wind farm	(i) Fairly accurate	(i) Much research not available	Applicability in power prediction and online monitoring to be explored
ANFIS	Actual data of wind farm	(i) Integrates best features of fuzzy systems and neural networks (ii) Accurate method (iii) Fewer parameters required in training therefore faster training compared to NN (iv) Tunable membership functions	(i) Computational complexity	Wind power assessment for sizing and power forecasting for energy trading Online monitoring applications, suitable for group of turbines
Copula model	Actual data of wind farm	(i) Considers joint probability distribution of wind speed and power (ii) Includes measures of uncertainty in performance estimates	(i) Needs advanced method for parameter estimation of marginals	Applications for condition monitoring to be investigated

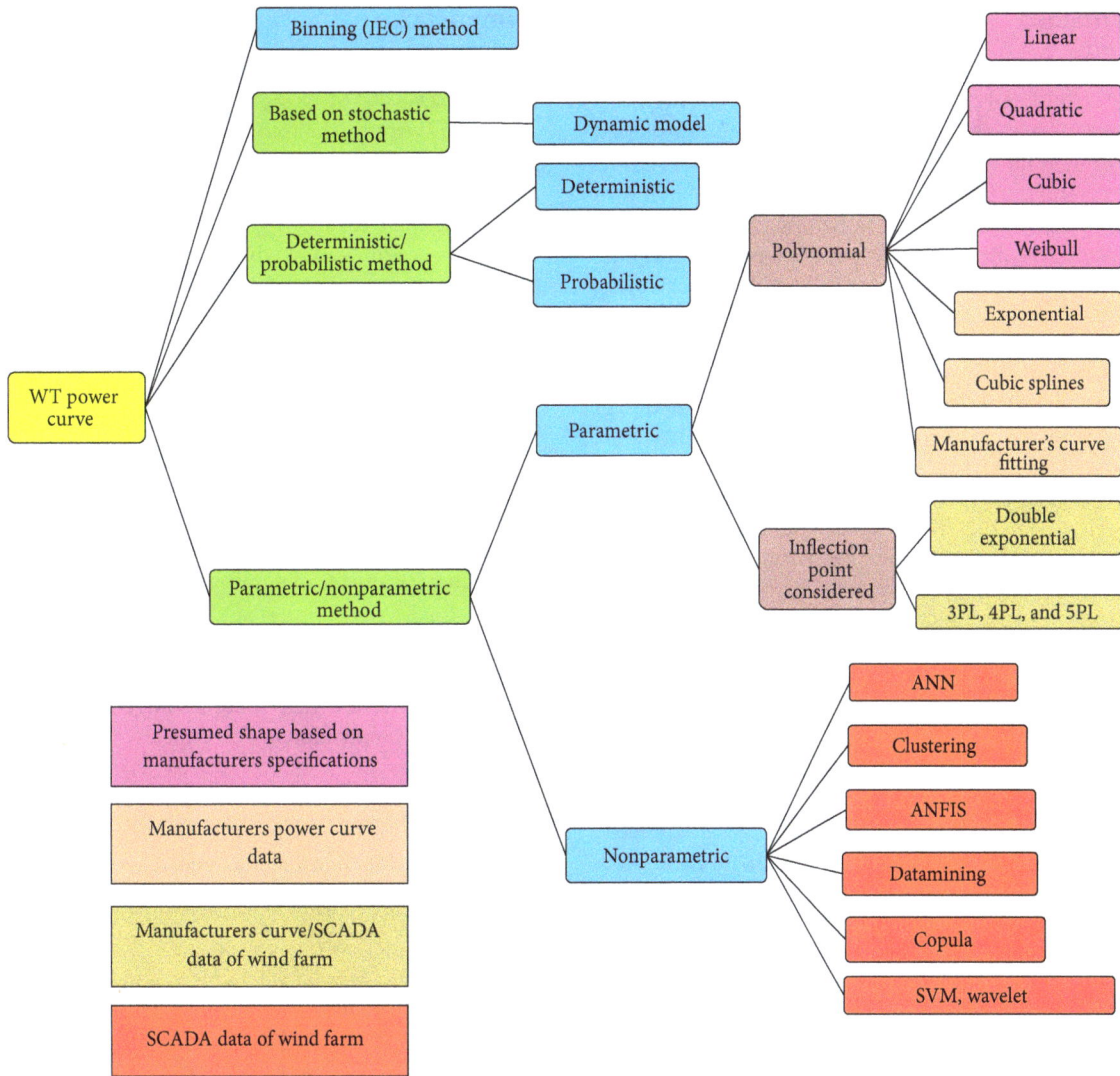

FIGURE 3: Wind turbine power curve models.

(viii) Appropriate evaluation of the developed models is a very important requirement in modelling. The right choice of an evaluation metric is important and depends on the data and analysis requirements. Models developed in the literature have used different performance metrics. The reviewed articles do not highlight the reason of preferring a particular criterion for evaluation. Different statistical measures can have different interpretations and appropriate selection of error metrics is crucial for analysis of the models. Moreover these models will ultimately be used in wind energy applications; therefore it is not appropriate to judge their suitability on the basis of goodness of fit parameters alone, but it should also be examined how successfully these models can be employed for the particular applications.

(ix) Power curves of wind turbines provided by the manufacturers are used for power prediction in most of the wind energy applications. These curves are developed under standard test conditions. The IEC 61400-12-1 is the most accepted standard for power curve measurement of single wind turbines. The wind conditions at the practical sites can be different from those at the test site. IEC curve may not always represent the wind conditions of other sites. Further research should focus on development of site specific power curves.

(x) The discrete model prescribed in IEC 61400-12 is simple, but a large amount of data is required to develop a reliable model.

(xi) The stochastic model that is proposed in some works is independent of turbulence intensity but does not include the effect of other influencing parameters.

(xii) Future works should also include the effect of various influencing parameters on the power curves.

9. Conclusions

Accurate modelling of WT power curves is crucial for successful design and operation of any wind energy conversion system. This paper presented an overview of different approaches used for modelling of wind turbine power curve. There are several methods of modelling which have their own advantages and disadvantages. Polynomial approximation based power curve models which have been used widely are simple to use and can be utilized for predicting power during initial resource assessment and designing of small systems. The four- and five-parameter logistic function based power curve models which consider the inflection point on these curves are promising methods which can help in reduction of power prediction errors and improved performance in online monitoring of these curves. Also most of the models in earlier works are developed from manufacturers curves of wind turbines. However, the manufacturer supplied curves are turbine specific and represent their behaviour under standard test conditions. They can be applied for power prediction of single turbines and for sites with steady winds. Improved models are required which can represent the conditions at large wind farms with a group of turbines installed and sites having complex terrains. Power curves derived from actual wind speed and output power data from wind farms can take into account various site specific factors resulting in better models. A wind farm's SCADA data is a valuable resource which can be exploited for this purpose. Nonparametric methods of power curve modelling used in the literature include neural networks, clustering, data mining, ANFIS, and copula models. The nonparametric methods are suitable for extracting models from large data. Moreover, these models can incorporate the effect of parameters other than wind speed on the power curves more easily than the parametric models. Literature survey reveals that ANN and ANFIS nonparametric methods perform well among other models for power prediction and online monitoring applications. Further research should focus on development of site specific power curves. Future models should be able to minimize the prediction errors and should be suitable for online monitoring of turbines. Methods which can quantify the power curve departures from expected values for identification of turbine faults should be explored. As the power output of wind turbines is strongly dependent on wind speed of a potential wind farm site, selection of appropriate wind speed model along with the power curve model is an important requirement for accurate prediction of wind farm output. Different wind speed modelling techniques have also been reviewed briefly in this paper. It can be concluded that selection of appropriate model, solution technique, and proper algorithms for a particular application is important for efficient modelling and can contribute significantly to developing reliable and efficient wind energy based power system.

Nomenclature

P: Power output of wind turbine (W)
v: Wind speed (m/s)
A_w: Area swept by the rotor blades (m^2)
ρ: Air density (kg/m^3)
λ: Tip speed ratio
β: Pitch angle (degree)
C_p: Power coefficient of wind turbine
k: Shape parameter Weibull PDF
c: Scale parameter Weibull PDF
v_c: Cut-in wind speed of turbine (m/s)
v_r: Rated wind speed of turbine (m/s)
v_f: Cut-off (or furling) wind speed of turbine (m/s)
P_r: Rated power of wind turbine (W)
η_w: The efficiency of WTG and the corresponding converter
θ: Vector parameter of parametric models.

Competing Interests

The authors declare that there is no conflict of interests regarding the publication of this paper.

References

[1] A. Kusiak, H. Zheng, and Z. Song, "On-line monitoring of power curves," *Renewable Energy*, vol. 34, no. 6, pp. 1487–1493, 2009.

[2] M. Schlechtingen, I. F. Santos, and S. Achiche, "Using data-mining approaches for wind turbine power curve monitoring: a comparative study," *IEEE Transactions on Sustainable Energy*, vol. 4, no. 3, pp. 671–679, 2013.

[3] S. Mathew, *Wind Energy-Fundamentals, Resource Analysis and Economics*, Springer, Berlin, Germany, 2006.

[4] A. Kusiak, H. Zheng, and Z. Song, "Models for monitoring wind farm power," *Renewable Energy*, vol. 34, no. 3, pp. 583–590, 2009.

[5] F. A. L. Jowder, "Wind power analysis and site matching of wind turbine generators in Kingdom of Bahrain," *Applied Energy*, vol. 86, no. 4, pp. 538–545, 2009.

[6] W. R. Powell, "An analytical expression for the average output power of a wind machine," *Solar Energy*, vol. 26, no. 1, pp. 77–80, 1981.

[7] T.-P. Chang, F.-J. Liu, H.-H. Ko, S.-P. Cheng, L.-C. Sun, and S.-C. Kuo, "Comparative analysis on power curve models of wind turbine generator in estimating capacity factor," *Energy*, vol. 73, pp. 88–95, 2014.

[8] T. Üstüntaş and A. D. Şahin, "Wind turbine power curve estimation based on cluster center fuzzy logic modeling," *Journal of Wind Engineering and Industrial Aerodynamics*, vol. 96, no. 5, pp. 611–620, 2008.

[9] J.-Y. Park, J.-K. Lee, K.-Y. Oh, and J.-S. Lee, "Development of a novel power curve monitoring method for wind turbines and its field tests," *IEEE Transactions on Energy Conversion*, vol. 29, no. 1, pp. 119–128, 2014.

[10] B. Stephen, S. J. Galloway, D. McMillan, D. C. Hill, and D. G. Infield, "A copula model of wind turbine performance," *IEEE Transactions on Power Systems*, vol. 26, no. 2, pp. 965–966, 2011.

[11] S. Gill, B. Stephen, and S. Galloway, "Wind turbine condition assessment through power curve copula modeling," *IEEE Transactions on Sustainable Energy*, vol. 3, no. 1, pp. 94–101, 2012.

[12] S. Diaf, M. Belhamel, M. Haddadi, and A. Louche, "Technical and economic assessment of hybrid photovoltaic/wind system with battery storage in Corsica island," *Energy Policy*, vol. 36, no. 2, pp. 743–754, 2008.

[13] L. Horváth, T. Panza, and N. Karadža, "The influence of high wind hysteresis effect on wind turbine power production at Bura-dominated site," in *Proceedings of the European Wind Energy Conference and Exhibition (EWEC '07)*, pp. 1017–1022, Milan, Italy, May 2007.

[14] T. Jin and Z. Tian, "Uncertainty analysis for wind energy production with dynamic power curves," in *Proceedings of the IEEE 11th International Conference on Probabilistic Methods Applied to Power Systems (PMAPS '10)*, pp. 745–750, IEEE, Singapore, June 2010.

[15] J. R. McLean, "WP2.6-Equivalent wind power curves," Tech. Rep. EIE/06/022/SI2.442659, 2008.

[16] E. Hedevang, "Wind turbine power curves incorporating turbulence intensity," *Wind Energy*, vol. 17, no. 2, pp. 173–195, 2014.

[17] S. Li, D. C. Wunsch, E. A. O'Hair, and M. G. Giesselmann, "Using neural networks to estimate wind turbine power generation," *IEEE Transactions on Energy Conversion*, vol. 16, no. 3, pp. 276–282, 2001.

[18] Z. O. Olaofe and K. A. Folly, "Wind energy analysis based on turbine and developed site power curves: a case-study of Darling City," *Renewable Energy*, vol. 53, pp. 306–318, 2013.

[19] International standard IEC 61400-12-1, Wind turbines- Power performance measurement of electricity producing wind turbines, 2005.

[20] H. Mellinghoff, *Development of Power Curve Measurement Standards*, DEWEK 2012: 11, Deutsche Windenergie-Konferenz, Bremen, Germany, 2012.

[21] S. A. Akdağ and A. Dinler, "A new method to estimate Weibull parameters for wind energy applications," *Energy Conversion and Management*, vol. 50, no. 7, pp. 1761–1766, 2009.

[22] J. A. Carta, P. Ramírez, and S. Velázquez, "A review of wind speed probability distributions used in wind energy analysis. Case studies in the Canary Islands," *Renewable and Sustainable Energy Reviews*, vol. 13, no. 5, pp. 933–955, 2009.

[23] V. Lo Brano, A. Orioli, G. Ciulla, and S. Culotta, "Quality of wind speed fitting distributions for the urban area of Palermo, Italy," *Renewable Energy*, vol. 36, no. 3, pp. 1026–1039, 2011.

[24] M. Lydia, A. I. Selvakumar, S. S. Kumar, and G. E. P. Kumar, "Advanced algorithms for wind turbine power curve modeling," *IEEE Transactions on Sustainable Energy*, vol. 4, no. 3, pp. 827–835, 2013.

[25] S. Diaf, G. Notton, M. Belhamel, M. Haddadi, and A. Louche, "Design and techno-economical optimization for hybrid PV/wind system under various meteorological conditions," *Applied Energy*, vol. 85, no. 10, pp. 968–987, 2008.

[26] L. Nand Kishore and E. Fernandez, "Reliability well-being assessment of PV-wind hybrid system using Monte Carlo simulation," in *Proceedings of the International Conference on Emerging Trends in Electrical and Computer Technology (ICE-TECT '11)*, pp. 63–68, Tamil Nadu, India, March 2011.

[27] P. Milan, M. Wachter, and J. Peinke, "Stochastic modeling and performance monitoring of wind farm power production," *Journal of Renewable and Sustainable Energy*, vol. 6, no. 3, Article ID 033119, 2014.

[28] E. Anahua, F. Böttcher, S. Barth, J. Peinke, and M. Lange, "Stochastic analysis of the power output for a wind turbine," in *Proceedings of the European Wind Energy Conference (EWEC '04)*, London, UK, November 2004.

[29] E. Anahua, S. Barth, and J. Peinke, "Markovian power curves for wind turbines," *Wind Energy*, vol. 11, no. 3, pp. 219–232, 2008.

[30] P. Giorsetto and K. F. Utsurogi, "Development of a new procedure for reliability modeling of wind turbine generators," *IEEE Transactions on Power Apparatus and Systems*, vol. 102, no. 1, pp. 134–143, 1983.

[31] M. K. Deshmukh and S. S. Deshmukh, "Modeling of hybrid renewable energy systems," *Renewable and Sustainable Energy Reviews*, vol. 12, no. 1, pp. 235–249, 2008.

[32] E. Sainz, A. Llombart, and J. J. Guerrero, "Robust filtering for the characterization of wind turbines: improving its operation and maintenance," *Energy Conversion and Management*, vol. 50, no. 9, pp. 2136–2147, 2009.

[33] D. Villanueva and A. E. Feijóo, "Reformulation of parameters of the logistic function applied to power curves of wind turbines," *Electric Power Systems Research*, vol. 137, pp. 51–58, 2016.

[34] D. Rodbard, "Statistical quality control and routine data processing for radioimmunoassays and immunoradiometric assays," *Clinical Chemistry*, vol. 20, no. 10, pp. 1255–1270, 1974.

[35] V. Sohoni, S. C. Gupta, and R. K. Nema, "A comparative analysis of wind speed probability distributions for wind power assessment of four sites," *Turkish Journal of Electrical Engineering and Computer Sciences*, 2016.

[36] M. Lydia, S. Suresh Kumar, A. Immanuel Selvakumar, and G. Edwin Prem Kumar, "Wind resource estimation using wind speed and power curve models," *Renewable Energy*, vol. 83, pp. 425–434, 2015.

[37] J. Zeng and W. Qiao, "Short-term wind power prediction using a wavelet support vector machine," *IEEE Transactions on Sustainable Energy*, vol. 3, no. 2, pp. 255–264, 2012.

[38] Curve fitting toolbox, for use with MATLAB, User's Guide, Version 1.The Mathworks, 2002.

[39] H. Yang, L. Lu, and W. Zhou, "A novel optimization sizing model for hybrid solar-wind power generation system," *Solar Energy*, vol. 81, no. 1, pp. 76–84, 2007.

[40] L. Xu, X. Ruan, C. Mao, B. Zhang, and Y. Luo, "An improved optimal sizing method for wind-solar-battery hybrid power system," *IEEE Transactions on Sustainable Energy*, vol. 4, no. 3, pp. 774–785, 2013.

[41] X. Liu and W. Xu, "Minimum emission dispatch constrained by stochastic wind power availability and cost," *IEEE Transactions on Power Systems*, vol. 25, no. 3, pp. 1705–1713, 2010.

[42] Y. M. Atwa, E. F. El-Saadany, M. M. A. Salama, R. Seethapathy, M. Assam, and S. Conti, "Adequacy evaluation of distribution system including wind/solar DG during different modes of operation," *IEEE Transactions on Power Systems*, vol. 26, no. 4, pp. 1945–1952, 2011.

[43] R. Karki, P. Hu, and R. Billinton, "Reliability evaluation considering wind and hydro power coordination," *IEEE Transactions on Power Systems*, vol. 25, no. 2, pp. 685–693, 2010.

[44] P. Wang, Z. Gao, and L. Bertling, "Operational adequacy studies of power systems with wind farms and energy storages," *IEEE Transactions on Power Systems*, vol. 27, no. 4, pp. 2377–2384, 2012.

[45] L. Lu, H. Yang, and J. Burnett, "Investigation on wind power potential on Hong Kong islands—an analysis of wind power and wind turbine characteristics," *Renewable Energy*, vol. 27, no. 1, pp. 1–12, 2002.

[46] R. Chedid, H. Akiki, and S. Rahman, "A decision support technique for the design of hybrid solar-wind power systems," *IEEE Transactions on Energy Conversion*, vol. 13, no. 1, pp. 76–83, 1998.

[47] L. Wang and C. Singh, "Adequacy-based design of a hybrid generating system including intermittent sources using constrained particle swarm optimization," *IEEE Transactions on Power Systems*, vol. 27, no. 4, pp. 2377–2383, 2007.

[48] A. K. Kaviani, G. H. Riahy, and S. M. Kouhsari, "Optimal design of a reliable hydrogen-based stand-alone wind/PV generating system, considering component outages," *Renewable Energy*, vol. 34, no. 11, pp. 2380–2390, 2009.

[49] B. S. Borowy and Z. M. Salameh, "Optimum photovoltaic array size for a hybrid wind/PV system," *IEEE Transactions on Energy Conversion*, vol. 9, no. 3, pp. 482–488, 1994.

[50] B. S. Borowy and Z. M. Salameh, "Methodology for optimally sizing the combination of a battery bank and PV array in a wind/PV hybrid system," *IEEE Transactions on Energy Conversion*, vol. 11, no. 2, pp. 367–375, 1996.

[51] M. G. Khalfallah and A. M. Koliub, "Suggestions for improving wind turbines power curves," *Desalination*, vol. 209, no. 1–3, pp. 221–229, 2007.

[52] C. Carrilo, A. F. O. Montaño, J. Cidras, and D. Dorado, "Review of power curve modelling for wind turbines," *Renewable and Sustainable Energy Reviews*, vol. 21, pp. 572–581, 2013.

[53] S. Shokrzadeh, M. Jafari Jozani, and E. Bibeau, "Wind turbine power curve modeling using advanced parametric and non-parametric methods," *IEEE Transactions on Sustainable Energy*, vol. 5, no. 4, pp. 1262–1269, 2014.

[54] X. Liu, "An improved interpolation method for wind power curves," *IEEE Transactions on Sustainable Energy*, vol. 3, no. 3, pp. 528–534, 2012.

[55] S. Diaf, D. Diaf, M. Belhamel, M. Haddadi, and A. Louche, "A methodology for optimal sizing of autonomous hybrid PV/wind system," *Energy Policy*, vol. 35, no. 11, pp. 5708–5718, 2007.

[56] F. O. Hocaoğlu, Ö. N. Gerek, and M. Kurban, "A novel hybrid (wind-photovoltaic) system sizing procedure," *Solar Energy*, vol. 83, no. 11, pp. 2019–2028, 2009.

[57] V. Thapar, G. Agnihotri, and V. K. Sethi, "Critical analysis of methods for mathematical modelling of wind turbines," *Renewable Energy*, vol. 36, no. 11, pp. 3166–3177, 2011.

[58] P. G. Gottschalk and J. R. Dunn, "The five-parameter logistic: a characterization and comparison with the four-parameter logistic," *Analytical Biochemistry*, vol. 343, no. 1, pp. 54–65, 2005.

[59] B. Ai, H. Yang, H. Shen, and X. Liao, "Computer-aided design of PV/wind hybrid system," *Renewable Energy*, vol. 28, no. 10, pp. 1491–1512, 2003.

[60] Y. A. Katsigiannis, P. S. Georgilakis, and E. S. Karapidakis, "Hybrid simulated annealing-tabu search method for optimal sizing of autonomous power systems with renewables," *IEEE Transactions on Sustainable Energy*, vol. 3, no. 3, pp. 330–338, 2012.

[61] S. Li, D. C. Wunsch, E. O'Hair, and M. G. Giesselmann, "Comparative analysis of regression and artificial neural network models for wind turbine power curve estimation," *Journal of Solar Energy Engineering*, vol. 123, no. 4, pp. 327–332, 2001.

[62] F. Pelletier, C. Masson, and A. Tahan, "Wind turbine power curve modelling using artificial neural network," *Renewable Energy*, vol. 89, pp. 207–214, 2016.

[63] M. Schlechtingen and I. Ferreira Santos, "Comparative analysis of neural network and regression based condition monitoring approaches for wind turbine fault detection," *Mechanical Systems and Signal Processing*, vol. 25, no. 5, pp. 1849–1875, 2011.

[64] Y. Wang, D. G. Infield, B. Stephen, and S. J. Galloway, "Copula-based model for wind turbine power curve outlier rejection," *Wind Energy*, vol. 17, no. 11, pp. 1677–1688, 2014.

[65] L. Zheng, W. Hu, and Y. Min, "Raw wind data preprocessing: a data-mining approach," *IEEE Transactions on Sustainable Energy*, vol. 6, no. 1, pp. 11–19, 2015.

[66] National Renewable Energy Laboratory (NREL) western data set, station ID 2 dataset year 2006, http://wind.nrel.gov/Web_nrel/.

[67] S. R. Jang, C.-T. Sun, and E. Mizutani, *Neuro-Fuzzy and Soft Computing-a Computational Approach to Learning and Machine Intelligence*, PHI ltd, New Delhi, India, 2003.

[68] S. R. Jang, C.-T. Sun, and E. Mizutani, *Neuro-Fuzzy and Soft Computing-A Computational Approach to Learning and Machine Intelligence*, PHI ltd, New Delhi, India, 2003.

[69] M. S. Mohan Raj, M. Alexander, and M. Lydia, "Modeling of wind turbine power curve," in *Proceedings of the IEEE PES Conference on Innovative Smart Grid Technologies—India*, Kollam, India, December 2011.

[70] S. L. Chiu, "Fuzzy model identification based on cluster estimation," *Journal of Intelligent and Fuzzy Systems*, vol. 2, pp. 267–278, 1994.

[71] J. Han and M. Kamber, *Data Mining: Concepts and Techniques*, Morgan Kaufmann, San Francisco, Calif, USA, 2006.

[72] A. M. Prasad, L. R. Iverson, and A. Liaw, "Newer classification and regression tree techniques: bagging and random forests for ecological prediction," *Ecosystems*, vol. 9, no. 2, pp. 181–199, 2006.

[73] I. H. Witten and E. Frank, *Data Mining: Practical Machine Learning Tools and Techniques*, Morgan Kaufmann, San Francisco, Calif, USA, 2nd edition, 2005.

[74] M. Lydia, S. S. Kumar, A. I. Selvakumar, and G. E. Prem Kumar, "A comprehensive review on wind turbine power curve modeling techniques," *Renewable and Sustainable Energy Reviews*, vol. 30, pp. 452–460, 2014.

[75] S. A. Akdağ and Ö. Gler, "A comparison of wind turbine power curve models," *Energy Sources, Part A: Recovery, Utilization and Environmental Effects*, vol. 33, no. 24, pp. 2257–2263, 2011.

13

Power Consumption: Base Stations of Telecommunication in Sahel Zone of Cameroon; Typology Based on the Power Consumption—Model and Energy Savings

Albert Ayang,[1] **Paul-Salomon Ngohe-Ekam,**[2] **Bossou Videme,**[3] **and Jean Temga**[4]

[1]*Higher Institute of the Sahel, Department of Renewable Energy, University of Maroua, P.O. Box 46, Maroua, Cameroon*
[2]*National Advanced School of Engineering, Energy and Automatic Laboratory, University of Yaounde I,*
 P.O. Box 8390, Yaounde, Cameroon
[3]*Higher Institute of the Sahel, Department of Information and Telecommunication, University of Maroua,*
 P.O. Box 46, Maroua, Cameroon
[4]*Ecole Polytechnique Montreal, Polygrames Laboratory, P.O. Box 2500, Chemin Montreal, M6106, Canada H3T1J4*

Correspondence should be addressed to Albert Ayang; ayabache@yahoo.fr

Academic Editor: Mattheos Santamouris

In this paper, the work consists of categorizing telecommunication base stations (BTS) for the Sahel area of Cameroon according to their power consumption per month. It consists also of proposing a model of a power consumption and finally proceeding to energy audits in each type of base station in order to outline the possibilities of realizing energy savings. Three types of telecommunication base stations (BTS) are found in the Sahel area of Cameroon. The energy model takes into account power consumption of all equipment located in base stations (BTS). The energy audits showed that mismanagement of lighting systems, and of air-conditioning systems, and the type of buildings increased the power consumption of the base station. By applying energy savings techniques proposed for base stations (BTS) in the Sahel zone, up to 17% of energy savings are realized in CRTV base stations, approximately 24.4% of energy are realized in the base station of Missinguileo, and approximately 14.5% of energy savings are realized in the base station of Maroua market.

1. Introduction

In order to cope with the development of the world, the requirements in telecommunication will continuously increase. In order to allow a vast and rapid communication (*i.e., to maximize the range of signals and the extent of the telephone and broadcast coverage*), telecommunication and broadcast companies (*namely, MTN, CAMTEL and ORANGE, CRTV with its transmitters, and other broadcast channels*) proceeded with the installation of pieces of equipment of telecommunications in several rural and urban areas in Cameroon, on the mountains and the buildings. These installations require a reliable electric power supply, being without interruption.

Unfortunately, many areas are electrically isolated because they are not supplied by the interconnected electrical networks (*according to [1] only about 14% of the 13000 villages have access to electricity in Cameroon*). In these particular areas, the installations of telecommunications witness a serious problem of electrical energy supply, despite the use of power generating units (*generating units use petrol or gas oil for fuel, which from the environmental point of view contribute to pollution effect of greenhouse, and consequently accelerate the phenomenon of global warming*). As for the urban areas, despite the presence of the interconnected electrical supply networks of AES-Sonel, telecommunication installations face serious problems of supply electric power in view of the important increase of telephone subscribers (*close to 10 million subscribers according to [2]*) and the recurrence of unballastings. Facing the difficulties of supply permanent and reliable energy, in spite of large investments (*according to a source close to the general management, in ten years of activity in Cameroon, MTN affirms that it has invested, in Cameroon, more than 137 249 519,62 USD for telecommunication equipment*), the companies of telecommunication know serious problems concerning the coverage of the network (*according to [2], 20% of coverage against 95% of network coverage imposed by Agency of Regulation of Telecommunications*).

According to [3], approximately 600 TWh or 3% of the world's electrical energy is consumed by the ICTs (*information and communication technologies*) causing approximately 2% of the CO_2 emissions in the whole world; 9% of this consumption of ICTs is caused by communication networks radio [4]. Within these radio communication networks, 10% of the energy is consumed by the users of terminals, while 90% is consumed by telecommunication base stations [5]. Thus, the increase in the number of base stations by the telephone and audiovisual companies in Cameroon implies an increase in the global energy consumption that is increase in energy costs, which has also an impact on global warming especially in the Sahelian areas of Cameroon where we often encounter high temperatures. The rest of the paper is organized as follows: in Section 2, we present the bibliographical approach or the existing state of the art on the power consumption of base stations and the various existing approaches to be able to save energy in the base stations of telecommunication. The classification of base stations and the description of some base stations of Sahelian zone of Cameroon are presented, respectively, in Sections 3 and 4. The results on energy audits carried out in the three base stations and the proposal of a power consumption model are, respectively, presented in Sections 5 and 6. In addition, some solutions of realizable energy saving are detailed, respectively, in Sections 7 and 8. Finally, applicable technical solutions, in telecommunication base stations in the Sahelian zone in order to increase the energy efficiency, are presented in Section 9.

2. State of the Art: Bibliographic Approaches of Energy Savings in Telecommunication Base Stations

The growing interest towards new and reliable services in the field of mobile telecommunication has led to an increase in installation of number of base stations in the whole world. Besides, the traditional concept of the deployment of base stations ensures a continuous operation in order to constantly guarantee a quality of the service of network in any place. According to [6], these two reasons have contributed synergetically during the last decade to an important increase in energy consumption of base stations belonging to mobile telephone network operators.

According to [7], the distribution of power consumption averages of the various components of base stations is recapitulated in Figure 1.

It shows the power consumption by component in a base station; the largest energy consumer in base stations is the radiofrequency equipment (power amplifier plus the transceivers and cables), which consumes approximately

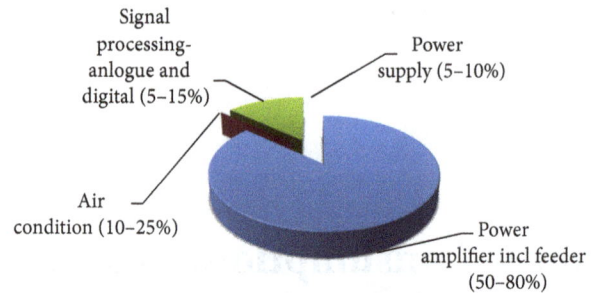

FIGURE 1: Energy consumption of the various components of the base stations [7].

65% of the total energy. Among the other components of the base station, the important energy consumers are air conditioning (17.5%), digital signal processor (10%), and the AC/DC converter (7.5%).

We notice that the radio operator equipment (the module of digital signal processing, the power amplifiers of transceivers, the radio frequencies, and connecting wires) and the systems of air cooling are the large-scale consumers of energy in telecommunication base stations. Emphasis must thus be laid on these components to reduce the total energy consumption of base stations.

In the zones of Sahel, the annual average temperatures are high; to reduce the energy consumption of base stations in these areas, an effort must be made on the control of the internal temperature of the room sheltering the equipment or on the system of air cooling.

To optimize energy consumption in a telecommunication base station, we answer three principal questions: optimization of energy consumption of BTS (*base transceiver stations*), energy optimization of the site sheltering the BTS (*base transceiver stations*), and the energy optimization of the network and radio frequency connection.

2.1. Power Optimization Consumption of BTS (Base Transceiver Stations). Research is focused on several components of the BTS to improve their energy efficiency. Research is more focused on the amelioration of the linearization and energy efficiency of the power amplifier. The energy efficiency can be improved by using an especially designed power amplifier containing special materials for the transistors of the power amplifier, like materials of high frequency such as Si, GaAs [9]. A numerical technique of predistortion can be used in the power amplifier to cancel the distortion of energy and so give a better linearity [10].

The power consumption of the digital signal processor can be reduced by using, for example, integrated circuits architectures like ASIC, FPGA, or DSP which are combined to obtain a better efficiency [11].

The AC/DC converter can be ameliorated by using converters that have a good efficiency in terms of energy being able to improve; thus, the total energy efficiency of the BTS, even in situations where the traffic load, is very bulky.

The power consumption caused by air conditioning can be reduced by lowering the operational temperature of base stations to the minimum or by using additional elements such as heat exchangers, membrane filters, and "smart" fans or heat modules [12].

2.2. Power Optimization of the Premises Accommodating the BTS. Energy savings in a base station can be obtained by putting into place the distributed architecture of a base station, where the radio frequency equipment is placed near the antenna in order to minimize losses in cables [13]. The possibilities to use renewable energies such as photovoltaic panels and wind energy on the sites of base stations are under study. By combining these two sources of renewable energies, one can reduce the potential of power consumption cost of a base station by 50% [14].

2.3. Optimize Power Consumption of the Network and Radio Frequency Connection. The potential of energy savings at the level of the connection is mostly found in the techniques of transmission by interface with air.

The level of contact considers the possible modes of sleep of certain components of the base station, when some of them can be switched off during certain times. In this case, the base station must provide a certain difference between the transmissions ordered by the traffic load in strong connection or in weak connection [15].

The energy efficiency in a base station obtained by the modes of sleep can be increased through the implementation of tedding techniques and flowering of cells. These techniques, used for the conception of base stations that have transitional states of sleep, consist of a progressive commutation of "switching off and on" a base station. It is shown that these transitional states are too short, which enable the base station to switch off and on for a short time which does not have a great significant reduction in the energy savings obtained through the approaches of sleep [16].

The system of 4G (4th generation of wireless networks) is envisaging the possibility of a dynamic allocation of the spectrum of frequencies as a function of the traffic load [17]. The cancellation of interferences in cellular networks by using the distributed antenna systems and algorithms, such as linear forcing to zero, the minimal error squared, and the cancellation of successive interferences, contributes also to the reduction of energy consumption [18].

At the level of the network, one of the most important approaches to reducing the consumption of energy is the dynamic management of network resources, which in fact enables the switching off of equipment of base stations at the time of weak traffic load. In such a scenario, neighboring base stations must ensure network coverage and take care of the network traffic of subscribers situated in the area where the base stations are not activated [15]. This can be combined with a dynamic selection of transmitter power, by tilting the antenna, by relaying multiskip, or by coordinating the transmission and reception in several points (multipoint) [19].

An important concept to reduce energy consumption of mobile telephone networks is presented in the context of the evaluation of energy efficiency, which includes several models that can ameliorate the energy efficiency on weak load

traffic [18]. These models are subdivided into small sample models for short-term and large sample models for long-term. The small sample models for short-term are power models of which the cartography of RF power (radio frequency) brings out of the antenna radiated power, and the latter is assimilated to the total power supply of the site of the base station. The large sample models for long-term include the traffic models that describe the variation of the traffic load during a day and the models of deployment at small samples existing in geographically large areas. The energy efficiency can also be ameliorated without switching off certain equipment of the base station, using a technique called attenuation of cellular networks. This method is built as a problem of optimization at multiperiod enabling the attenuation frequency to switch off certain frequencies of the canals of the base station. The attenuation frequency can be combined with the service of attenuation, which stops certain services at high rate of data on the permitted frequencies during the periods when the traffic load demand is weak [20].

In the case of network architecture of heterogeneous mobile telephone, the network itself represents the potential of reducing energy consumption. In this type of network, the macrocells are completed with cells of weak transmission power such as micro-, Pico-, and femtocells [17]. The macrocells ensure permanent network coverage; meanwhile, the putting into service and out of service of small cells depend on the traffic load present. The possibility of applying techniques such as the zooming of cells, where the cell can adjust its size depending on the situation of the traffic load, is also explored in [21].

The potential of reduction of the energy consumption of networks is also found at the level of its planning and its functioning. One of the models proposed is the *Traffic-Aware Network Planning cadre* and *Green Operation (TANGO) Framework*, which seems to be an implementation of the future, being capable of increasing the energy efficiency of mobile telephone networks while conserving the quality of the service at a satisfactory level [22]. Besides, certain initiatives are based on the possibility of making energy savings through a cooperation between competing operators that offer the same services in the same area of coverage (generally in towns). The fact is that one of the operators can completely switch off its base stations during a weak traffic load, while the base stations of the second operator accept the subscribers from the two operators. According to the authors of [23], such an approach can offer reductions in energy consumption by 20%. In [8], the authors propose several solutions that can ameliorate the energy efficiency of base stations of 4G (4th generation of wireless networks). These solutions can be observed from time, the frequency, and spatial domain, and the most promising solutions are hybrid solutions that combine the solutions in the different domains to adapt the energy consumption of the site of the base station to the conditions of the traffic load. In fact, the simultaneous use of most of the approaches mentioned will have a synergetic effect that leads completely to an energy efficiency of mobile telephone networks.

It turns out that base stations are greatest energy consumers in the mobile communication chain. This energy

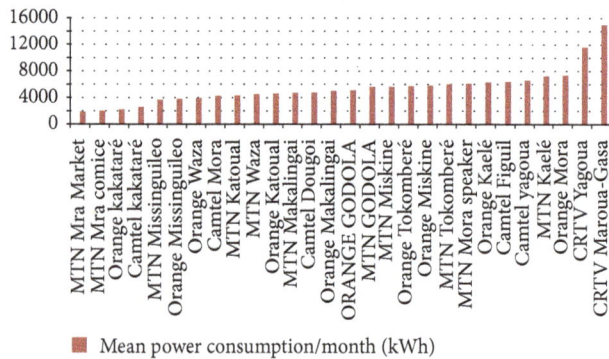

FIGURE 2: Energy consumption of some telecommunications base stations in the Sahel area of Cameroon.

TABLE 1: Equipment of the TV room.

Equipment	Power	Other characteristics and working hours
TV emitter	2,5 KW	Working 24 h/24
Air conditioning system	2900 W	Working 24 h/24 at 20°C
Hot air extractor	2850 W	Working 24 h/24
Seven lighting lamps	36 W × 7	Working 24 h/24
Communication equipment of the national gendarmerie	Power 20 W, supply 48 V	Working 24 h/24

consumption is better controlled in western countries than in sub-Saharan countries (where the average temperatures are very high); thus, we have to find how to control the power consumption of base stations situated in the Sahel area of Cameroon (the annual average temperatures are high) by proposing solutions of energy savings. The approaches of energy efficiency in base stations in general are recent and do not reveal the specific case of base stations situated in Sahel areas (where one witnesses average temperatures variations).

3. Power Consumption and Classification of Base Stations in the Sahel Zone of Cameroon

The visit to several sites accommodating base stations allowed to count base stations in terms of their energy consumption (see Figure 2).

It turns out that according to the classification of base stations defined in [24], one finds two types which include base stations of whose equipment is placed on the ground and indoors (encountered fields in town) and the base stations of distributed architecture and outdoors (encountered where there are buildings).

Besides these two types, there are also the relay stations of audiovisual companies such as CRTV. In sum, in the Sahel zone, there are three types of base stations:

(i) Traditional base stations indoors (the equipment is placed on the ground).

(ii) Base stations of distributed architecture and outdoors.

(iii) Base stations of audiovisual companies in which the specificity is the presence of several radio transceivers and TV (television).

It turns out from the above figure that the base stations of distributed architecture and outdoors (from MTN Maroua Market to Camtel kakataré) consume less power than the traditional base stations (from MTN Missinguileo to orange Mora). The latter in their turn consume less energy than the audiovisual stations (CRTV Maroua and CRTV Yagoua). Thus, there are three ranges of power consumption:

(i) Nonordinary base stations which have large power consumption (more than 10 000 kWh/month), it is

the case of the station CRTV Maroua (has several transmitters of high power and air conditioning system of high power).

(ii) Traditional base stations (equipment placed on the ground in a premises) which have medium power consumption (consumption ranging from 3 000 kWh/month and 10 000 kWh/month), it is the case of the station MTN Missinguileo (indoor station using air conditioners).

(iii) Base stations of distributed architecture and outdoors which have small power consumption (consumption less than 3 000 kWh/month), it is the case of the station MTN Maroua Market (outdoor station and with distributed architecture).

Some energy audits were carried out on the sites of CRTV Maroua, MTN Missinguileo, and MTN Maroua Market.

4. Characteristics of Base Stations of CRTV Maroua, MTN Missinguileo, and MTN Maroua Market

4.1. CRTV Maroua Site.
CRTV Maroua, like all other audiovisual companies, is endowed with several radio transceivers and television enabling the service in radio/television communication in Far North Region.

The equipment ensuring the transmission and the reception are accommodated in GASA neighborhood. They are distributed in three rooms (TV room, FM room, and RFI room). The electric power delivered by AES-Sonel on the site is about 160 kVa with a power factor of 0.8.

The transceivers (TX) of all rooms, and others equipment in this site, are represented in Supplementary Annexes 1 (In Supplementary Material available online at http://dx.doi.org/10.1155/2016/3161060).

4.1.1. TV Room.
The TV room is the one accommodating the TV emitter. The equipment found in this room is listed in Table 1.

The TV transceiver enables collecting television signals coming from National TV, amplifying, and supplying almost all the Far North Region. The hot air extractor enables emptying the room, at every moment when the temperature of the room is high and the air conditioning system is out of service.

TABLE 2: Equipment of the FM room.

Equipment	Power	Other characteristics and working hours
2 FM TX 10 KW (national program and local program)	2 × 22.5 KVA	Working 24 h/24; frequencies of emission: 98.10 MHz for FM National and 94.80 MHz for FM Regional
Dummy load	1 KVA	Working 24 h/24
PIE RACK	1 KVA	Working 24 h/24
Measuring equipment	1 KVA	Working 24 h/24
Dehydrator	1,5 KVA	Serial number: 35651, Sumiden Opcom, Ltd, NP: 1604, working 24 h/24
Two air conditioners	2 × 12 KVA	Working 24 h/24, adjusted at 19°C
Spare	0,5 KVA	Working 24 h/24
Transformer isolator	80 KVA	Working 24 h/24
Voltage regulator	75 KVA	AVR and PDB, work 24 h/24
Two lighting lamps	2 × 36 W	Working 24 h/24

TABLE 3: Equipment of the RFI room.

Equipment	Power	Other characteristics
TX/RX RFI north Cameroon	1 KW (2 × 500 W)	FMTX Model T213SJ, DRX 3200, QPSK receiver
TX/RX 900 GSM et 1800 GSM ORANGE Cameroon	Max 3600 W × 2	Alcatel 02G 89450736652, ABIS1TXRX ABIS 1TXRX
Energy bay ELTEC (charger of batteries made rectifiers AC/DC)		Output voltage 53,4 V
16 batteries	12 V-160 Ah	Monolite 12 FFT 160, 12 scpe CF 150 12 V 160 Ah/10 hr to 1,80 V/cell at 20°C 12 V 160 Ah/8 hr to 1,75 V/cell at 77°C Float voltage 13,62 V at 20°C/13,56 V at 77°C
Five security lamps	11 W × 5	220 V-50 Hz
Three lighting lamps	36 W × 3	240 V-50 Hz, work at 24 h/24
Two split air conditioning systems	3800 W	Maintain the internal temperature of the room averagely at 24°C. Work 24 h/24 and adjusted at 20°C.

4.1.2. FM Room. The FM room is the one accommodating the regional and national FM transceivers.

The equipment located in this room is listed in Table 2.

The two national and regional transceivers enable supplying the entire region by radio waves of CRTV Maroua and the national station broadcasted from Yaounde; they also have a range of 200 Km.

The two air conditioning systems enable cooling the FM room, because

(i) the FM transceivers in their internal working procedure produce heat;

(ii) the room in which those pieces of equipment are accommodated is built out of concrete (and thus absorbs the heat of the day and rejects it into the room) and has only one opening (door);

(iii) the external temperature in Sahel zone is high.

The dehydrator dehydrates the air of the room in order to maintain the hygrometry around 50%; the values of temperatures are given by the temperature sensor.

The frequency selector switch enables conditioning the different frequencies in order to move towards the antennas; it also enables equally switching from regional FM to national FM.

4.1.3. RFI Room. The equipment found in this room is listed in Table 3.

Apart from the equipment of power consumption, found in the three rooms (TV, FM, and RFI), there are a TV set (77 W, 260 V) functioning on average 18 h/24 (from 6:00 am to 12:00 midnight), 4 lighting lamps in the corridor of 36 W each (working time: whole day), and 8 external lighting lamps of 36 W each (working time: whole day).

4.2. Missinguileo Site. The base station of Missinguileo neighborhood is a station of emission and reception. It is consisted of the radio antennas (the *"drums" or point to point antenna*), antennas of cover (*antenna with small range in the form of "stick"*), and also the antennas WIMAX emission (used for the internet network). This station is located on a mountain of approximately 230 meters high. The pylon in question has a height of 55 m. It is an indoor station (*in an iron container*). The temperature in the room containing the machines is maintained at 21°C by the cooling systems. Equipment characteristics' are listed in Table 4 and pictures of these equipment are represented in Supplementary Annexes 2.

4.3. Site of Maroua Market Site. The site sheltering the base station of MTN Maroua Market is set on a building of three

TABLE 4: Equipment of MTN Missinguileo site.

Equipment	Electrical characteristics	Other characteristics
AES-Sonel supply	Amperage 10–30 A, voltage 220 V/400 V, cos φ = 0,8 Power supplied: 15 kVa	Couple C = 2,5 wh/tr
2 TX/RX 900 GSM (DRU) et TX/RX 1800 GSM (DRU)	Max 3600 W × 3 Supply 48 V DC	ERICSON 02G 89450736652; work at 24 h/24
Energy bay Emerson Network Power (charger of batteries made of rectifiers AC/DC)	Redresseurs 220 V–24 V Puissance minimale 10 KW	Works 24 h/24
Eight accumulator batteries	Power sonic PG-6V 220B; 6 volts 226 A.H	6 V-200 A; autonomy 8 h; temperature 27°C.
Power system ERICSON	DC output: 24 V ou 48 V DC/200A/6,4 KW	Works 24 h/24
Standby generating set	Apparent power 20 KVA; power factor 0.8; output 16 KW	Works automatically in case of AES-Sonel power cut
Two air conditioning systems	Cooling input power: 2410 W Rated voltage: 220 V–230 V	Only one system works and is adjusted at 23°C, 24 h/24. The second is under failure
Five buoying lamps of red coulor	11 W × 2	220 V-50 Hz, work from 6:00 pm to 6:00 am
Four lighting lamps	36 W × 4	240 V-50 Hz, work at 24 h/24
One radio equipment Airmux 200	Supply 48 V DC 3 W	Serial LIU STM-1/STS-3; E1/T1 way side A B. P1-Direction KATOUAL CH1 P2-Direction KATOUAL CHP
WIMAX equipment	Supply 48 V DC 14 W	Supplies six directorates in internet network: (i) Lycée Kaelé (Centre multimédia) (ii) Camair-co (iii) Régionale AMCHIDI (iv) Direction Sonel Maroua (v) CDC Bomtock (vi) Afriland First Bank Kousseri
Multiplexer OSN 2500 Digital multiplexer 2/34	400 W 15 W	Of trademark *HUAWEI*, work 24 h/24 SMU 16 × 2; MMU 34 + 2

levels. The BTS system (excluding the aerials) is set on the 2nd level and the aerials are set on the roof of the 3rd level. The equipment is outdoors or are set on one shelter (shelter open to the free air). This site is located near the market of Maroua.

The three BTS enable covering in network, all the population of the market, and the neighboring area. In this zone of coverage, the traffic is dense considering the density of the population and the flow of people in the market. This type of site does not require air conditioning system (air conditioner) as in the indoor sites.

Equipment characteristics are listed in Table 5 and images of these pieces of equipment are represented in Supplementary Annexes 2 showing equipment set on the shelter and the aerials set on the roof.

5. Energy Audits in Base Stations

5.1. Site of CRTV Maroua

5.1.1. Remarks and Investigations Made on the Site of CRTV Maroua

(i) Lighting is insufficient in this day (small openings or windows and they are less wider).

(ii) The buildings are constructed in concrete cement and the roof is paved out of concrete.

(iii) The air conditioners of the FM room equally air-condition the TV room and sometimes the RFI room since the air conditioners of these last two rooms are regularly broken down.

(iv) The lamps are lit constantly. The lamps located outside and in the corridors of the buildings are sometimes not switched off during the day, which also increase the power consumption of the site.

(v) The site has only one source of energy, and the power generating unit used as standby is out of service for more than five years.

5.1.2. The Power Consumption of the CRTV Site from December 10 to July 12. By observing Figure 3, from December 10 to July 12, the energy consumption of the CRTV site is varied. During the months of December 10, January 11, February 11, August 11, September 11, November 11, January 12, and February 12, the energy consumption is lower than the average consumption (14750 kWh). This low consumption compared to the others is due, on the one hand for January, February, and November, to the low temperature (the less the temperature is, the less

TABLE 5: Equipment of the MTN Maroua Market site.

Equipment	Electrical characteristics	Other characteristics and working hours
AES-Sonel supply	Amperage 10–30 A, voltage 220 V/400 V, cos φ = 0.8 Power supplied: 15 kVa	Couple C = 2,5 wh/tr
2 TX/RX 900 GSM (DRU) et TX/RX 1800 GSM (DRU)	Max 3600 W × 3 Supply 48 DC	Ericsson 02G 89450736652 work at 24 h/24
Two energy bays Ericsson (charger of batteries made of AC/DC rectifier)	Rectifier 220 V–24 V Power 10 KW	Each bay possesses an evacuation system of the internal heat to the casket, enabling keeping a fairly low temperature
16 accumulator batteries	Power sonic PG-6V 220B; 6 volts 226 A.H	6 V-200 A; autonomy 7.8 h; temperature 27°C.
Power system Ericsson	DC output: 24 V or 48 V DC/200 A/6,4 KW	Works 24 h/24
Two lighting lamps	36 W × 2	240 V-50 Hz, work from 6:00 pm to 6:00 am
Two equipment radios Winlink 2000	Supply 48 V DC 2 × 2,4 W	Serial LIU STM-1/STS-3; E1/T1 way side A B. P1-Direction Maroua Dougoi CH1 P2-Direction Maroua Dougoi CHP P3-Direction Maroua Comice CH2 P4-Direction Maroua Comice CHP Work at 24 h/24

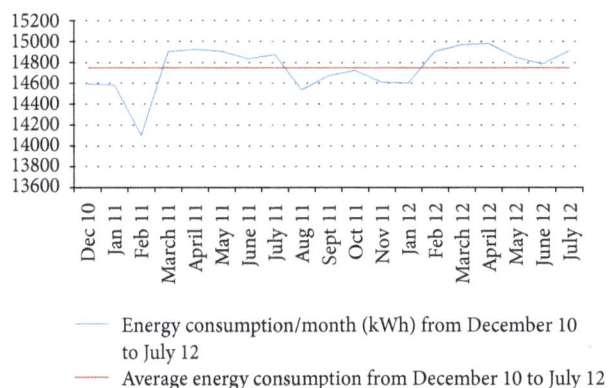

— Energy consumption/month (kWh) from December 10 to July 12

— Average energy consumption from December 10 to July 12

FIGURE 3: Monthly energy consumption curve, CRTV site from December 10 to July 12.

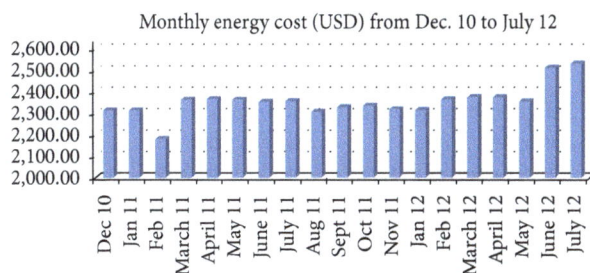

FIGURE 4: Estimation costs curve for monthly energy consumption, CRTV site from December 10 to July 12.

the air conditioners function); On the other hand, the drops of consumption in energy for August and September are due to the multiple power cuts because of heavy rains. As for the consumption higher than the average, it is due to the high temperatures during the months concerned.

Thus it arises from the graph that low consumption is recorded during the rainy and cold seasons, and the highest consumption takes place during the dry season (March, April, and May).

5.1.3. Estimation of Costs Related to the Monthly and Annual Energy Consumption.

As represented in Figure 4, the cost related to the energy consumption of the CRTV site Maroua varies; the months when consumption in energy is high also have a high cost of energy. For the year 2011, the cumulative of expenditure on the power consumption is approximately 27 908,409 USD.

We notice that in 2012, although consumption of energy in March (14 969 kWh) and April (14 981 kWh) exceeds the consumption of the months of June (14783 kWh) and July (14 910 kWh), the energy costs related to the months of June (2 511,077 USD) and July (2 532,508 USD) are higher than the costs of energy related to the months of March (2374,593 USD) and April (2376,855 USD); this is explained by the new grid of tariffing of electricity of AES-Sonel whose price of the kWh rise from 0,157896 USD to 0,169910 USD.

5.2. Site of MTN Missinguileo

5.2.1. Remarks and Investigations Carried Out on the Site of MTN Missinguileo

(i) The site of MTN Missinguileo being located at a high altitude, that is why, on this site, the temperature being relatively low;

(ii) The container, in which the equipment is placed, being made out of iron on a surface of approximately 50 m^2 and a 3,5 m height;

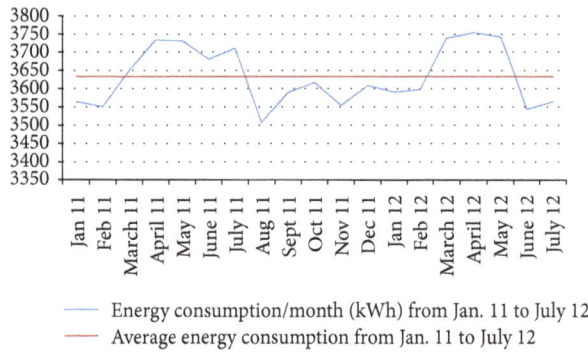

FIGURE 5: Monthly energy consumption curve, Missinguileo site from January 11 to July 12.

(iii) Lack of openings along the wall for the exchange of air with the outside;

(iv) External lighting being lit on 24 h/24;

(v) Only one air conditioner ensuring the air conditioning system;

(vi) The source of emergency power supply being a power generating unit, consuming on average 2000 liters of gas oil per year.

5.2.2. Existing Energy Consumption MTN Missinguileo 2011. By observing Figure 5, it turns out that the mean of the monthly consumption, from January 11 to July 12, is estimated at 3 633 kWh.

We notice that throughout the year 2011, the greatest energy consumption is recorded at the time of April (3733 kWh) and of May (3730 kWh) which belong to, according to the seasons of the Sahel zone, the hottest months. Once again, we say that the rise of consumption in energy is related to the rise of the temperatures on the sites of the base stations. August records lowest consumption in energy (3507 kWh); this low consumption is once more due to the inopportune and intense power cuts in the town of Maroua during the rainy season; this month is equally one of the least hot months of the year in the Sahel region of Cameroon. The annual energy consumption is approximately 43496 MWh. As for the year 2012, we notice that the hottest months of the year, March (3740 kWh), April (3755 kWh), and May (3743 kWh), are the most energy "consumers."

As for monthly load factors of the site, they are not set out for insufficient data; the maximum demands in power are difficult to establish since the only material enabling the records is the AES-Sonel meter who only indicates instantaneously the total energy consumed on the site.

5.2.3. Estimation of Costs Related to the Monthly and Annual Energy Consumption. As represented in Figure 6, the variation of energy costs related to the consumption of electrical energy from AES-Sonel origin on the MTN Missinguileo site. The most significant costs are those of the months of August11 (lowest energy cost because of low energy consumption) and June 12 and July12 (highest energy costs because of the new

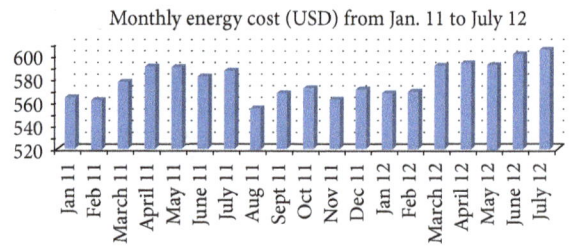

FIGURE 6: Estimation costs curve for monthly energy consumption, Missinguileo site from January 11 to July 12.

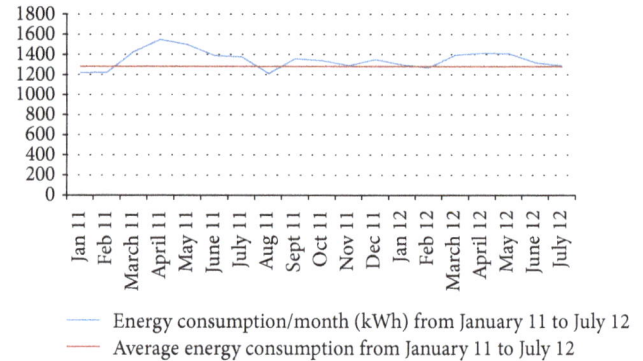

FIGURE 7: Monthly energy consumption curve, MTN Maroua Market site from January 11 to July 12.

grid of tariffing of electricity). The cumulative cost related to the energy consumption from source AES-Sonel for the year 2011 is estimated at 6879,99 USD.

The cost of one liter gas oil is 0,8918 USD (*source*: *CSPH*; Calculations SIE-Cameroun); thus, the annual cost 2011 related to the gas oil is 2000 L × 0,8918 USD = 1783,6 USD/year.

The total cost for expenditure 2011 on the energy consumption is thus 8 663,59 USD.

5.3. Site of MTN Maroua Market

5.3.1. Remarks and Investigations Carried Out on the Site of MTN Maroua Market

(i) The site of MTN Maroua Market is set on the roof of a building.

(ii) Outdoor station does not require air conditioning.

(iii) The lamps are lit on 24 h/24.

5.3.2. Existing Power Consumption MTN Maroua Market from January 11 to July 12. From Figure 7, it appears that the average of the monthly consumption in energy from January 11 to July 12 turns around 1 281 kWh. We notice that, for the year 2011, the energy consumption is higher in month of April (1551 kWh) and lower in August (1221 kWh). For the same reasons explained above, in the case of the base station of Missinguileo, during the hottest months, there is more consumption in energy.

Monthly energy cost (USD) from Jan. 11 to July 12

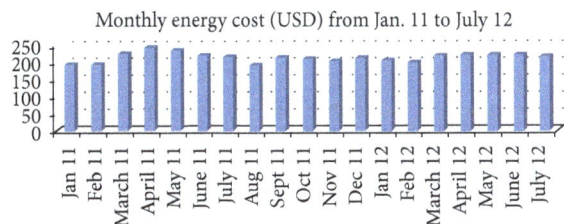

FIGURE 8: Estimation costs curve for monthly energy consumption, MTN Maroua Market site from January 11 to July 12.

During the year 2012, we note that in February the energy consumption (1268 kWh) remains lowest of the 07 recorded first. High consumption is known, once more during the hot periods of the year, which are March (1395 kWh), April (1413 kWh), and May (1409 kWh).

As for the monthly load factors of the site, they are not set out for insufficient data; the maximum demands for power are difficult to establish since the only material allowing the records remains the AES-Sonel meter, only indicating instantaneously the energy consumption of the site.

5.3.3. Estimation of Costs Related to the Monthly and Annual Energy Consumption. The variation of electrical energy costs from AES-Sonel on the site of MTN Maroua Market is represented in Figure 8. These costs vary around a monthly average of 214,88 USD. The cumulative cost of the year 2011 related to the energy consumption is estimated at 2578,713 USD.

6. Proposal of a Model of Energy Consumption of Base Stations

Faruk et al. [25] propose the following model of energy consumption for a BTS (1): the total power of a BTS and total energy consumed are P_{BTS} and E_{BTS}:

$$P_{\mathrm{BTS}} = P_{\mathrm{DP}} + P_{\mathrm{Ampl}} + P_{\mathrm{RU}} + P_{\mathrm{cov}} + \sum_i^m P_{\mathrm{AC}_i} + \sum_j^m P_{\mathrm{LB}_j}, \quad (1)$$

$$E_{\mathrm{BTS}} = P_{\mathrm{BTS}} \cdot t.$$

$P_{\mathrm{DP}}, P_{\mathrm{Ampl}}, P_{\mathrm{RU}}, P_{\mathrm{cov}}, P_{\mathrm{AC}_i}$, and P_{LB_j}, respectively, are power of digital signal processing, power of amplifier, power of radio unit, power of AC/DC converter, power of air conditioner i, and power of lamp j, and t is the operating time.

To have an expression for the total energy consumption of a telecommunication base station (MTN, ORANGE, CAMTEL, and CRTV) in the Sahel zone of Cameroon, we considered the following:

(i) The energy consumption of the various BTS (BTS GSM900, BTS GSM1800,...).

(ii) The energy consumption of technologies of transmission for internet (WIMAX,...). and/or radio transmission at long distances point to point such as the Airmux 200 and IRT 2000.

(iii) The energy consumption of the air conditioners (if possible).

(iv) The energy consumption of the lighting lamps.

(v) Losses of energies caused by the cables connecting the equipment of transmission and the antennas.

The energy consumption is defined as the product of the power supplied and the working time. It is given as follows:

$$\begin{aligned} E_{\mathrm{BTS}} &= P_{\mathrm{BTS}} \cdot t \\ &= \sum_{i=1}^n \sum_{j=1}^n P_{\mathrm{bts}_i} \cdot t_j \\ &+ \sum_{i=1}^m \sum_{j=1}^m P_{\text{another trans equip}_i(\text{WIMAWX},...)} \cdot t_j \\ &+ \sum_{i=1}^k \sum_{j=1}^k P_{\text{multipl equip}_i} \cdot t_j + \sum_{i=1}^l \sum_{j=1}^l P_{\mathrm{clim}_i} \cdot t_j \\ &+ \sum_{i=1}^g \sum_{j=1}^g P_{\mathrm{lamp}_i} \cdot t_j + \sum_{i=1}^f \sum_{j=1}^f P_{\text{losses in cables}_i} \cdot t_j. \end{aligned} \quad (2)$$

With P_{bts_i}, $P_{\text{another trans equip}_i(\text{WIMAWX},...)}$, $P_{\text{multipl equip}_i}$, P_{clim_i}, P_{lamp_i}, $P_{\text{losses in cables}_i}$, respectively, being power of each equipment bts_i of transmission, power of other transmission equipment such as the WIMAX, Airmux 200, IRT 2000, and VSAT (very small aperture terminal), power of the equipment of the different multiplexers and others as well, power of the air conditioner i, power of lamp j, and power lost in the cable i and t is the operating time.

E_{BTS} represents the total energy consumption of the site accommodating the base station.

Thus, the meters of energy found on the different sites of the base stations record at constantly cumulated E_{BTS}.

7. Energy Saving Realized on the Sites of the Base Stations after Recommendations of Solutions in Energy Saving

The solutions for energy savings were proposed in the three base stations. They are based firstly on the lighting and air conditioning system.

7.1. Solutions of Energy Saving on the Site of CRTV Maroua. We have proposed to increase the range of air conditioning of the air conditioners of the various rooms and to reduce the operating time of several lighting lamps (see Table 6).

(i) Increase of Range of Air Conditioning. The range of air conditioning of the FM and RFI rooms is increased by 2°C and 3°C in the night; thus, the air conditioners of the FM rooms are regulated to 21°C instead of 19°C in the day (8:00 am to 10:00 pm) and 22°C in the night (10:00 pm to 8:00 am). In the same way, the air conditioners of the RFI are regulated to 22°C instead of 20°C in the day (8:00 am to 10:00 pm) and 23°C in the night (10.00 pm to 8:00 am). This work was carried out from the 5th to 7th June 2012.

TABLE 6: Proposal of energy savings on the lighting aspect and air conditioning of the CRTV Maroua site.

Period	Characteristics of the site
8:00 am–6:00 pm	Widening of the climatic range by 2°C of the FM and RFI rooms, the lighting lamps are switched off
6:00 am–10:00 pm	Widening of the climatic range by 2°C of the FM and RFI rooms, the lighting lamps are switched on
10:00 pm–6:00 am	Widening of the climatic range by 3°C of the FM and RFI rooms, the lighting lamps are switched on
6:00 am–8:00 am	Widening of the climatic range by 3°C of the FM and RFI rooms, the lighting lamps are switched off

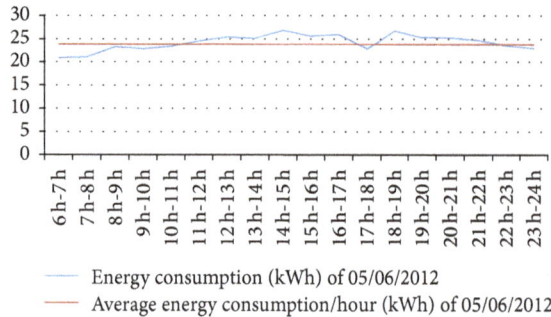

FIGURE 9: Hourly energy consumption curve of 05/06/2012.

(ii) Reduction of Operating Time of the Lamps. We have seen that the lighting lamps are lit only at the night that means from 6:00 pm to 6:00 am, and the records were carried out from 5th to 7th June 2012.

We have in total 23 lamps, of 36 W each, which light the site.

The periods in Table 6 are considered for energy consumption.

7.2. Application of Energy Savings Proposed. To evaluate the changes on the level of the energy consumption and the temperatures in the various rooms of the station, we followed the layout of power consumption time and the recorded temperatures time of the day of the 05/06/2012. On that day, all the air conditioners are under operation as at the date of the 27/05/2012, except for change on the climatic conditions and the lighting as indicated in the table of proposal for an energy savings (see Figures 9 and 10). The curves of variation of the energy consumption of the site of CRTV in the date of 05/06/2012 (see Figure 9) show an increasing time power consumption in the morning and decreasing in the evening. The time average of power consumption is 23,7955 kWh lower than that recorded in the date of 27/05/2012 (28,60 kWh).

7.3. Energy Savings and Saving Costs Realized. To establish the difference between the energy consumption of the 27/05/2012, where the proposals for an energy saving (lighting and system of air conditioning) are not applied, and the energy saving from 5th to 7th June 2012, when the proposals of energy saving (lighting and system of air conditioning) are applied, we recapitulated the energy consumption of these days in Table 7 and Figure 11.

We notice that the power consumption, at various periods considered of the 27/05/2012 are higher than those of the

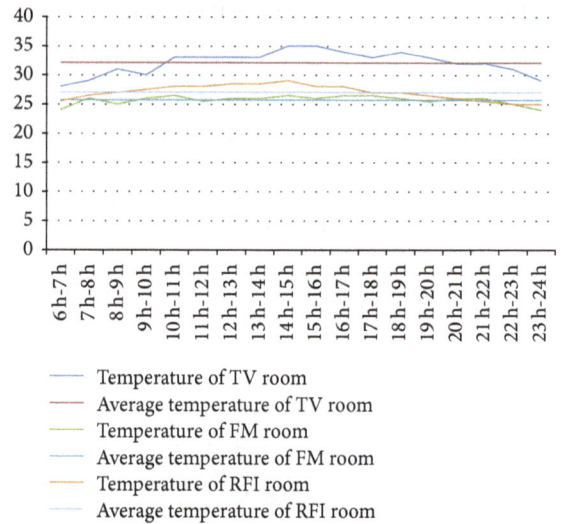

FIGURE 10: Hourly temperature curve of 05/06/2012.

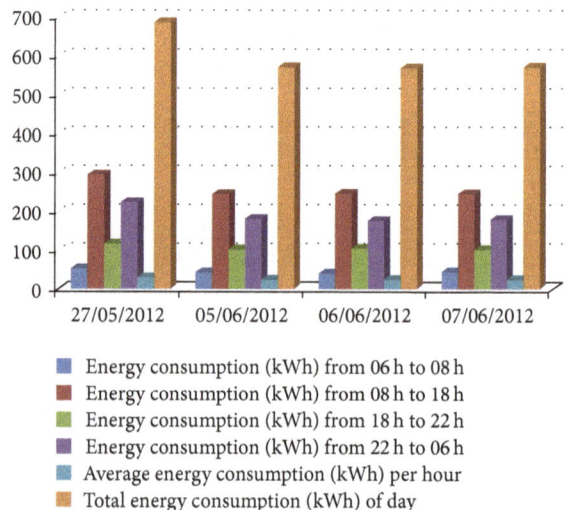

FIGURE 11: Comparison between energy consumption of the site, case of a day not applying the proposals of energy savings (27/05/2012), case of a day for which the air conditioner was under failure, and case of days applying the proposals of energy savings (05, 06, and 07 June 2012).

days of 5th, 6th, and 7th June 2012. The daily power consumption average from 5th to 7th June 2012 is approximately 570 kWh below 687 kWh (that of the 27/05/2012) that is to say a realizable energy saving of 117 kWh/day or 17% energy savings. By projecting measurements of energy saving over

TABLE 7: Recapitulative of energy consumption of the 27/05/2012 and of from the 5th to 7th June 2012.

Dates	Energy consumption (kWh) of the site 6:00 am–8:00 am	Energy consumption (kWh) of the site 8:00 am–6:00 pm	Energy consumption (kWh) of the site 6:00 pm–10:00 pm	Energy consumption (kWh) of the site 10:00 pm–6:00 am	Mean energy consumption (kWh) per hour	Total energy consumption (kWh) of the day	Energy savings realized % to the reference (27/05/2012)
27/05/2012	52.28	295	116	223.64	28.625	687	—
05/06/2012	41.872	245,42	102	181.8	23.7955	571.092	116 kWh which is 16.88%
06/06/2012	40.569	247,78	104.23	176.92	23.7291	569.499	118 kWh which is 17.17%
07/06/2012	44.532	246.34	101.45	179.63	23.8313	571.952	115 kWh which is 16.73%

TABLE 8: Proposals of energy savings on the lighting and air conditioning aspects of MTN sites.

Period	Characteristics of the site of MTN Missinguileo	Characteristics of the site of MTN Maroua Market
6:00 am–6:00 pm	No widening of the range of air conditioning, lamps switched off	Lamps switched off
6:00 pm–6:00 am	No widening of the range of air conditioning, external lamps switched on	Lamps switched on

one month or a year, one will have approximately energy saving 3510 kWh/month or 42120 kWh/year; In terms of cost, the realizable energy savings are 595,94 USD/month or 7151,34 USD/year. This realizable energy saving can supply an indoor base station like that of MTN Missinguileo (whose average consumption is 114 kWh/day) or two outdoor base stations (absence of air conditioner) such as the site of MTN Maroua Market (of which the mean consumption is 54 kWh/day).

8. Solutions of Energy Savings on the Sites of Missinguileo and Maroua Market

On the MTN sites of studied, we have proposed the reduction of the operating time of lighting lamps. Theses lamps can thus work from 6:00 pm to 6:00 am instead of allowing them to work throughout the day. As for the air conditioning system of the site of MTN Missinguileo, we have equally proposed an increase of the range of air conditioning at certain times of the day. Table 8 details proposals of energy savings on the lighting systems and air conditioning aspects of MTN sites.

By applying the proposals for an energy saving presented above on the sites of MTN Maroua Market et MTN Missinguileo, we plotted the curves of follow-up of the power consumption during three different days (from 16th to 18th July 2012).

8.1. Application of the Proposals of Energy Savings on the Missinguileo Site. Table 9 and Figure 12 recapitulate the power consumption of the MTN Missinguileo site from 16th to 18th July 2012, and the consumption of the 06/07/2012 being used as comparison with other consumption.

By comparing the energy consumption at the time of 16th, 17th, and 18th July 2012 to those of 6th July 2012, we notice that the energy consumption (consumption from 6:00 am to 6:00 pm, 6:00 pm–6:00 am, and the total daily consumption) of the 6th July 2012 is higher than those of the 16th, 17th, and 18th July 2012. The daily power consumption average from 16th to 18th July is approximately 93 kWh lower than 123 kWh (that of the 6th July 2012), which is 30 kWh/day, or 24.40% of realizable energy savings per day, or 900 kWh/month and 10 800 kWh/year. In terms of costs, this energy saving can generate each month 152,80 USD (900 kWh × 0.169 USD) or each year 1 883,6 USD. A good management of the system of lighting can generate energy saving and costs in the base station Missinguileo.

8.2. Application of Proposals of Energy Savings on the Maroua Market Site. While applying the proposals of an energy savings proposed, we followed the variation of the total energy consumption of the site during the days from the 16th to 18th July 2012. The results obtained are recapitulated in Table 10 and the histogram is represented in Figure 13, establishing the comparison between the day of Monday 9th July 2012 (where the proposals of energy savings were not yet applied) and days from the 16th to 18th July 2012 (application of proposals for an energy saving). This histogram (Figure 13) shows that the energy consumption (consumption from 6:00 am to 6:00 pm, consumption from 6:00 pm to 6:00 am, average consumption per hour, and total average consumption per day) recorded on the 09/07/2012 is higher than that recorded from the 16th to 18th July 2012.

The average power consumption from the 16th to 18th July 2012 is 53 kWh/day.

TABLE 9: Comparison between energy consumption of the site, case of a day not applying the proposals of energy savings (06/07/2012), case of a day for which the air conditioner was under failure, and case of days applying the proposals of energy savings (16, 17, and 18 July 2012): Missinguileo site.

Dates	Energy consumption (kWh) from 6 am to 6 pm	Energy consumption (kWh) from 6 pm to 6 am	Average energy consumption (kWh) per hour	Total energy consumption (kWh) daily	Energy savings realized, by the reference date (06/07/2012)
06/07/2012	70	53	5,125	123	—
11/07/2012	35,6	34	2,9	69,6	53 kWh or 43,41%
16/07/2012	46,2	46,6	3,86	92,8	30,2 kWh or 24,55%
17/07/2012	45,3	46,8	3,84	92,1	30,9 kWh or 25,12%
18/07/2012	46,9	46,3	3,88	93,2	29,8 kWh or 24,22%

TABLE 10: Comparison between the energy consumption of the site, case of a day not applying the proposals of energy savings (09/07/2012) and case of days applying the proposals of energy savings (16, 17, and 18 July 2012): Maroua Market site.

Dates	Energy consumption (kWh) from 6 am to 6 pm	Energy consumption (kWh) from 6 pm to 6 am	Average energy consumption (kWh) per hour	Total energy consumption (kWh) daily	Energy savings realized, by the reference date (09/07/2012)
09/07/2012	33	29	2,58	62	—
16/07/2012	28,4	25,6	2,25	54	8 kWh or 13%
17/07/2012	28,1	25,2	2,22	53,3	8,7 kWh or 14%
18/07/2012	27,8	25	2,2	52,8	9 kWh or 14,83%

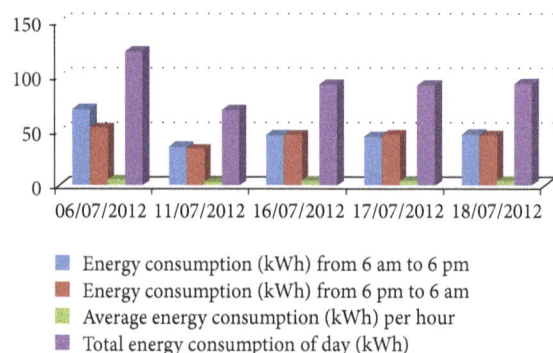

FIGURE 12: Comparison between energy consumption of the site, case of a day not applying the proposals of energy savings (06/07/2012), case of a day for which the air conditioner was under failure and case of days applying the proposals of energy savings (16, 17, and 18 July 2012): Missinguileo site.

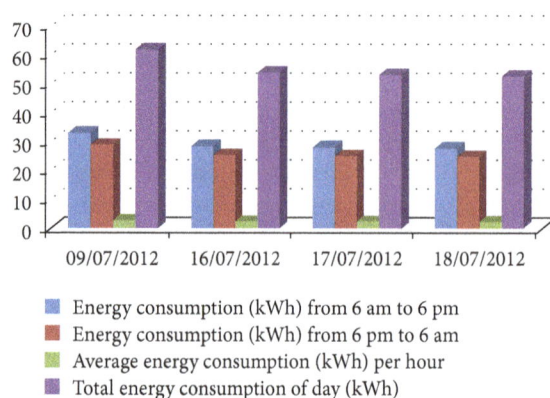

FIGURE 13: Comparison between the energy consumption of the site, case of a day not applying the proposals of energy savings (09/07/2012) and case of days applying the proposals of energy savings (16, 17, and 18 July 2012): Maroua Market site.

It is lower than that of the day of the 09/07/2012 (62 kWh/day), which is 9 kWh/day, or 270 kWh/month, or 3240 kWh/year, or 14.5% *of energy saving realizable*. In terms of saving costs, it is 45,84 USD/month or 550,10 USD/year.

TABLE 11: Techniques of solutions of energy efficiency.

Base station equipment, energy savings potential	Technical proposals of energy savings
Lighting systems	(i) Having the light level detectors in areas containing the base stations; (ii) Using more efficient lighting lamps to contribute to the reduction of cooling load; (iii) Replacing nonfunctional lighting lamps to ensure visual comfort of the user.
Air conditioner system	(i) Better air conditioner system adapting to the energy demand; (ii) Delay the start of the air conditioning units to avoid current peaks; (iii) Set the appropriate temperature and humidity inside the rooms, from 19°C to 30°C and 75% to 52% (thermal comfort in the Sahel [8]); (iv) Have the switcher time, on air conditioners lines (especially during cold periods, running from 9 am to 5 pm, meaning that air conditioners operate from 9 am to 5 pm during cold periods); (v) Avoid placing the air conditioner condensers in direct sunlight (vi) If possible, instead of air conditioners, have fans (with variable speed) which can recover heat or having outdoor BS, thus deleting conditioners and ventilators.
Locations of windows' rooms	(i) Windows must be in good seal; (ii) The windows should be wide so we can take advantage of wide daylight to reduce energy consumption due to lighting lamps; (iii) The windows locations are important on the internal temperature of the rooms (better South or North), if they are facing east or west they contribute to either rising temperature of the room (the sun hitting the windows) or cooling the room (shading phenomenon).
Base transmitter station and others equipment	(i) Having BTS and transmission equipment which are energy efficient; (ii) Having the BTS and transmission equipment that can switch off during periods of low traffic; (iii) The BTS and other transmission equipment being as near as possible to the transmitting antennas (the compact BS or distributed architecture BS); (iv) Equipment placing in the open air (outdoor base station) being less energy consumers; (v) Using the AC/DC converters and current stabilizer having an effectiveness about 95%.
Cables connections	The energy losses from cables must be less as possible.
Power supply	(i) The better power supply adapting to the energy load; (ii) Control and monitor energy consumption using "smart meters."
Maintenance	Ensure strict monitoring of the preventive maintenance program.
Energy management	Engineers and technicians in energy must manage the energy consumption.
Sensitization	Educate staff in energy savings.

The applications of the proposals of an energy saving which we carried out in the three sites of the base stations show us that energy saving is realizable (in a few days) when we operate on the lighting system and air conditioning system. If we apply those proposals in for one month or a year, the results will be more visible.

To reduce indeed the power consumption in the base station of telecommunication, we proposed a technical description of solutions being able to vigorously make the base stations located in the Sahel zone more effective.

8.3. Technical Description of Solutions of Energy Efficiency (Energy Savings) Proposed in Base Stations Situated in the Sahel Zone of Cameroon. We describe in Table 11 a set of recommendations to be able to realize energy savings in a base station situated in the Sahel zone.

9. Conclusions

In this paper, we have presented some approaches of energy savings and power consumption on the sites of the base stations of Telecommunication recently encountered in literature review of research based on the energy efficiency of the mobile communication. Moreover, in the Sahel zone, according to the average power consumption per/month, we met three categories of the base stations of telecommunication for which we have carried out energy audits. These audits reveal wasting of energy, and the proposals and application of some techniques of energy saving have enabled recording energy saving and considerable costs. The energy model proposed is more complete since it takes into account all the equipment consuming energy of the base station. To further reduce the expenditure in energy consumption of telecommunication base stations in the Sahel zone and the emission of greenhouse gases, in addition to the installation of measurements of energy efficiency, it is necessary to conduct a comparative study to know if the use of renewable energy sources is technically and economically profitable.

Nomenclature

ICTs: Information and communication technologies
AC: Alternative current
E: Power consumption
TV: Television
RFI: Radio France International

ADM: Add drop multiplexing
AES-Sonel: Apply Energy Services-Sonel (société nationale d'électricité)
ASIC: Application-specific integrated circuit
BSC: Base station controller
BTS: Base transceiver station
CAMTEL: Cameroon telecommunication
CDMA: Code division multiple Access
USD: United states Dollar
CO_2: Dioxyde of carbone
CRTV: Cameroon radio and television
DC: Direct current
DSP: Digital Signal Processor
EDGE: Enhanced Data rates for GSM Evolution
EGPRS: Enhanced GPRS
FPGA: Field-programmable gate array
GaAs: Gallium arsenide
GES: Gaz à effet de serre
GPRS: General Packet Radio Service
GSM: Global System for Mobile Communications
HSCSD: High Speed Circuit Switched Data
ISO: Organisation internationale de normalisation
LTE: Long-Term Evolution
MSC: Mobile services switching centres
MTN: Mobile telephone network
PSU: Power supply unit
RU: Radio unit
RBU: Radio base units
RF: Radio fréquence
RX: Receiver
Si: Silicium
TIC: Technologie de l'information et de la communication
TRI: Temps sur retour d'investissement propre
TWh: Tera watt hour
TX: Transmitter
UIT: Union internationale des télécommunications
UMTS: Universal Mobile Télécommunications System
VSAT: Very small aperture terminal
WIMAX: Worldwide Interoperability for Microwave Access.

Competing Interests

The authors declare that there are no competing interests regarding the publication of this paper.

Acknowledgments

The authors express their kindly acknowledgment of mobile communication and audiovisuals companies (CAMTEL, ORANGE, MTN, and CRTV) represented in the northern part of Cameroon.

References

[1] Fond d'Energie Rurale, 2010, http://www.se4all.org/sites/default/files/Cameroon_RAGA_FR_Released.pdf.

[2] http://www.afriquinfos.com/articles/2012/5/26.

[3] I. Humar, X. Ge, L. Xiang, M. Jo, M. Chen, and J. Zhang, "Rethinking energy efficiency models of cellular networks with embodied energy," *IEEE Network*, vol. 25, no. 2, pp. 40–49, 2011.

[4] Ericsson, "Sustainable energy use in mobile communications," Ericsson White Paper, 2007.

[5] Le rapport TIC et Développement Durable, 2008, http://www.ladocumentationfrancaise.fr/var/storage/rapports-publics/094000118.pdf.

[6] A. Conte, "Power consumption of base stations," in *Proceedings of the TREND Plenary Meeting*, Ghent, Belgium, February 2012.

[7] Study Energy Efficient Radio Acess Network Technologies, "Alcatel-lucent/TU dresden vodafone chair mobile communications systems," ATIS Report on Wireless Network Energy Efficiency, 2009.

[8] T. Chen, Y. Yang, H. Zhang, H. Kim, and K. Horneman, "Network energy saving technologies for green wireless access networks," *IEEE Wireless Communications*, vol. 18, no. 5, pp. 30–38, 2011.

[9] J. Lorincz, T. Garma, and G. Petrovic, "Measurements and modelling of base station power consumption under real traffic loads," *Sensors*, vol. 12, no. 4, pp. 4281–4310, 2012.

[10] H. Hirata, K. Totani, T. Maehata et al., "Development of high efficiency amplifier for cellular base stations," *SEI Technical Review*, no. 70, pp. 47–52, 2010.

[11] S. Zoican, "The role of programmable digital signal processors (dsp) for 3 g mobile communication systems," *Acta Technica Napocensis*, vol. 49, pp. 49–56, 2008.

[12] S. N. Roy, "Energy logic: a road map to reducing energy consumption in telecom munications networks," in *Proceedings of the 30th International Telecommunications Energy Conference (INTELEC '08)*, pp. 1–9, IEEE, San Diego, Calif, USA, September 2008.

[13] M. Etoh, T. Ohya, and Y. Nakayama, "Energy consumption issues on mobile network systems," in *Proceedings of the International Symposium on Applications and the Internet (SAINT '08)*, pp. 365–368, Turku, Finland, August 2008.

[14] J. Lorincz, T. Garma, and G. Petrovic, "Measurements and modelling of base station power consumption under real traffic loads," *Sensors*, vol. 12, no. 4, pp. 4281–4310, 2012.

[15] L. M. Correia, D. Zeller, O. Blume et al., "Challenges and enabling technologies for energy aware mobile radio networks," *IEEE Communications Magazine*, vol. 48, no. 11, pp. 66–72, 2010.

[16] A. Conte, A. Feki, L. Chiaraviglio, D. Ciullo, M. Meo, and M. A. Marsan, "Cell wilting and blossoming for energy efficiency," *IEEE Wireless Communications*, vol. 18, no. 5, pp. 50–57, 2011.

[17] O. Blume, H. Eckhardt, S. Klein, E. Kuehn, and W. M. Wajda, "Energy savings in mobile networks based on adaptation to traffic statistics," *Bell Labs Technical Journal*, vol. 15, no. 2, pp. 77–94, 2010.

[18] C. Han, T. Harrold, S. Armour et al., "Green radio: radio techniques to enable energy-efficient wireless networks," *IEEE Communications Magazine*, vol. 49, no. 6, pp. 46–54, 2011.

[19] E. Oh, B. Krishnamachari, X. Liu, and Z. Niu, "Toward dynamic energy-efficient operation of cellular network infrastructure," *IEEE Communications Magazine*, vol. 49, no. 6, pp. 56–61, 2011.

[20] D. Tripper, A. Rezgui, P. Krishnamurthy, and P. Pacharintankul, "Dimming cellular networks," in *Proceedings of the IEEE Global Telecommunications Conference (GLOBECOM '10)*, pp. 1–6, Pittsburgh, Pa, USA, December 2010.

Power Consumption: Base Stations of Telecommunication in Sahel Zone of Cameroon; Typology...

161

[21] Z. Niu, Y. Wu, J. Gong, and Z. Yang, "Cell zooming for cost-efficient green cellular networks," *IEEE Communications Magazine*, vol. 48, no. 11, pp. 74–79, 2010.

[22] Z. Niu, "TANGO: traffic-aware network planning and green operation," *IEEE Wireless Communications*, vol. 18, no. 5, pp. 25–29, 2011.

[23] M. A. Marsan and M. Meo, "Energy efficient management of two cellular access networks," *ACM SIGMETRICS Performance Evaluation Review*, vol. 37, no. 4, pp. 69–73, 2010.

[24] Senza Fili Consulting, "White Paper 2010," http://www.senzafil-iconsulting.com.

[25] N. Faruk, A. A. Ayeni, M. Y. Muhammad et al., "Powering cell sites for mobile cellular systems using solar power," *International Journal of Engineering and Technology*, vol. 2, no. 5, 2012.

14

Development of a Cost-Effective Solar/Diesel Independent Power Plant for a Remote Station

Okeolu Samuel Omogoye, Ayoade Benson Ogundare, and Ibrahim Olawale Akanji

Electrical and Electronics Engineering Department, Lagos State Polytechnic, Ikorodu Campus, PMB 21606, Lagos, Nigeria

Correspondence should be addressed to Okeolu Samuel Omogoye; samelect2003@yahoo.com

Academic Editor: Mattheos Santamouris

The paper discusses the design, simulation, and optimization of a solar/diesel hybrid power supply system for a remote station. The design involves determination of the station total energy demand as well as obtaining the station solar radiation data. This information was used to size the components of the hybrid power supply system (HPSS) and to determine its configuration. Specifically, an appropriate software package, HOMER, was used to determine the number of solar panels, deep-cycle batteries, and rating of the inverter that comprise the solar section of the HPSS. A suitable diesel generator was also selected for the HPSS after careful technical and cost analysis of those available in the market. The designed system was simulated using the HOMER software package and the simulation results were used to carry out the optimization of the system. The final design adequately meets the station energy requirement. Based on a life expectancy of twenty-five years, a cost-benefit analysis of the HPSS was carried out. This analysis shows that the HPSS has a lower cost as compared to a conventional diesel generator power supply, thus recommending the HPSS as a more cost-effective solution for this application.

1. Introduction

The term remote station as used in this paper refers to a remote weather station, an espionage listening post, or a telecommunication repeater tower, and so forth, that is located where public utilities have not yet been made available. Such stations require electricity to operate the installed data communication and control equipment. In the past, before the prevalence of solar supply system, diesel generators were used to power the stations necessitating the stationing of human operators nearby to regularly visit the stations to carry out necessary maintenance services on the diesel generators, a situation similar to what is obtained for Nigerian global system for mobile (GSM) base stations. With the current preference for renewable energy sources, these remote station power supplies are being redesigned to take advantage of available technologies in renewable energy. This paper discusses the design of a hybrid power supply system (HPSS) comprising both solar and diesel complementary power plants taking into account the cost-effectiveness of the design system for a remote station. The design was carried out on the HOMER computer software platform. Section 1 discusses the introduction. Section 2 discusses the details of the design materials and methods. Section 3 discusses the simulation results, optimization, and cost-benefit analysis of HPSS while Section 4 discusses the conclusion and recommendation.

2. Materials and Methods

2.1. Station Power Demand Assessment. From the data obtained from the station, a profile is developed in Tables 1 and 2. The daily average load variation is depicted in Figure 1 and tabulated in Table 2. Assume that the load is constant for everyday of the year. The daily peak load of 98.86 kW is observed between 9:00 AM and 10:00 AM with 716.224 kWh d^{-1} energy consumption.

2.2. Solar and Wind Data for the Station Location

2.2.1. Photovoltaic Energy. Solar energy is one of the inexhaustible energy sources for renewable energy implementation of

TABLE 1: Remote station daily load demand.

S/N	Power consumption	Power (W)	Quantity	Load (W)	Load (kW)	Hour/day	On-time (time in use)
1	Air condition	2000	18	36000	36	8	08:00–16:00
2	Fans	56	78	4368	4.368	8	08:00–16:00
3	Lighting	40	152	6080	6.08	8	08:00–16:00
4	Printers	800	11	8800	8.8	8	08:00–16:00
5	Desktop computer	264	14	3696	3.696	8	08:00–16:00
6	Security light	60	9	540	0.54	8	08:00–16:00
7	Refrigerator	100	12	1200	1.2	8	08:00–16:00
8	Pumping machine	1119	1	1119	1.119	2	10:00–12:00
9	Television	60	12	720	0.72	4	12:00–16:00
10	Tachometer	16.1	1	16.1	0.016	3	13:00–16:00
11	Transformer trainer	600	1	600	0.6	3	13:00–16:00
12	Feedback power tester	675	1	675	0.675	3	13:00–16:00
13	Battery charger	4400	1	4400	4.4	6	10:00–16:00
14	System trainer	4.56	1	4.56	0.00456	3	13:00–16:00
15	Projector	210	2	420	0.42	6	10:00–16:00
16	Thurlby PL(320)	480	1	480	0.48	3	13:00–16:00
17	Function generator	500	1	500	0.5	3	13:00–16:00
18	Photocopy machine	1100	12	13200	13.2	6	10:00–16:00
19	Soldering iron	480	8	3840	3.84	7	09:00–16:00
20	DC power supply	375	1	375	0.375	3	13:00–16:00
21	Electric water heater	2990	12	35880	35.88	2	08:00–10:00
22	Oscilloscope	255	3	765	0.765	3	13:00–16:00
23	Signal generator	1000	2	2000	2	3	13:00–16:00
24	Feedback amplifier	155.9	1	155.9	0.1559	3	13:00–16:00

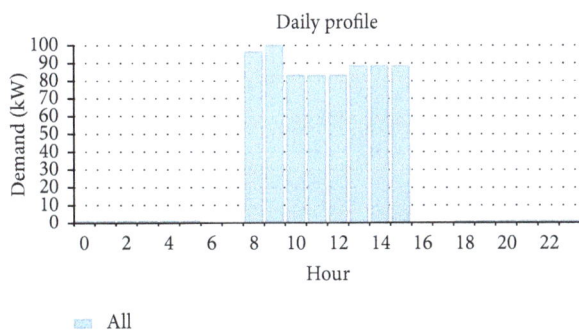

FIGURE 1: Station average daily load variation, Ikorodu, Lagos [7].

solar photovoltaic energy with a readily available standalone diesel generator [1]. Nigeria enjoys average daily sunshine hours of 6.25 hd^{-1} ranging between about 3.5 kWh m^{-2}d^{-1} in coastal areas and 9.0 kWh m^{-2}d^{-1} at far northern boundary [2]. In Nigeria, two main seasons are usually experienced because of the variation of climate from tropical to subtropical. The seasons are dry, which normally spans between October and March, and wet season, which usually spans between April and October. Variation in the stipulated months is usually experienced in the northern part of the country where the weather is very hot and dry. In the region, the raining season may vary between April and September, while in the southern part of the country, where the weather

is usually hot and humid, rainy season may occur between March and December [3]. The coastal area rarely produces temperature above 32°C while the temperature in the north may range between 32°C and 42°C with average humidity of about 95° Fahrenheit [4]. Ikorodu town in Lagos state has temperature of about 22°C, wind speed at 16 kmh^{-1} and 78% humidity [5].

2.2.2. Wind Resources. The wind speed in Ikorodu, Lagos, is very low and not suitable for a hybrid power system. This speed is averagely about 16 kmh^{-1} throughout the year. Because of this reason, wind turbine is not considered in this work. In this work, solar renewable energy resources are used. The solar data for clearness index and radiation index were obtained from NASA surface meteorology [6]. Ikorodu is a town and a Local Government Area in Lagos state, Nigeria, located along the Lagos lagoon. It shares boundary with Ogun state. The weather condition is 31°C, wind speed at 16 kmh^{-1}, 78% humidity at location latitude of 6.6000°N and longitude of 3.5000°E. Table 3 shows the solar resource profile for Ikorodu, Lagos, Nigeria. The latitude 6.6 and longitude 3.5 were chosen to run the HOMER data file from NASA surface meteorology of solar energy. The annual average monthly radiation incident on a horizontal surface is obtained as 4.74 kWh m^{-2}d^{-1} for year 2015.

Table 3 and Figure 2 reveal that February 2015 is the sunniest month of the year with solar energy of

TABLE 2: The electrical load (daily load demands) data for remote station.

Time	1	2	3	4	5	6	7	8	9	10	11	12	13	14	15	16	17	18	19	20	21	22	23	24	Total (watt)	Total (kwatt)
00:00–01:00						540																			540	0.54
1:00–2:00						540																			540	0.54
2:00–3:00						**540**																			540	0.54
3:00–4:00						540																			540	0.54
4:00–5:00						540																			540	0.54
5:00–6:00						540																			540	0.54
6:00–7:00																									0	0
7:00–8:00																									0	0
8:00–9:00	36000	4368	6080	8800	3696		1200														35880				96024	96.024
9:00–10:00	36000	4368	6080	**8800**	3696		1200												3840		35880				99864	99.864
10:00–11:00	36000	4368	6080	8800	3696		1200	1119					4400		420			13200	3840						83123	83.123
11:00–12:00	36000	4368	6080	8800	3696		1200	1119					4400		420			13200	3840						83123	83.123
12:00–13:00	36000	4368	6080	8800	3696		1200		720				4400		420			13200	3840						82724	82.724
13:00–14:00	36000	4368	6080	8800	3696		1200		720	16.1	600	675	4400	4.56	420	480	500	13200	3840	375		765	2000	155.9	88295.56	88.29556
14:00–15:00	36000	4368	6080	8800	3696		1200		720	16.1	600	675	4400	4.56	420	480	500	13200	3840	375		765	2000	155.9	88295.56	88.29556
15:00–16:00	36000	4368	6080	8800	3696		1200		720	16.1	600	675	4400	4.56	420	480	500	13200	3840	375		765	2000	155.9	88295.56	88.29556
16:00–17:00																									0	0
17:00–18:00																									0	0
18:00–19:00						540																			540	0.54
19:00–20:00						540																			540	0.54
20:00–21:00						540																			540	0.54
21:00–22:00						540																			540	0.54
22:00–23:00						540																			540	0.54
23:00–00:00						540																			540	0.54
																									716224.7	716.2247

TABLE 3: Daily averaged solar incident on a horizontal surface.

Month	Clearness index	Daily radiation (kWh/m²/day)
January	0.628	5.280
February	0.596	5.490
March	0.543	5.460
April	0.494	5.210
May	0.448	4.760
June	0.383	4.040
July	0.374	3.950
August	0.378	3.980
September	0.402	4.090
October	0.483	4.550
November	0.578	4.950
December	0.636	5.170

Annual average monthly radiation (kWh m^{-2}d^{-1}: 4.74) [6, 7].

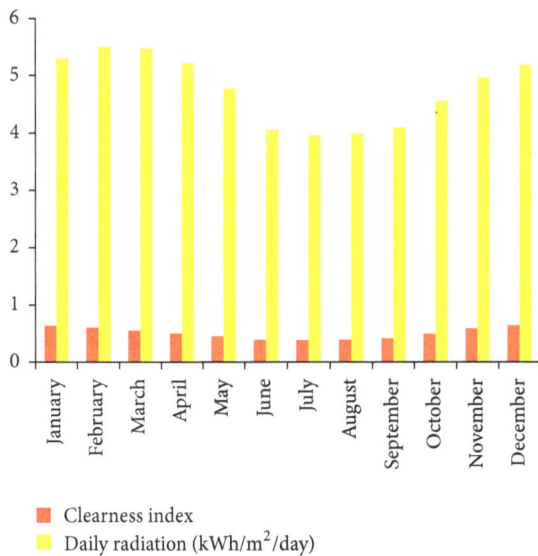

FIGURE 2: Station graphic display for solar radiation of Ikorodu, Lagos [7].

5.49 kWh m^{-2}d^{-1}, while July gave the lowest sunny month of the year with solar energy resource of 3.95 kWh m^{-2}d^{-1}. However, starting from the month of August to February, the solar radiation gained solar energy with differences from month to month cycle as (0.03), (0.11), (0.46), (0.40), (0.22), (0.11), and (0.21), respectively, whereas there are decreases in solar radiation from the months of March, April, May, June, and July with these values of solar (0.03), (0.25), (0.45), and (0.72), respectively. During these former months when the solar radiation dropped drastically, the standby diesel generator can stand in to compensate for the energy needed because the whole system setup is optimized.

2.2.3. Design of Station Solar Power Subsystem. The software used for the design and simulation of the PV/Diesel hybrid power system is HOMER optimization software. The input data used to run the HOMER program include

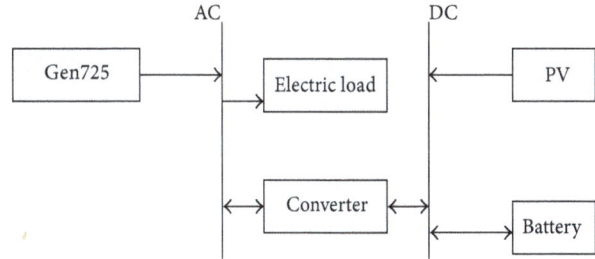

FIGURE 3: Hybrid renewable energy system architectural design [7].

(a) hourly load demand profile,

(b) sample of monthly solar radiation value of year 2005 PV system,

(c) the initial cost of each component of renewable energy, photo voltaic panel system, backup diesel generator, converter, and battery bank,

(d) the cost of AGO and yearly real interest rate of the project life cycle.

The hourly load demand data were gotten from load consumption of the station; the sample of monthly solar radiation value was gotten from NASA online data. 2005 data were the latest available data at June 1, 2015 [6, 7].

The initial capital cost of PV panel, diesel generator, battery, and converter was sought from Nigeria venture in Naira and later converted to dollars in order to be used in HOMER software [8]. HOMER software recognizes US dollars only. The current price of AGO was gotten from NNPC marketers for this simulation. Excel spreadsheet program is used to compute hourly load profile. HOMER is used for simulation of the system operation for 25 years. It reveals the energy balance calculation for every hour in each year. The configuration is sorted to be based on the life cycle cost (LCC) of the system. This (LCC) could be referred to as total net present cost (TNCP) [9]. The calculation here gives information about any cost which could occur during the stipulated project lifetime cycle of 25 years. These include initial setup costs, component replacement cost, maintenance cost, and fuel energy cost.

Designing a hybrid system demands correct components selection and sizing with appropriate operational strategy [10, 11]. In this work, solar energy is used with a diesel generator. The hybrid components include electric load demand, solar panels, battery, and converter. The architectural design of Figure 3 is obtained from HOMER V2.68.

For this work, the sensitivity variables were chosen based on monthly solar radiation and diesel fuel price. The simulation was based on life cycle cost (LCC) which reflected the total net present cost (TNPC).

HOMER was used to perform the optimization of the selective variables. This optimization gave the best hybrid renewable energy system size.

From the discussion so far, the four main components in PV/Diesel hybrid system are the PV panels, diesel generator, batteries, and converters. In order to determine the economic advantage of a hybrid system, one must determine the

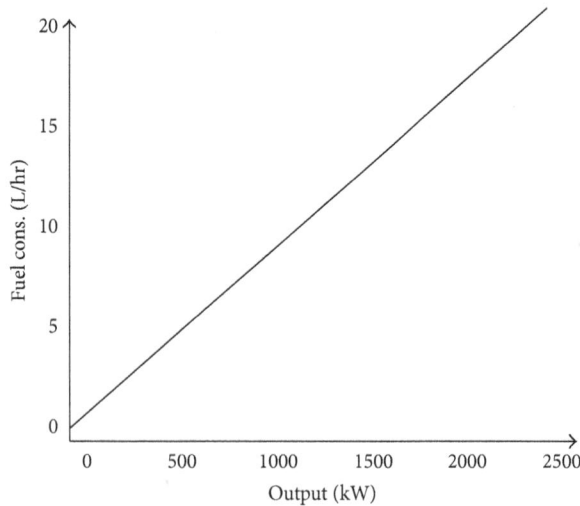

FIGURE 4: Fuel curve for generator set [7].

number of the above components units to be used, capital costs, replacement cost, and operation and maintenance cost. Operating hours must be stated in order for HOMER software to give accurate simulation.

2.3. HOMER Input Data

2.3.1. Solar Photovoltaic (PV). The photovoltaic module needed to power the station load of $716.224\,\mathrm{kWh\,d^{-1}}$ is estimated to cost $116,609 with the installation charges. The modules are expected to last for 25 years. The parameters used for the simulation of solar photovoltaic (PV) are suggested by HOMER.

2.3.2. Battery. The commonly used battery in hybrid system is the surrette 4kS25P.

2.3.3. Converter. 800 kW converter was used as input data for converter in this design.

2.3.4. Diesel Generator. A linear curve characterized by slope and intercept at no load is used to model the generator fuel consumption in HOMER. The graph of Figure 4 displays the load slope and intercept of 0.2372 (L/hr/kW output) and 0.0391 (L/hr/kW rated), respectively.

2.4. Design of the HPSS Components

2.4.1. Selection of Diesel Generator. The generator rated 725 kW was chosen on the HOMER software since the load requirement for the design is $716.224\,\mathrm{kWh\,d^{-1}}$. The site specific input of diesel generator was analysed as follows.

(1) Lifetime (Hours). This is the number of hours in which the generator is expected to provide service before replacement.

For the generator operating on minimum of 4 hours daily on weekdays, its lifetime in hours is calculated as follows:

(a) Hours/week = 4×5 = 20 hours/week.

(b) Hours/month = 20×4 = 960 hours/month.

(c) Hours/year = 80×12 = 960 hours/year.

(d) Lifetime hours (25 years) = 960×25 = 24000 hours/25 years.

For the generator working on weekly (7 days) basis, its lifetime in hours is calculated as follows:

(a) Hours/week = 4×7 = 28 hours/week.

(b) Hours/month = 28×4 = 112 hours/month.

(c) Hours/year = 112×12 = 1344 hours/year.

(d) Lifetime hours (25 years) = 1344×25 = 33600 hours/lifetime.

(2) Minimum Runtime (Minutes). This is when the dispatch starts the generator.

For the generator working on weekdays only, its minimum runtime in minutes is calculated as follows:

(a) Minutes/day = 4×60 = 240 minutes/day.

(b) Minutes/week = 240×5 = 1200 minutes/week.

(c) Minutes/month = 1200×4 = 4800 minutes/month.

(d) Minutes/year = 4800×12 = 57600 minutes/month.

(e) Lifetime minutes (25 years) = 57600×25 = 144000 minutes/25 years.

For the generator operating a minimum of 4 hrs daily on weekly basis, its minimum runtime in minutes is calculated as follows:

(a) Minutes/day = 4×60 = 240 minutes/day.

(b) Minutes/week = 240×7 = 1680 minutes/week.

(c) Minutes/month = 1680×4 = 6720 minutes/month.

(d) Minutes/year = 6720×12 = 80640 minutes/year.

(e) Lifetime minutes (25 years) = 80640×25 = 201600 minutes/25 years.

(3) Minimum Load Ratio (%). This is the minimum allowable load on the generator expressed as a percentage of its capacity. For minimum capacity of 0.54 kW, 82.724 kW, and 83.123 kW, the minimum load ratio is calculated as follows:

$$0.54\,\mathrm{kW} = \frac{0.54}{100} = 0.0054\%,$$

$$82.74\,\mathrm{kW} = \frac{82.72}{100} = 0.827\%, \tag{1}$$

$$83.12\,\mathrm{kW} = \frac{83.12}{100} = 0.831\%.$$

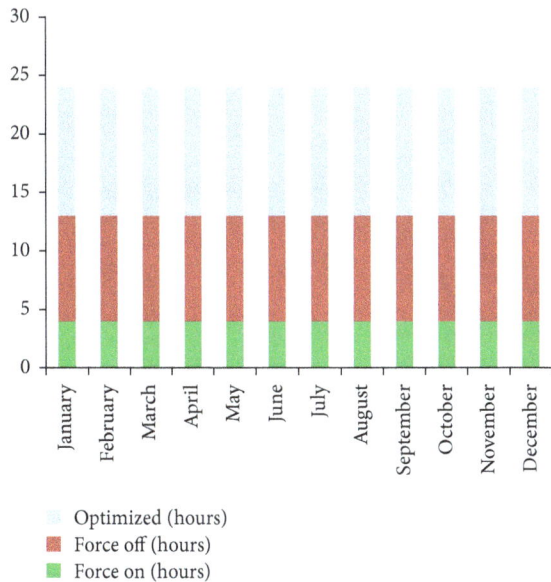

FIGURE 5: Generator operation schedule chart for diesel generator [7].

Legend for Figure 5:
- Optimized (hours)
- Force off (hours)
- Force on (hours)

(4) Fuel Curve. This provides assistance in calculating the two fuel curve inputs on the generator window.

(5) Schedule. HOMER schedule decides each time step. It operates and control the generator operation based on the electrical power demand. The operational schedule plan is shown in Figure 5.

2.4.2. Solar Photovoltaic. A generic flat plate of 350 W rating was chosen for the design. PV capacity and cost are analysed as follows:

Total load of 716.224 kWh d^{-1} divided by hours of sunlight per day (8 hrs) = 716.22/8 = 89.5275 kWd^{-1}.

For 0.35 kW panel capacity, this will give 89.5275/0.35 = 256 panels.

The price cost of 0.35 kW panels in dollars is $455 [12]. Therefore, 256 panels will cost 256 × 455 = $116480.

2.4.3. Battery. The CELLCUBE FB 200–800 battery of 800 kW capacity was chosen.

2.4.4. Power Converter. The generic system AC/DC inverter of 800 kW capacity was chosen for the design of this work.

3. Results and Discussions

3.1. Simulation Results. The simulation displayed information on the economic cost, electricity production, and environmental characteristic of each system component. The results obtained are presented in Tables 4, 5, and 6, respectively. The simulation reveals the optimized sizes of the solar

photovoltaic panel, battery bank, converter, and diesel generator as used in this work. This led to the design specification of the system components. Since the input data are the load requirement of the station, the hourly solar radiation and ambient temperature of the station obtained from NASA website were used. The analysed solar radiation gave the best optimized tilt angle of 48.65° for the PV panel. The result of the simulation reveals that hybrid system of 750 kV PV array, 3 units of 800 kWh battery, and 800KW AC/DC inverter would generate electricity of 1077343 kWh yr^{-1} while 725 kW diesel generator generates 24321 kWh yr^{-1}, making total power generation of 1101664 kWh yr^{-1} altogether at the cost of $62,050 (#12,410,000) [13].

3.2. Optimization Results. From the optimization results, Tables 5 and 6 showcase the optimization results with overall option. These results displayed the comparable results of the system configuration. The results analysed here are the system components sizes and numbers, initial capital cost and operating cost, NPC and COE, renewable fraction, and fuel consumption [13]. Tables 5 and 6 displayed the best optimal combination of energy system components. (one 725 kW diesel generator and 256 solar panels were used to assemble one 750 kW solar PV array, three 800 kW cell cube battery cycles charging, and one 800 kW converter). This hybrid system provides the lowest total net present cost when compared to existing diesel standalone generator as displayed in Table 7, with enough excess energy to meet the remote station energy demand if compared with the standalone diesel system as shown in Table 9. The presence of storage battery raises the initial capital cost of the system but battery storage reduces the operating hours of diesel generator in a system referring to the operational schedule in Figure 5 and therefore saves the world from global warming caused by high emission of toxic substances from generators and therefore reduced fuel consumption.

3.3. Cost-Benefit Analysis of HPSS

3.3.1. Existing System/Proposed Hybrid System. The existing diesel generator has lower initial capital cost, higher operating cost, and higher total net present cost for the whole project as shown in Table 7 and as illustrated on chart of Figure 6. This system emits more carbon monoxide (CO) and NO_2 as a result of fuel combustion of a lot of AGO. The hybrid solar PV/Diesel system can supply renewable energy corresponding to 98% of the daily energy demand to the station. The hybrid solar PV/Diesel system has reduced total net present cost as a result of less fuel consumption as shown in Table 7 and displayed in chart of Figure 7.

3.3.2. Economic Cost. The NPC involved in the two cases of standalone diesel generator and hybrid systems is displayed in Table 7. Subtracting the hybrid NPC from Standalone diesel NPC, the standalone system is $2,899,661 (#579,862,200) costlier if compared with the newly designed hybrid system NPC. An hybrid system saves cost and allows the station to

TABLE 4: HOMER simulation result showing sensitivity case only.

PV (kW)	Gen725 (kW)	FB 200–800	Converter (kW)	Dispatch	COE ($)	NPC ($)	Operating cost ($)	Initial capital ($)	Ren Frac (%)	Fuel (L)	Hours
		Architecture					Cost		System	Gen725	
750		2	400	CC	0.489	1,651,149	5,346	1,582,035	100		
750	725	1	400	CC	1.057	3,572,806	144,519	1,704,535	91	88,454	2,919
	725	2	400	CC	1.212	4,096,615	241,893	969,535	0	214,852	2,919

TABLE 5: HOMER simulation result showing optimization case with categorized option.

PV (kW)	Gen725 (kW)	FB 200–800	Converter (kW)	Dispatch	COE ($)	NPC ($)	Operating cost ($)	Initial capital ($)	Ren. Frac (%)	Fuel price ($/L)	Hours
		Architecture					Cost		System	Gen725	
750		2	400	CC	0.489	1,651,149	5,346	1,582,035	100	0.75	

TABLE 6: HOMER simulation result showing optimization case with overall option.

PV (kW)	Gen725 (kW)	FB 200–800	Converter (kW)	Dispatch	COE ($)	NPC ($)	Operating cost ($)	Initial capital ($)	Ren Frac (%)	Fuel (L)	Hours
		Architecture					Cost		System	Gen725	
89.6	725	3	400	CC	1.190	4,021,437	208,503	1,326,015	0	166,871	2,919
750	725	3	400	CC	1.203	4,064,293	145,408	2,184,535	100	82,734	2,919
0.4	725	2	400	CC	1.212	4,095,415	241,746	970,235	0	214,656	2,919
	725	2	400	CC	1.212	4,096,615	241,893	969,535	0	214,852	2,919
750	725	3	400	CC	1.219	4,119,949	149,713	2,184,535	91	88,474	2,919
89.6	725	3	800	CC	1.227	4,147,168	208,559	1,451,015	0	166,734	2,919
750	725	3	800	LF	1.240	4,189,552	145,428	2,309,535	100	82,734	2,919
0.4	725	2	800	CC	1.249	4,220,674	241,766	1,095,235	0	214,656	2,919

TABLE 7: Comparison simulation of economic results between existing system and a hybrid system.

Parameter	Existing diesel-only system		Proposed hybrid (PV/Diesel) system	
Value	Dollar ($)	Naira (#)	Dollar ($)	Naira (#)
Initial cost	1,574,535	314,907,000	2,184,535	436,907,000
Operating cost	416,895	83,379,000	145,408	29,081,600
Total NPC	6,963,954	1,392,790,800	4,064,293	812,858,600

enjoy optimum economic conditions. Moreover, the operational life of diesel only is low (2 years) as predicted by HOMER software in Table 8, while in Hybrid system its operational life is extended (5 years) as shown in Table 8.

3.4. Electricity Production. The standalone diesel generator set produces 321,686 kWh yr^{-1} (100%) of the total electricity with a capacity factor of 12% compared to the proposed hybrid system that will produce 1,077,343 kWh yr^{-1} (98%) from solar PV array and 24,317 kWh yr^{-1} (2%) from diesel generator with a capacity factor of 3% making a total of 1,101,660 kWh yr^{-1} (100%). The load demand is

261,422 kWh yr^{-1}, while excess electricity from the existing system is 26 kWh yr^{-1}; the proposed project has excess electricity of 761,261 kWh yr^{-1} as shown in Table 9. This information is displayed graphically in Figures 8 and 9, respectively:

$$\text{Excess energy} = \text{Total energy production}$$
$$- \text{Total energy consumption,}$$
$$\text{Excess energy} = (761{,}261 - 26) \text{ kWh yr}^{-1}$$
$$= 761{,}235 \text{ kWh yr}^{-1}.$$

(2)

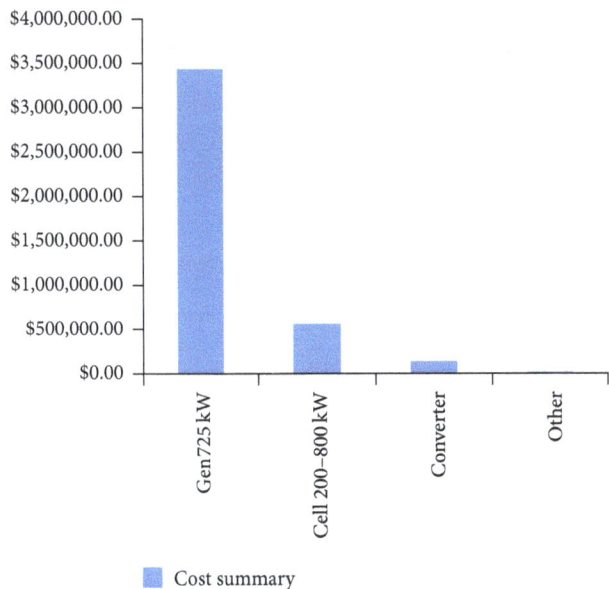

FIGURE 6: Cash flow analysis of NPC of existing diesel-only generator set [7].

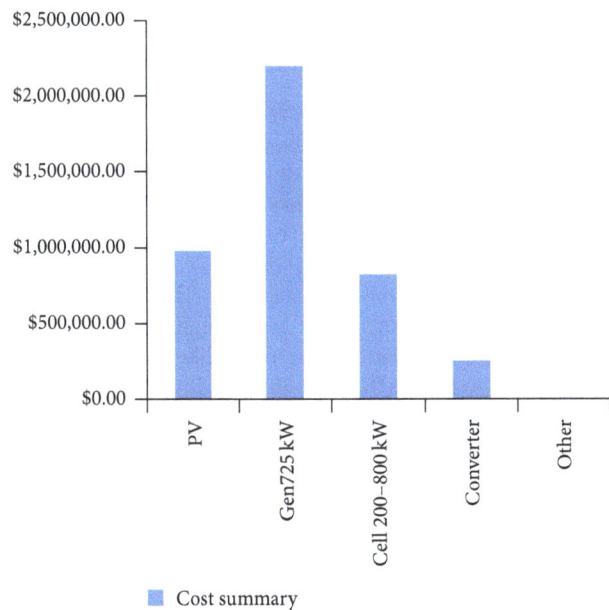

FIGURE 7: Cash flow analysis of NPC of optimized hybrid energy system [7].

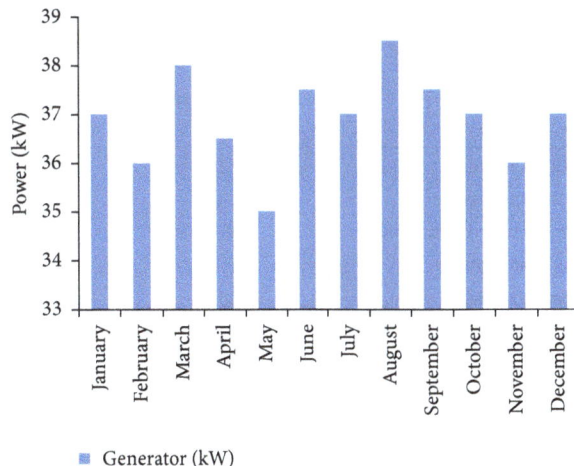

FIGURE 8: Monthly average of electrical power production of diesel-only generator set [7].

TABLE 8: Comparison simulation of existing system (diesel only) and proposed hybrid system.

Quantity	Diesel only		Diesel in hybrid system	
	Value	Units	Value	Units
Operational life	2.0	yr	5.0	yr
Capacity factor	12	%	3	%
Hours of operation	7,008	$h\,yr^{-1}$	2919	$h\,yr^{-1}$
Fuel consumption	274,877	$L\,yr^{-1}$	88,474	$L\,yr^{-1}$

TABLE 9: Comparison of simulation results of electricity production ($kWh\,yr^{-1}$).

Quantity	Diesel only		Hybrid system (Solar PV/Diesel)	
	$kWh\,yr^{-1}$%	%	$kWh\,yr^{-1}$%	%
Load consumption				
AC primary load	261,422	100	261,422	100
Production				
PV array	None	None	1,077,343	98
Diesel generator	321,686	100	24,317	2
Total energy	321,686	100	1,101,660	100
Excess electricity	26		761261	

3.5. Environmental Pollution. The standalone diesel generator set operates for 7,008 h annum^{-1}, with total fuel consumption of 74,877 L annum^{-1}. It generates 721.420 tonnes of CO_2, 3.026 tonnes of CO, 0.346 tonnes of UHC, 0.087 tonnes of PM, 1.485 tonnes of SO_2, and 3.024 tonnes of NO_2 as shown in Table 10. In contrast, in hybrid PV/Diesel system, the diesel generator operates for 2,919 h annum^{-1} and has a fuel consumption of 88,474 L annum^{-1} as shown in Table 10. This system emits 232.201 tonnes of CO_2, 0.97 tonnes of CO, 0.111 tonnes of UHC, 0.028 tonnes of PM, 0.478 tonnes of

SO_2, and 0.973 tonnes of NO_2 annually. Considering the environmental hazard, the higher the operational hours of a diesel generator, the higher the pollutant emission, and vice versa. Therefore, standalone generating set poses more danger to the environment if compared with a hybrid system.

3.6. Economics and Constraints. The project lifetime is fixed to be 25 years at annual interest rate of 5.88%. The safety margin of the operating reserve ensures the reliability of the power supply irrespective of the load variation. No capacity shortage was noted. The operating reserve as expressed in percentage of hourly load was 10%. Meanwhile, the operating reserve as a percentage of solar power output is fixed at 25%.

TABLE 10: Comparison of simulation results of emissions from standalone diesel generator set and proposed system.

| Pollutant | Emissions (kg yr^{-1}) | | | | |
| | Existing diesel only | | Proposed hybrid system | | Difference |
	(kg yr^{-1})	(ton yr^{-1})	(kg yr^{-1})	(ton yr^{-1})	(ton yr^{-1})
Carbon dioxide	721,420	721.420	232,201	232.201	489.219
Carbon monoxide	3024	3.024	973	0.973	2.054
Unburned hydrocarbon	346	0.346	111	0.111	0.235
Particular matter	87	0.087	28	0.028	0.059
Sulphur dioxide	1485	1.485	478	0.478	1.007
Nitrogen oxide	3024	3.024	973	0.973	2.051

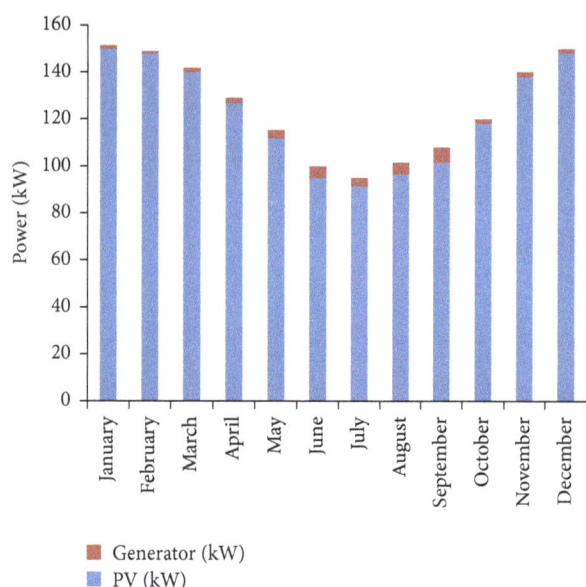

FIGURE 9: Monthly average of electrical power production of hybrid energy system [7].

3.6.1. System Economics. The capital costs for all the system components prices as quoted in this paper were sorted from local suppliers in Nigeria and outside Nigeria [8]. The costs estimates used in this paper were obtained from series of internet search. These prices may slightly defer from the actual prices. This is due to fluctuation of market prices in Nigeria. The replacement cost is assumed to be the same as the initial cost in this paper. The system components maintenance costs are estimates based on approximate time of 25 years required for the station. One dollar conversion to naira at the time that this work was done is N200 of Nigerian currency [14].

4. Conclusion

The analysis shows that the optimal hybrid system discussed in this paper is economically profitable for use in a remote station compared to using a diesel-only power generating set. This hybrid system reduces the fuel consumption and carbon dioxide (CO_2) by 32.18%, from Tables 8 and 10, respectively. From the environmental point of view, global warming could be controlled if we all embrace the advantages of this renewable energy technology. The simulation results analysis made it known that if battery is used with a hybrid PV/Diesel system, a high reduction in NPC and emission of toxic substances is also possible. In the same vein, this system meets the annual load demand of the remote station at a reduced cost and it is reliable with cheap maintenance cost. The objectives of this work are to provide reliable, clean, and environmentally friendly alternative power supply to the station at a much reduced cost. The analysis shows that a photovoltaic/diesel hybrid power system is reliable and economically viable for use at this proposed site.

4.1. Recommendation. These few recommendations are, however, worth noting. The excess energy generated in this paper could be used to power the neighbouring community but in the case of a station where no one lives around, multiple generators are used instead of one generator with higher capacity to reduce the amount of excess electricity generated in order to minimize energy wastage [15]. The emission from diesel generators in both cases is too high and this causes global warming; therefore equipment like carbon capture could be used to reduce the emission. During the design, each component must be stated accurately with respect to load demand and each component size must be at least 10% greater than the load demand. The input voltages of the converter and the battery must be the same. Many sensitivity variables required a lot of simulation time to run; therefore too many sensitivity variables may be avoided.

Conflict of Interests

The authors declare that there is no conflict of interests regarding the publication of this paper.

References

[1] A. V. Anayochukwu and E. A. Nnene, "Simulation and optimization of photovoltaic/diesel hybrid power system for health services facilities in rural environment," *Electronic Journal of Energy and Environment*, vol. 1, no. 1, pp. 57–70, 2013.

[2] E. J. Bala, J. O. Ojisu, and I. H. Umar, "Government policies and programmes on the development of solar PV sub-sector in Nigeria," *Nigeria Journal of Renewable Energy*, vol. 8, no. 1-2, pp. 1–6, 2000.

[3] O. Ojo, *Fundamentals of Physical and Dynamics Dimalology*, SEDEC Publication, Lagos, Nigeria, 1st edition, 2000.

[4] T. S. Falade, "Solving housing problem in lokoja," B.Sc. Research Report, Department of Architecture, A.B.U. Zaria, 1995.

[5] Ikorodu Location, October 2015, http://www.accuweather.com/en/ng/ikorodu/253770/weather-forecast/253770.

[6] Ikorodu solar radiation data, October 2015, http://eosweb.larc.nasa.gov/cgi-bin/sse/homer.cgi.

[7] HOMER V2.68. National Renewable Energy Laboratory (NREL), USA, 2010.

[8] Renewable energy shop in Nigeria, 2015, http://www.solarshop-nigeria.com/.

[9] S. M. Hussain and D. K. Sharma, "Techno-economic analysis of solar PV/diesel hybrid energy system for electrification of television substation. A case study of Nepal television substation at Ilam," in *Proceedings of the IOE Graduate Conference*, pp. 420–428, October 2014.

[10] B. S. Borowy and Z. M. Salameh, "Optimum photovoltaic array size for a hybrid wind/pv system," *IEEE Transactions on Energy Conversion*, vol. 9, no. 3, pp. 482–488, 1994.

[11] R. Dufo-López and J. L. Bernal-Agustín, "Design and control strategies of PV-diesel systems using genetic algorithms," *Solar Energy*, vol. 79, no. 1, pp. 33–46, 2005.

[12] Renewable solar panel price, http://www.pvpower.com/.

[13] G. Bekele and G. Tadesse, "Feasibility study of small Hydro/PV/Wind hybrid system for off-grid rural electrification in Ethiopia," *Applied Energy*, vol. 97, pp. 5–15, 2012.

[14] WebCrawler, Exchange Rate in Nigeria, September 2015, http://www.money.webcrawler.com/.

[15] M. Nour and G. Rohani, "Prospect of stand-alone PV-diesel hybrid power system for rural electrification in UAE," *International Journal of Renewable Energy Research*, vol. 4, no. 3, 2014.

15

The Economics of Renewable Energy Sources into Electricity Generation in Tanzania

Baraka Kichonge,[1] Iddi S. N. Mkilaha,[2] Geoffrey R. John,[2] and Sameer Hameer[3]

[1]*Mechanical Engineering Department, Arusha Technical College (ATC), P.O. Box 296, Arusha, Tanzania*
[2]*College of Engineering and Technology (CoET), University of Dar es Salaam (UDSM), P.O. Box 35131, Dar es Salaam, Tanzania*
[3]*Nelson Mandela African Institution of Science and Technology (NM-AIST), P.O. Box 447, Arusha, Tanzania*

Correspondence should be addressed to Baraka Kichonge; kichongeb@nm-aist.ac.tz

Academic Editor: Soteris Kalogirou

The study analyzes the economics of renewable energy sources into electricity generation in Tanzania. Business as usual (BAU) scenario and renewable energy (RE) scenario which enforce a mandatory penetration of renewable energy sources shares into electricity generations were analyzed. The results show total investment cost for the BAU scenario is much lower as compared to RE scenario while operating and maintenance variable costs are higher in BAU scenario. Primary energy supply in BAU scenario is higher tied with less investment costs as compared to RE scenario. Furthermore, the share of renewable energy sources in BAU scenario is insignificant as compared to RE scenario due to mandatory penetration policy imposed. Analysis concludes that there are much higher investments costs in RE scenario accompanied with less operating and variable costs and lower primary energy supply. Sensitivity analysis carried out suggests that regardless of changes in investments cost of coal and CCGT power plants, the penetration of renewable energy technologies was still insignificant. Notwithstanding the weaknesses of renewable energy technologies in terms of the associated higher investments costs, an interesting result is that it is possible to meet future electricity demand based on domestic resources including renewables.

1. Introduction

Energy is an essential and dominant component in achieving the interrelated economic and sustainable development of any country. Global energy demand is increasing at an exponential rate as a result of the exponential growth of world population [1]. Increases in global energy demand combined with fossil fuel depletion and the concern over environmental degradation put renewable energy sources as future energy supply [2, 3]. The energy consumption status of Tanzania is dominated by biomass which accounts for approximately 90% of total primary energy supply [4, 5]. Renewable energy sources available are biomass, hydro, geothermal, biogas, wind, and solar [6–8]. Geothermal potential is approximated at 650 MW with resource assessment still under preliminary surface studies [7, 9, 10]. Biomass estimated energy potential was at 12 million TOE in 2010 mainly from agriculture wastes, plantation forests, and natural forests [11].

Hydropower potential of Tanzania is estimated at 4700 MW of which only 553 MW has been realized at macro level and 12.8 MW at micro level generations [9]. Tanzania experiences between 2800 and 3500 sunshine hours per annum with solar irradiation between average values of 4 and 7 kWh/m^2 per day across the country [7, 12, 13]. Proven wind potential of 200 MW has been identified so far in Singida with studies going on in other areas of the country [7].

The technologies mix in the Tanzanian electricity sector comprises mainly hydro and thermal power plants specifically gas-fired and heavy fuel oil (HFO) [5]. Dependence on higher shares of electricity generated from hydropower plants has previously affected the security of its supply due to changing weather patterns [14]. A high dependence on hydropower resulted into power cuts and rationing caused by severe drought conditions as experienced in the past decades [9, 14]. The challenges of security of supply due to changing weather patterns forced a shift to more thermal generation to compensate for the hydroelectricity shortage [7, 9]. The incorporation of thermal power plants into electricity generations helped to boot out challenges of security of supply with changed generation mix [15]. In contrast, a shift to more thermal power plants has considerably increased the level of greenhouse emissions and other pollutants from power sector [16]. With the exclusion of hydropower plants, the share of renewable energy sources in the portfolio of technologies generating electricity in the country is insignificant [5, 9]. The participation of renewable energy sources predominantly wind, solar PV, and thermal is uncertain and thus calls for a balanced and diversified range of electricity generation technologies with less level of greenhouse emissions and other pollutants. Renewable energy sources are less vulnerable to climate unpredictability and are generally clean energy [1, 17, 18].

The country's dependence on hydropower [15], the vulnerability of the hydropower generation to extreme weather conditions, the unpredictability of HFO and natural gas prices, and the global growth in demand for energy sources are all drivers encouraging the pursuit for alternative energy sources in electricity generation in Tanzania. Renewable energy sources are attractive to complement the country's primary energy sources mix of supply for electricity generation. Renewable energy technologies are commercially available and have shown promising cost reduction due to their increased global uses. Literatures shows that penetration of renewable energy sources into electricity generation decreases generation and transmission costs owing to learning effects and increasing fossil fuel costs [19]. Despites their promising benefits, the country's potential in terms of renewable energy sources is yet to be fully utilized for electricity generation primarily due to the limited policy interest and investment levels [18]. The contribution of renewable energy sources with the exclusion of hydropower was only 0.55% in 2013, which is a small proportion as related to nonrenewable sources [20].

The country's approach for assured sustainable energy future for electricity generation is on the use of renewable energy sources through the adoption of renewable energy technologies [9]. However, there is a need to increase renewable energy sources utilization as a means of diversifying energy mix for the country. But then again, electricity generation through the use of renewable energy sources requires sufficient information on their economics convenience. This study is therefore focused on the analysis of economic convenience in renewable energy sources penetration into electricity generation in Tanzania using MESSAGE model. Modelling results will open up knowledge for long-term electricity sector planning and provides additional information and facts for policy and decision makers.

2. Methodology

The methodology applied in this study is centered on the plausible scenarios optimization representing expansion of the electricity generation system with an objective of meeting projected demand. The optimization was done in the long-term basis with a planning horizon from 2010 to 2040 adopted for this study.

2.1. Modelling Tool. MESSAGE is the analytical tool that formulates and evaluates alternative energy supply strategies consonant with user-defined constraints on new investment limits, market penetration rates for new technologies, fuel availability and trade, and environmental emissions, among many others [21–23]. The choice of MESSAGE was based on its ability to offer the essential engineering methodology to optimize electricity demand in the long-term horizon [24]. MESSAGE is equipped with features which provides the opportunity to define constraints between all types of technology-related variables. MESSAGE allows the user to limit one technology in relation to some other technologies such as maximum share of a certain technology that can be handled in electricity network or define further constraints between production and installed capacity. Extreme flexibility of the model can be used to analyse energy and electricity markets and climate change issues [22, 23].

MESSAGE works on the principle of reference energy system which allows representation of the entire energy network including existing and future technologies [25–28]. MESSAGE modelling procedure is based on building the energy flows network to describe the whole energy system, starting from level of domestic energy passing through primary and secondary level and ending by the given demand at a final level. Final demand level is distributed according to the types of consumption, for example, electricity [21, 25, 28]. In MESSAGE specific technology performance is compared with its alternatives on a life cycle analysis basis. When energy consumption is to be met by various options, MESSAGE selects the optimal solution from the most appropriate option considering the calculated discounted cost of the delivered energy unit while taking account of the whole technology investment cost, operation and maintenance (O&M) costs, and fuel cost at a constant price of the base year. MESSAGE approach allows the realistic evaluation of the long-term role of an energy supply option under competitive conditions [21, 28, 29]. For the case of Tanzania, MESSAGE has been applied to model energy supply options for electricity generation [30] with the electricity demand based on the Model for Analysis of Energy Demand (MAED) [22, 23, 31].

2.2. Modelling Framework. Modelling the economics of renewable energy sources into electricity generation in Tanzania consists of the optimization of the energy supply system. The conceptual modelling framework as applied in this study follows an approach depicted in Figure 1. The framework includes formulation of plausible scenarios using MESSAGE model to optimize energy supply options considering energy resources and technologies constraints.

2.3. Electricity Demand Forecasts. The electricity demands in this study are based on the official projections as given in Power System Master Plan (PSMP) 2012 update [5]. The electricity demand as depicted in Figure 2 represents short-, medium-, and long-term forecasts of interconnected power

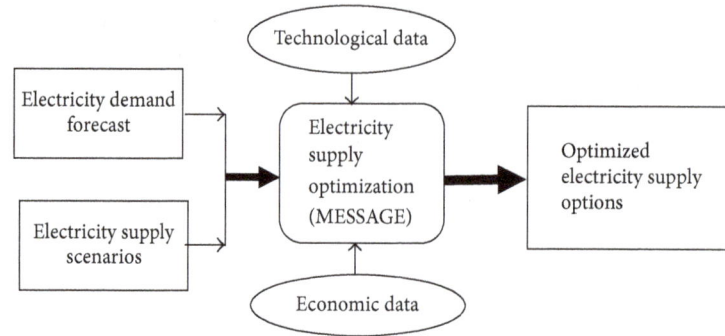

FIGURE 1: Conceptual modelling framework.

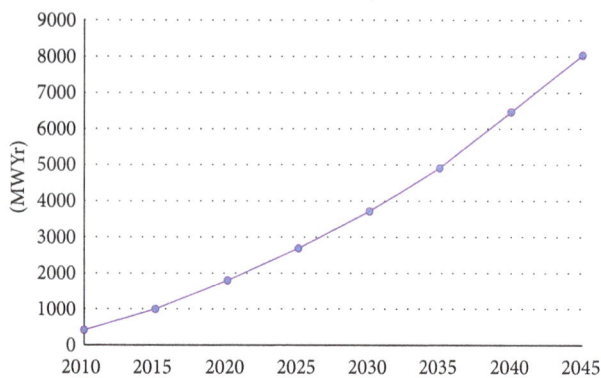

FIGURE 2: Electricity demand forecasts.

system and the isolated systems in the country. As opposed to MAED forecasts, the PSMP applies the trend methodology which is not consistent with the forecast of economic development [15].

2.4. Modelling Scenarios and Technologies. Two plausible modelling scenarios named as business as usual (BAU) and renewable energy (RE) were chosen in modelling the economics of renewable energy sources into electricity generations. BAU scenario is an overall electricity generation scenario (a reference case) which intends to illustrate how the electricity generation mix would take into account the renewable energy sources into power generations. BAU scenario follows official planned power system technologies capacities additions based on conventional energy sources such as natural gas, coal, HFO, and hydropower [5]. In BAU scenario, primary electricity conversion technologies which include hydro, coal, and natural gas power plants continue to dominate generation capacity without limitations in their expansion. Thermal power plants fired by natural gas continue their current domination with later entry of more coal fired power plants in the thermal portfolio for electricity generation. The BAU scenario also considers hydropower potential of 4700 MW to be unchanged over the study period due to the firm water sources availability as a result of rain and dry seasons being within the known and acceptable limits. The seasonality of hydropower in BAU scenario is accounted for by thermal generation primarily from natural gas and coal.

Renewable and nonrenewable energy sources are allowed to compete equally for the share in electricity generation. The main feature of BAU is to consider the growth of energy systems to minimize total discounted energy costs based on the technology and resource cost as inputs to the model.

As the objective of this study is to model the economics of renewable energy sources into electricity generation, an alternative RE scenario is developed to cater for that purpose. It is a known fact that electricity generated from coal, hydropower, and natural gas is characterized with low prices as compared to those from renewable energy sources. For example, cost of generating electricity using solar energy is quite higher than that of thermal generation such as combined cycle gas turbine (CCGT) [32]. Thermal generation technologies such as CCGT have the ability to infiltrate the market easily as opposed to renewable energy technologies [32, 33]. Without government support through internalization of external costs or providing incentives, penetration of renewable energy sources for electricity generation is difficult. However, the RE scenario is developed from BAU scenario to include a mandatory penetration of renewable energy sources into electricity generation. All technoeconomic inputs to the RE scenario are exactly the same as in BAU scenario. This makes RE scenario in effect as BAU scenario with additions of the mandatory penetration of renewable energy sources into electricity generation. The RE scenario introduces gradual increase of the renewable energy sources into electricity generation by imposing constraints in the model to allow their penetration. The imposed mandatory penetration of renewable energy sources requires a 15% share of wind, solar PV, geothermal, biomass, and solar thermal (added together) sources of the total electricity generation by 2040 starting from 5% in 2020 and progressively increasing to 10% in 2025 and 15% in 2030 through 2040.

For the purpose of this study, the definition of renewable energy is limited to geothermal, biomass, solar thermal, solar PV, and wind sources and excludes hydro because it is a matured energy source that has been competitively used in Tanzania. Hydro is the renewable energy source which is commercially viable on a large scale level because of its least costly way of keeping large amount of energy in the form of electricity but constrained owing to societal and environmental barriers [33, 34]. Furthermore, hydro produces negligible amounts of greenhouse gases and can easily adjust the amount of electricity produced with regard to demands. The choice of the mandatory penetration of renewable energy sources shares was stirred by descriptive scenarios which depicts global shares for electricity generation ranging within 25%–70% by 2050 [32, 35].

Modelling employed technologies using natural gas, biomass, geothermal, solar, wind, coal, and imported oil products (HFO) as fuels for the optimization of power generations. Nonrenewable technologies that were employed included natural gas technologies, hydro, HFO, and coal power plants. Natural gas technologies were bounded on gas turbine (GT) and combined cycle gas turbine (CCGT) power plants. MESSAGE model data collection and entry for optimization purposes were preceded by a quantitative analysis, specifically projection of PSMP 2012 electricity demand beyond 2035 and future fuel and technology costs. Data for the study consisting of electricity demand [5], technologies, technological constraints and efficiencies, technology's lifetime, investments, fixed costs, and capacity boundary were derived from a number of similar studies.

3. Results and Discussions

BAU and RE scenarios, as previously defined in methodology section, have been optimized using the MESSAGE model to decide the optimal supply options for Tanzania electricity generation from 2010 to 2040. The optimal electricity generations mix for BAU and RE scenarios considering electricity generation mix, operating and variable costs, primary energy production, and investment costs is presented and discussed in the next section.

3.1. Least-Cost Electricity Generation Mix. It can be seen graphically from Figures 3 and 4 that there are four main energy supply sources for electricity generation throughout the study period distinguished by hydro, natural gas, geothermal energy, and coal. The plots present only the most relevant energy sources (those that can be easily seen in the plots, excluding nonrelevant sources). The distribution of energy supply mix for electricity generation in both scenarios has been characterized by a substantial increase in hydro, natural gas, and geothermal and coal shares mostly after the year 2020. In both scenarios a total of 11,291 GWh will be generated in 2015 as compared to 54,981 GWh in 2035. The results match official projections of which a total of 11,246 GWh and 47,724 GWh were projected as demands in the years 2015 and 2035, respectively [5, 9]. According to the results, BAU scenario is dominated by natural gas with a share of 4968.6 GWh of the total electricity generated in 2015, followed by coal and hydro at 3586 GWh and 2722.2 GWh, respectively. Optimization results suggest hydropower will continue having a weighty part in terms of the capacity and production as it is the least-cost technology with the greatest potential in the country. Furthermore, optimization results in BAU scenario suggest geothermal as the most promising renewable energy technology which is able to compete without a mandatory introduction policy of renewable energy. Geothermal technology achieves competitive costs for electricity generation in the BAU scenario where other

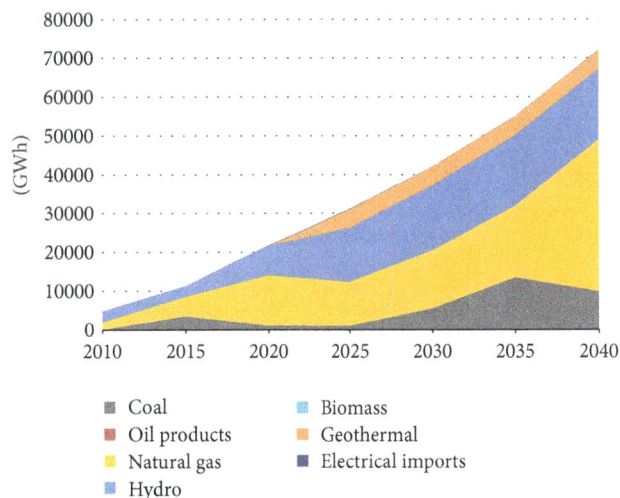

FIGURE 3: Electricity production by energy source, BAU scenario.

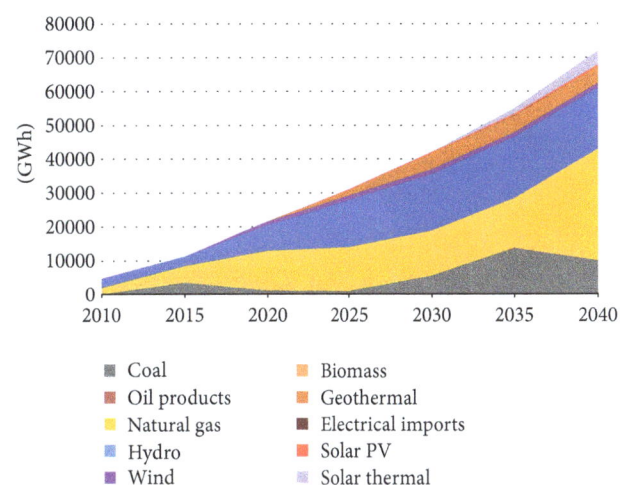

FIGURE 4: Electricity production by energy source, RE scenario.

renewable energy technologies did not. The competitiveness of geothermal technology is attributed to a combination of lesser investment costs and the assumed technology maximum operation time of 85% as compared to 25% and 35% for solar PV and wind technologies, respectively.

Penetration of renewable energy sources into electricity generations in BAU is very small since optimization was based on meeting demand at a least-cost composition of energy sources and technologies. RE scenario represents an increase in the share of renewable energy from 1,124 GWh representing 5% in the year 2020 to 3,155 GWh (10%) in 2025, 6499 GWh (15%) in 2030, and 10,876 GWh in the year 2040. The increase follows a mandatory introduction of renewable energy into the scenario from 2020. Electricity generation in RE scenario is dominated by natural gas power plants which contributes 45.6% of the total electricity generated in the year 2040. RE scenario optimization results suggest wind as the most promising renewable energy technology. Wind technology achieves competitive costs in electricity generation earlier than other renewable energy technologies.

The competitiveness of wind technology is attributed to a combination of lesser investment costs and the assumed technology maximum operation time of 35% as compared to the adopted 25% for solar PV. Comparison between the scenarios suggests a decrease of nonrenewable sources from 49,023 GWh to 42,987 GWh with the difference being replaced by renewable energy sources. Optimization results in RE scenario suggest that a rising share of renewable energy technologies over the years will be able to replace a big portion of nonrenewable energy sources. Notwithstanding an important decrease of nonrenewable energy sources for electricity generation over the study period in RE scenario, the country's power system will however require these sources as least-cost solution.

Hydro and geothermal shares are limited in the electricity generation mix due to energy potential constraint of 4700 MW and 650 MW, respectively, even though they have low operating cost advantages [7, 36, 37]. Cost profiles for most of the renewable energy technologies are high in terms of capital investment though they have low running costs [38]. Geothermal energy was the only renewable energy source able to penetrate into electricity generations mix in 2020 due to the fact that it is characterized by high availability capable of providing base load power for 24 hours a day and lower operating costs as compared to other renewable energy technologies [18, 38, 39].

3.2. Economics of Scenarios. The total investments costs for BAU and RE scenarios are presented in Figure 5. It can be observed that there is a huge difference from an economic point of view in both BAU and RE scenarios. The total investment costs of RE scenario differ by a margin of 1818 million US$ as compared to BAU scenario. There is an observed marginal difference of investment costs in BAU and RE scenarios for the period from 2015 to 2020. This is because the constraints requiring the mandatory introduction of renewable energy were not yet imposed in that period leaving the model to choose the least-cost supply options. With mandatory inclusion of renewable energy imposed in the year 2020, the investment costs for RE scenario increased gradually and reached 2616.8 million US$ in 2035 as compared to 1624 million US$ for BAU in the same year. There were no investments in the year 2040 as was the last year of study.

Comparisons in terms of investments cost, operating and maintenance fixed costs (O&M), and operating and maintenance variable costs (O&M) are presented in Figure 6. It can be observed from data that the main difference among BAU and RE scenarios is in the variable O&M costs (which includes fuel costs) and investment cost. The RE scenario entails higher total fixed O&M costs which are 1187.35 million US$ but again lower variable O&M costs of 1318 million US$ as compared to BAU scenario. The collective effect of the total O&M costs for RE scenario is 2505.44 million US$ higher than those for the BAU scenario which stand at 2210 million US$. However, in terms of the total costs (which include investment and O&M costs), BAU scenario is characterized with lower total costs than those of RE scenario. RE scenario demonstrated less operating and maintenance variable costs for the years 2020 to 2040 as compared to BAU scenario

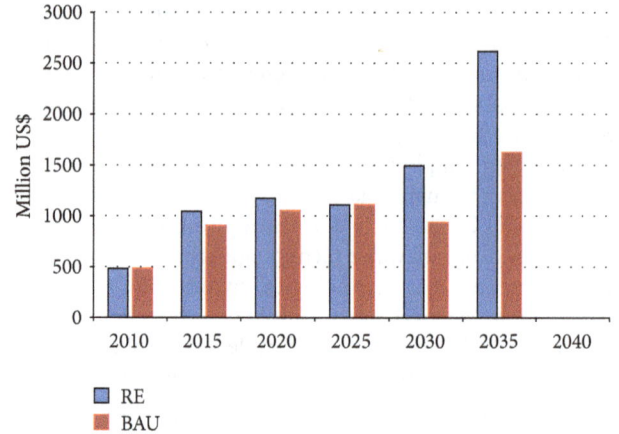

FIGURE 5: Investment costs comparison between BAU and RE scenarios.

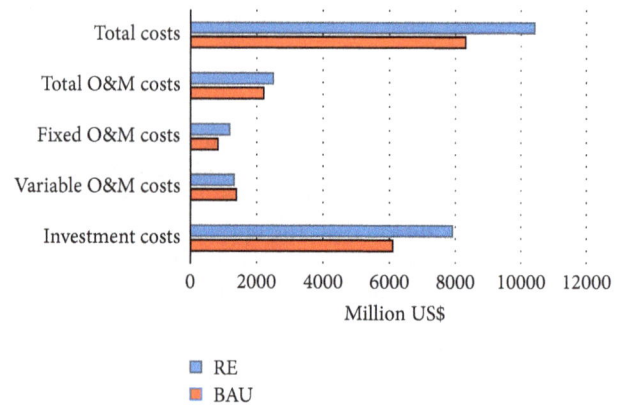

FIGURE 6: BAU and RE scenarios costs comparison.

except for the year 2015 where there was no penetration of renewable energy sources. Utilization of renewable technologies in electricity generation has a significant impact on operating and maintenance variable costs. This is due to less frequency of maintenance of parts particularly for wind and solar PV as compared to conventional technologies used in nonrenewable generations. Result depicts operating and maintenance variable costs for BAU are high as compared to RE scenario. However, the marginal differences in operating and maintenance variable costs are attributed to the fact that the renewable energy penetration accounts for a small proportion in share at only 5% in 2020 to 15% in 2040 of the total electricity generation. The differences in operating and maintenance variable costs would be higher if the share of renewable energy technologies is increased. As the technology in the production of renewable energy technologies improves, it is expected that these costs will be reduced [40, 41].

3.3. Primary Energy Production. The primary energy supply in BAU scenario is expected to increase from approximately 21,138.2 GWh in 2015 to approximately 109,799 GWh in 2040. If the proposed RE scenario is implemented in place of BAU from 2020 to 2040, the primary energy supply will be reduced to 96,742 GWh. On comparing primary energy supplies of the two scenarios, it is observed that 86,396 GWh of coal will be needed in RE scenario as compared to 87,380 GWh of BAU scenario for a period from 2015 to 2040. Similar observation is noted for the case of natural gas consumption in which only 209086 GWh will be required for RE scenario as compared to 240276 GWh for BAU scenario. Additionally, the reduced consumption of fossil fuel (coal and natural gas) translates into small variable O&M cost in RE scenario. The small proportions in the reduction of fossil fuel consumptions are attributed to marginal share of the renewable energy penetration which amounts to only 15%. The reduction in fossil fuel consumption would be much higher if the share of renewable energy technologies is increased in the generation of electricity. The reduction in primary energy supply in RE scenario as compared to BAU scenario translates into massive reduction in CO_2 emissions.

3.4. Sensitivity Analysis. In the sensitivity analysis carried out in this section, the technical parameters for all technologies in the model were not altered as they are the results of literature review and experience facts. Therefore, throughout the sensitivity analysis workout, the study assumed the uncertainty is anticipated to be a result of economic variables. In view of that, the sensitivity analysis adopted for this study consisted of the variations in renewable energy shares and the variations in investments and fuel costs.

3.4.1. Economic Effects of Variations in Renewable Energy Shares. Along with the BAU and RE scenarios, additional scenario abbreviated as RE1 was developed and modelled in order to evaluate the economic consequences with respect to the renewable energy share changes. RE1 scenario technoeconomic inputs are exactly as those of RE except for the mandatory shares of renewable energy sources penetration into electricity generation. RE1 scenario requires a 25% share of wind, solar PV, geothermal, biomass, and solar thermal (summed together) energy of the total electricity generation by 2040 starting from 10% in 2020 and progressively increasing the share to 15% in 2025 and 20% in 2030 and thereafter to 25% in 2035 through 2040. The effect of mandatory penetration of renewable sources to the model depicts the displacement of generation from fossil fuel sources, particularly natural gas and coal energy sources. As the mandatory contribution of renewable energy sources reaches the capacity set in the RE and RE1 scenarios and it is not increased, other least-cost sources occupy the generation system. Still thermal and hydropower generations support the system but the generation from these technologies decreases as renewable energy technologies specifically solar PV, solar thermal, wind, and geothermal energy enter the system as visible from 2020 throughout 2040.

Results show renewable energy shares increases in RE1 scenario upsurge the total investments cost to 10426 million US$ in comparison to 7918 million US$ and 6099.9 million US$ for RE and BAU scenarios, respectively. The general conclusion with regard to an increased share of renewable energy is that the more the shares of renewable energy sources are included in the electricity generation the more the impact

of increased investments costs observed is. However, the increased shares of renewable energy entail reduced O&M costs and the reduced CO_2 emissions as compared to that of BAU scenario. With increased renewable shares, there is substantial opportunity for Tanzania to meet its future electricity demand sustainably and thus economic growth through renewable energy sources. Since electricity supply in the country is dominated by fossil fuels and hydropower, political will and policies tailor-made to the promotion of other renewable energy sources are necessary to accomplish their penetration. RE and RE1 are sustainability scenarios as they are being provided for with sources that are likely to continuously be at disposal opposite to fossil fuels sources which are depletable and endure heavy environmental costs in terms of CO_2 emissions into the atmosphere.

3.4.2. Economic Effects of Variations in Investments and Fuel Costs. The investments cost of coal and natural gas power plant specifically CCGT was changed to observe the change in the installed capacity of the whole system and its impacts on the penetration of renewable energy technologies without mandatory penetration policy. The choice of these technologies was based on the fact that they have dominated the entire least-cost generation system as observed in the results. What is more, these technologies have been the priority for development of the existing generation capacity in PSMP as hydropower plants capacity is limited at 4700 MW [8, 9]. Sensitivity analysis on the investment cost was carried out by step increase of coal power plant investment cost from study adopted value of 1900 US$/kW to 2400 US$/kW. Similarly, the investment cost of coal power plant was decreased to 1700 US$/kW. During sensitivity analysis, the value of discount rate was 10% while the investments costs of other technologies including natural gas were kept constant as in BAU scenario.

The observed changes on the installed capacity shares of coal power plant as compared to that of CCGT power plant from the year 2020 are depicted in Figures 7 and 8. Coal power plants shares increase as their investment cost decreases while at the same time there is an observed marginal decrease in the shares of CCGT power plants. Similarly, as the investment cost of coal power plant increases there is a similar decrease of the technology shares as compared to that of CCGT power plant. In general, the sensitivity analysis on investment costs of coal and CCGT power plants suggests the rise of CCGT power plant investment cost above 1950 US$/kWh while keeping coal at 2300 US$/kWh or less, favouring coal power plants installed capacity shares growths. The rise of coal power plant investment cost to 2300 US$/kWh without similar rise of the investment cost for CCGT power plant to above 2000 US$/kWh will have negative impact on the coal power plant shares. In terms of coal and natural gas fuel variations, the observation on the sensitivity analysis suggests that gas fuel cost is less sensitive than that of coal. Furthermore, the variations of investment cost of coal and natural gas technologies do not influence renewable energy technologies penetration into electricity generation. The sensitivity analysis further reinterprets the influence of the market environment and the challenges on which

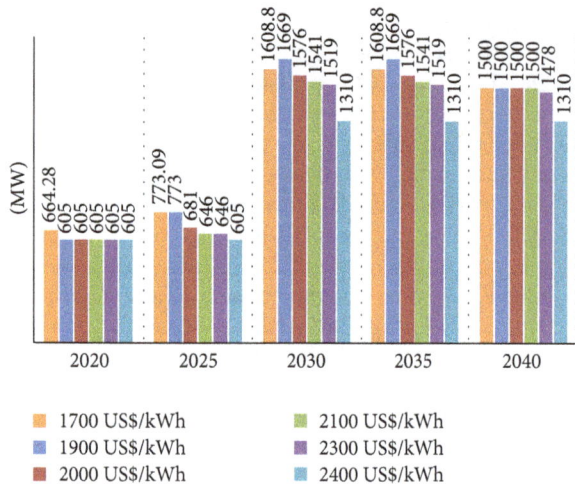

FIGURE 7: The installed capacity shares changes for coal power plants.

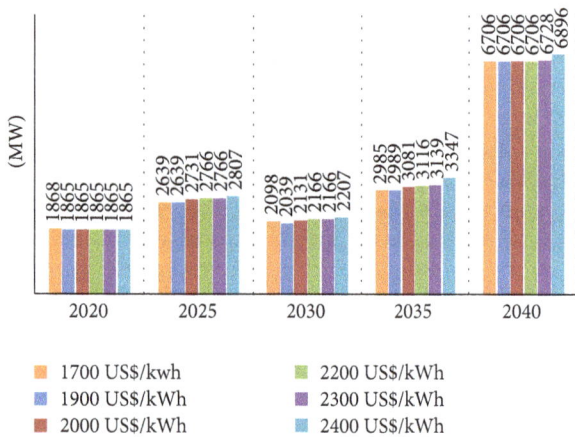

FIGURE 8: The installed capacity shares changes for CCGT power plants.

renewable energy technologies run into when trying to penetrate the generation mix.

4. Conclusion

The study presented a detailed analysis of the economics of renewable energy sources mixing into electricity generation in Tanzania. MESSAGE least-cost optimization results suggest that renewable energy sources and thus their technologies require compulsory policy measures to penetrate into the country's electricity system. Renewable energy technologies failures to fit in into the country's electricity system as suggested in the results were mainly due to technoeconomic competitiveness shown by conventional technologies under least-cost basis. Furthermore, the least-cost optimization results as considered in adopted scenarios of this study reflect the impact of the market environment and the challenges of renewable energy technologies explicitly wind and solar encounter in actuality. Mandatory penetration of renewable energy sources into electricity generation allows for realizing

substantial reduction in primary energy supply, O&M costs, and CO_2 emissions. However, RE scenario is still more expensive than BAU scenario in terms of investment cost. It is concluded that economic feasibility of renewable energy sources into electricity generation depends much on research and development (R&D) of renewable energy technologies that should allow for the investment cost decline coupled with efficiency improvement. Moreover, the use of renewable energy sources increases country's energy security from disruptions of supply. The weaknesses of renewable energy incorporation into electricity generations are on the high cost of its implementation. Renewable energy mandatory incorporation into electricity generation as was shown in RE scenario requires more investments than the case in BAU scenario. An interesting result is that the RE scenarios support the view that it is possible to meet future electricity demand based on domestic resources in spite of the associated higher investments costs.

Competing Interests

The authors declare that they have no competing interests.

Acknowledgments

The authors would like to thank Arusha Technical College (ATC), Nelson Mandela African Institution of Science and Technology (NM-AIST), Tanzania Atomic Energy Commission (TAEC), and College of Engineering and Technology (CoET) of the University of Dar es Salaam (UDSM) for their enabling environment that allowed successful completion of this work.

References

[1] A. Demirbas, A. Sahin-Demirbas, and A. Hilal Demirbas, "Global energy sources, energy usage, and future developments," *Energy Sources*, vol. 26, no. 3, pp. 191–204, 2004.

[2] M. Balat and G. Ayar, "Biomass energy in the world, use of biomass and potential trends," *Energy Sources*, vol. 27, no. 10, pp. 931–940, 2005.

[3] H. Garg and G. Datta, "Global status on renewable energy," in *Solar Energy Heating and Cooling Methods in Building, International Workshop: Iran University of Science and Technology*, pp. 19–20, 1998.

[4] K. T. Kabaka and F. Gwang'ombe, "Challenges in small hydropower development in Tanzania: rural electrification perspective," in *Proceedings of the International Conference on Small Hydropower-Hydro*, pp. 22–24, Kandy, Sri Lanka, October 2007.

[5] MEM, *Power System Master Plan 2012 Update*, Ministry of Energy and Minerals, Dar es Salaam, Tanzania, 2012.

[6] T. Alfstad, *Development of a Least Cost Energy Supply Model for the SADC Region*, University of Cape Town, Cape Town, South Africa, 2004.

[7] S. Kihwele, K. Hur, and A. Kyaruzi, "Visions, scenarios and action plans towards next generation tanzania power system," *Energies*, vol. 5, no. 10, pp. 3908–3927, 2012.

[8] MEM, *The Energy Sector Overview*, vol. 2013, Ministry of Energy and Mineral-Tanzania, 2013.

[9] MEM, *Scaling Up Renewable Energy Programme (SREP)—Investment Plan for Tanzania*, vol. 2013, Ministry of Energy and Minerals, Dar es Salaam, Tanzania, 2013.

[10] T. Mnjokava, Geothermal exploration in Tanzania-Status report. Presentation by Geological, 2008.

[11] L. Wilson, *Biomass energy systems and resources in tropical Tanzania [thesis]*, Royal Institute of Technology KTH, Stockholm, Sweden, 2010.

[12] D. Casmiri, *Energy Systems: Vulnerability—Adaptation—Resilience (VAR)*, vol. 2013, HELIO International, 2009.

[13] F. Mramba, Renewable energies in Tanzania: Opportunities and Challenges, 2013, http://www.malekigroup.com/images/ABW2013_2/Felchesmi_Mramba.pdf.

[14] S. Loisulie, *Vulnerability of the Tanzanian Hydropower Production to Extreme Weather Events*, vol. 2010, Sokoine University of Agriculture Faculty of Science, Morogoro, Tanzania, 2010.

[15] MEM, *Joint Energy Sector Review (JESR) 2012/13—Tanzania*, Ministry of Energy and Minerals, Dar es Salaam, Tanzania, 2013.

[16] IEA, "IEA statistics: CO_2 emissions from fuel combustion highlights," in *IEA Statistics*, vol. 2014, International Energy Agency (IEA), Paris, France, 2014.

[17] K. A. Hossain, "Global energy consumption pattern and GDP," *International Journal of Renewable Energy Technology Research*, vol. 1, pp. 23–29, 2012.

[18] S. Karekezi and W. Kithyoma, "Renewable energy development," in *Proceedings of the Workshop on African Energy Experts on Operationalizing the NEPAD Energy Initiative*, pp. 2–4, Dakar, Senegal, June 2003.

[19] O. Hohmeyer, *IPCC Scoping Meeting on Renewable Energy Sources: Proceedings*, 2008, https://www.ipcc.ch/pdf/supporting-material/proc-renewables-lubeck.pdf.

[20] IEA, *Energy Balances of Non-OECD Countries 2014*, International Energy Agency, Paris, France, 2014.

[21] A. Hainoun, M. Seif Aldin, and S. Almoustafa, "Formulating an optimal long-term energy supply strategy for Syria using MESSAGE model," *Energy Policy*, vol. 38, no. 4, pp. 1701–1714, 2010.

[22] IAEA, *Assessing Policy Options for Increasing the Use of Renewable Energy for Sustainable Development: Modelling Energy Scenarios for Ghana*, International Atomic Energy Agency, Vienna, Austria, 2006.

[23] IAEA, "Model for Analysis of Energy Demand (MAED-2)," in *Computer Manual Series No. 18*, vol. 2012, International Atomic Energy Agency, Vienna, Austria, 2006.

[24] D. F. Mora Alvarez, *Large scale integration of renewable energy sources for power generation in Colombia: a sensible alternative to conventional energy sources; scenario 2010–2050 [Ph.D. thesis]*, University of Flensburg, Flensburg, Germany, 2012.

[25] T. Pinthong and W. Wongsapai, "Evaluation of energy demand and supply under electricity generation scenarios of Thailand," in *Proceedings of the World Renewable Energy Congress*, Bangkok, Thailand, May 2009.

[26] L. Rečka, "Electricity system optimization: a case of the Czech electricity system—application of model MESSAGE," in *Proceedings of the 5th International Days of Statistics and Economics*, Prague, Czech Republic, September 2011.

[27] S. Selvakkumaran and B. Limmeechokchai, "Assessment of Thailand's energy policies on energy security," in *Proceedings of the International Conference and Utility Exhibition on Power and Energy Systems: Issues & Prospects for Asia (ICUE '11)*, pp. 1–6, IEEE, Pattaya City, Thailand, September 2011.

[28] N. Van Beeck, *Classification of Energy Models*, 1999.

[29] IAEA, "Guidance for the application of an assessment methodology for innovative nuclear energy systems," in *INPRO Manual—Overview of the Methodology*, vol. 1, International Atomic Energy Agency, Vienna, Austria, 2008, IAEA-TECDOC-1575 Rev. 1.

[30] B. Kichonge, G. R. John, and I. S. N. Mkilaha, "Modelling energy supply options for electricity generations in Tanzania," *Journal of Energy in Southern Africa*, vol. 26, no. 3, pp. 41–57, 2015.

[31] B. Kichonge, G. R. John, I. S. Mkilaha, and H. Sameer, "Modelling of future energy demand for Tanzania," *Journal of Energy Technologies and Policy*, vol. 4, no. 7, pp. 16–31, 2014.

[32] N. L. Panwar, S. C. Kaushik, and S. Kothari, "Role of renewable energy sources in environmental protection: a review," *Renewable and Sustainable Energy Reviews*, vol. 15, no. 3, pp. 1513–1524, 2011.

[33] J. Sathaye, O. Lucon, A. Rahman et al., "Renewable energy in the context of sustainable development," in *Renewable Energy Sources and Climate Change Mitigation*, chapter 9, pp. 707–790, Cambridge University Press, Cambridge, UK, 2011.

[34] IEA, "Structure of operation and maintenance training programmes," IEA Technical Report, International Energy Agency (IEA), Paris, France, 2000.

[35] J. Hamrin, H. Hummel, and R. Canapa, *Review of the Role of Renewable Energy in Global Energy Scenarios*, IEA, Paris, France, 2007.

[36] I. Dincer, "Renewable energy and sustainable development: a crucial review," *Renewable and Sustainable Energy Reviews*, vol. 4, no. 2, pp. 157–175, 2000.

[37] M. A. Kusekwa, *Biomass Conversion to Energy in Tanzania: A Critique*, InTech, Rijeka, Croatia, 2013.

[38] A. Evans, V. Strezov, and T. J. Evans, "Assessment of sustainability indicators for renewable energy technologies," *Renewable and Sustainable Energy Reviews*, vol. 13, no. 5, pp. 1082–1088, 2009.

[39] R. E. H. Sims, H.-H. Rogner, and K. Gregory, "Carbon emission and mitigation cost comparisons between fossil fuel, nuclear and renewable energy resources for electricity generation," *Energy Policy*, vol. 31, no. 13, pp. 1315–1326, 2003.

[40] P. Hearps and D. McConnell, *Renewable Energy Technology Cost Review*, Technical Paper Series, Melbourne Energy Institute, 2011.

[41] IRENA, *Renewable Power Generation Costs*, International Renewable Energy Agency, 2012.

16

The Links between Energy Consumption, Financial Development, and Economic Growth in Lebanon: Evidence from Cointegration with Unknown Structural Breaks

Salah Abosedra,[1] Muhammad Shahbaz,[2] and Rashid Sbia[3,4]

[1]College of Business Administration, American University in the Emirates, Academic City, Dubai, UAE
[2]Department of Management Sciences, COMSATS Institute of Information Technology, 1.5 km Defence Road, off Raiwind Road, Defence Road, Lahore 54000, Pakistan
[3]Ministry of Finance, P.O. Box 333, Manama, Bahrain
[4]Department of Applied Economics, Free University of Brussels, Avenue F. Roosevelt 50, CP 140, 1050 Brussels, Belgium

Correspondence should be addressed to Muhammad Shahbaz; shahbazmohd@live.com

Academic Editor: Jin-Li Hu

We investigate the relation between financial development, energy consumption, and economic growth in the economy of Lebanon over the period $2000M_2$–$2010M_{12}$. Our findings confirm the existence of cointegration among the variables. The results indicate that financial development and energy consumption contribute to economic growth in Lebanon. The impact of energy consumption on economic growth is positive showing the significance of energy as a main stimulant of economic growth. Financial development is also found to play a vital role in enhancing economic growth. Financial development and economic growth also result in further increase in energy consumption. We offer some policy implications specific to Lebanon considering the recent discovery of large oil and gas reserves in the country and the historical importance of its banking sector which remains a center of Lebanon's service-oriented economy.

1. Introduction

Energy economics has drawn substantial attention from academicians in recent times. A number of studies have investigated the causal relationship between energy consumption and economic growth. This issue is important because energy drives the wheels of economic growth since it is a key factor of production, along with capital, and labor. In addition, the higher the GDP per capita, the more the energy demand, which is a relation that is intuitively appealing. The pioneering study by J. Kraft and A. Kraft [1] confirms this by providing evidence of unidirectional causality from GNP to energy use for the US over the period 1947–1974.

Much of the research aimed at exploring the long run relationship and the direction of causality between economic growth and energy use has included several other variables, for example, population, urbanization, financial development, and so forth, to better understand the underlying dynamics

of the relation. Lee et al. [2] included capital stock and labor to explain energy use for some Asian nations. They found that the positive link between economic growth and energy demand gets stronger as relevant variables are included. Ang [3] explored the dynamic causal relationships between GDP and energy consumption in France. He found that economic growth influences energy consumption (and pollution) in the long run, but the relation reverses in the short run in case of France. Apergis and Payne [4–6] and Wolde-Rufael [7] argued that rise in energy demand in African countries is closely linked to income. Population growth creates pressure on rural resources, forces people to move to urban areas, and thus increases energy demand. For sustained economic growth, the increased energy demand over a long period must be met from new sources, or by developing cost-effective alternative energy. Using both bi- and multivariate models for New Zealand, Bartleet and Gounder [8] found causality running from real GDP to energy use. They also found that real GDP and employment exert significant impact on capital formation, where capital stock plays an important role in determining the direction of causality.

The objective of this paper is to examine the existence of long run relationship among energy consumption ($\ln E_t$), financial development ($\ln F_t$), and economic growth ($\ln Y_t$) for Lebanon using monthly data over the period of $2000M_2$–$2010M_{12}$. The ARDL bounds testing approach to cointegration is used to examine a long run relation among the series. The innovative accounting technique (variance decomposition approach and impulse response function) (the methodological framework is based on the studies by Shan [9] and Shahbaz et al. [10]) shows the response of dependent variable to shocks arising in independent variables. We use this method to examine the short run dynamics and the direction of causality. This method is helpful in determining the relative strength of causal relation beyond the chosen time frame [9, 11] and the magnitude of the feedback among the series.

The rest of the paper is organized as follows. Section 2 reviews the relevant literature with focus on the link between economic growth and energy consumption, economic growth and financial development, and financial development and energy consumption. Section 3 discusses the motivation of this study. Section 4 describes the data and estimation strategy. Results of the study are reported in Section 5. Conclusion and policy implications are reported in Section 6.

2. Literature Review

The question of whether energy consumption and/or financial development sway the rate of economic growth of a country or a region has shaped an important query among economists in the literature for some time. This interest is driven primarily from the important policy implications that can be obtained from such studies in relation to the desired action(s) that can accelerate the rate of economic growth and prosperity of a country. Empirical studies in this regard, however, provided conflicting results so economists' views on this issue have not been unanimous. Below, we will provide a brief review of the studies that addressed the impacts of these variables on economic growth.

2.1. Economic Growth and Energy Consumption. Energy forms the lifeblood of the world economy as it is an essential input to producing almost all of the goods and services of the modern global economy. It contributes to economic growth directly as it creates jobs and value associated with extracting, transforming, and distributing of energy. Furthermore and more importantly, this sector's activities relate to and strengthen the rest of the economy as energy forms an input for almost all production processes of goods and services. Supply interruptions of many sources of energy are known to have a great impact as they can harshly impact the economies of almost all countries. In addition, stable and lower energy prices are known to help stimulate the growth rate of any economy. This is because lower energy prices result in increasing disposable income for consumers and lowering

costs for firms. The resulting improved profit margins for firms and higher disposable income for consumers provide incentives for accelerated rates of growth.

One can distinguish between four different hypotheses explaining relationship between the energy consumption and economic growth based on the type of the relationship between both variables [12].

(i) Neutrality hypothesis implies no causality between energy consumption and economic growth. This means that energy conservation may not adversely affect economic growth.

(ii) Conservation hypothesis advocates for an implementation of conservative energy policy, as economic growth would not be slowed down by conservation. Economically, this means that causality is unidirectional running from economic growth to energy consumption.

(iii) Growth hypothesis is supported when unidirectional causality is running from energy consumption to economic growth. The latter indicates that energy conservation may reduce investment and negatively influence economic growth.

(iv) Feedback hypothesis confirms the interdependence between energy consumption and economic growth since both variables affect each other. This encourages the implementation of energy expansionary policies for long run sustainable economic growth.

The energy growth nexus has been extensively investigated since the pioneer work of J. Kraft and A. Kraft [1]. The authors examined the relationship between energy consumption and economic growth in case of the United States. The empirical results revealed that economic growth Granger causes energy consumption. Later on, a large number of empirical studies using different approaches, time periods, and proxy variables have tested this causal relationship in a number of countries. Abosedra and Baghestani [13], Cheng and Lai [14], Soytas and Sari [15], Oh and Lee [16], Jumbe, [12], Fatai et al. [17], Lee et al. [2, 18–20], Chontanawat et al. [21], Narayan and Smyth [22], Apergis and Payne [4], and Bowden and Payne [23] among others have examined this issue for different countries and over various sample periods. Lines of evidence from these empirical studies are still mixed at best and controversial results in terms of the direction of causality and the strength of impact of energy use on economic growth are reported. Some papers documented unidirectional causality from energy consumption to economic growth (growth hypothesis). Narayan and Smyth [22], Apergis and Payne [4], Odhiambo [24], Bowden and Payne [23], Tsani [25], Wang et al. [26], and Yazdan and Hossein [27] found evidence supporting this view.

Other researchers have documented unidirectional causality running from economic growth to energy consumption (conservation hypothesis). This hypothesis is supported by Lise and van Montfort [28], Erdal et al. [29], Huang et al. [30], Mallick [31], Sa'ad [32], Binh [33], Qazi et al. [34], and Soile [35]. Next, the feedback hypothesis is supported if bidirectional causality between energy consumption and growth is found. This is supported by Glasure [36], Lee and Chang [20], Belke et al. [37], Eggoh et al. [38], Marques et al. [39], Kaplan et al. [40], Shahbaz et al. [41], Fuinhas and Marques [42], Zeshan [43], and Shahbaz et al. [10]. The final hypothesis is that of neutrality, where no causality between energy consumption and growth is found, and it is supported by Soytas et al. [44] and Gross [45]. In the case of Lebanon, Dagher and Yacoubian [46] applied the cointegration developed by Johansen [47] and Granger causality by Toda and Yamamoto [48] as well as the VECM Granger causality to examine cointegration and causality relationship between energy consumption and economic growth. Their results indicated that long run relationship exists between the variables and energy consumption and economic growth are bidirectional Granger cause.

2.2. Financial Development and Economic Growth. Since pioneering works of Schumpeter [49], of Goldsmith [50], and recently of McKinnon [51] and Shaw [52], the relation between financial development and economic growth has attracted interest of both theorists and practitioners. Demirguc-Kunt and Levine [53] and Levine [54] provided an extensive literature survey on this topic. Güryay et al. [55], Wolde-Rufael [7, 56], and Shahbaz [57], among others, have used both cross-country and time series data to investigate the relationship between financial development and economic growth.

Financial development has a positive effect on economic growth as it may increase the efficiency of capital accumulation [50] and/or augment the level of saving and consequently investment [51, 52]. By increasing the size of savings and improving the efficiency of investment, financial development leads to higher economic growth [54, 58, 59]. Furthermore, financial development would support financial innovation and promote the adoption of advanced technology [57, 60]. In the related literature, this is known as "supply-leading." In other words, financial development causes economic growth in the sense of Granger. Ibrahim [61] observed that financial development stimulates economic growth in Malaysia. Jalil and Ma [62] found that financial development contributes to economic growth by increasing capital formation in Pakistan and China. Shahbaz [57] showed the same results in case of Pakistan. Coccorese [63] reported that economic growth is Granger cause of financial development. Masih et al. [64] applied long run structural modeling (LRSM) to examine the causality between financial development and economic growth in Saudi Arabia. They validated the existence of supply-side hypothesis. Kar and Mandal [65] also noted that financial development promotes economic growth by enhancing capitalization in India.

"Demand-following" hypothesis suggests that financial development follows economic growth. Patrick [66] pointed out that "demand-following" relationship indicates that real economic activity Granger causes financial development by generating demand for financial services as an economy develops. Liang and Teng [67] investigated the relationship between financial development and economic growth in China using VAR approach. They noted cointegration between the variables and economic growth Granger causes

financial development. Fung [68] finds that positive impact of economic growth on financial development is due to productivity boost. Chukwu and Agu [69] also supported demand-side hypothesis in case of Nigeria. Amarathunga [70] also confirmed the presence of demand-side hypothesis in case of Sri Lanka. In case of South Africa, Odhiambo [71] investigated the relationship between financial development and economic growth by applying the ARDL bounds testing and the VECM Granger causality approaches. The results indicated that the variables are cointegration and financial development is Granger cause of economic growth. Odhiambo [72, 73] investigated the causality between financial development, foreign capital inflows, and economic growth for Tunisia. The empirical results revealed that financial development follows economic growth and that financial development and foreign capital inflows are interdependent. Hassan et al. [74] also reported unidirectional causality running from economic growth to financial development in developing economies. Odhiambo [72, 73] used trivariate model to examine causality between financial development and economic growth by incorporating foreign capital inflows. The empirical evidence validated the existence of demand-side hypothesis in Tanzania.

The feedback effect between financial development and economic growth is also found in the existing literature. For example, Ilhan [75] applied cointegration and error correction method for the relationship between financial development and economic growth. He noted that cointegration exists and that the feedback effect is validated in case of South Africa. Zheng et al. [76] used bivariate framework model and reported that financial development and economic growth are complementary in case of China. The same exercise was conducted by Al-Malkawi et al. [77] in case of UAE and results found a feedback effect between financial development and economic growth. Eslamloueyan and Sakhaei [78] applied generalized least square (GLS) method with cross-section seemingly unrelated regression (SUR) to probe the relationship between financial development and economic growth and confirmed the findings of Husam-Aldin et al. [79]. On the contrary, Bakhouche [80] reported that financial development does not promote economic growth and that economic growth does not contribute to financial development, that is, neutral hypothesis. Gantman and Dabós [81] also supported the view by Bakhouche [80] that financial development and economic growth are independent.

2.3. Financial Development and Energy Consumption. Love and Zicchino [82] reported that financial development impacts on real interest can possibly result in an increase in investment. This in turn can promote economic growth and generate employment opportunities which further increase income. Such impact will increase purchases by consumers especially of durable items [83, 84] which add further to energy use. This shows that the linkages between energy use and economic growth are better understood when we go beyond a simple bivariate framework. Karanfil [85] suggested using stock market capitalization, liquid liabilities, and domestic credit to the private sector, each as share of GDP among the financial variables. Yandan and Lijun [86] examined the impact of financial development on primary energy consumption in Guangdong (China). Their study finds Granger causality running from energy consumption to financial development, while the reverse is insignificant. Sadorsky [83] examined 22 emerging economies (1990–2006) using different indicators of financial development. This included FDI, bank deposits as share of GDP, stock market capitalization as share of GDP, stock market turnover ratio, and total stock market value traded as a share of GDP. His results confirmed that energy consumption is positively linked to economic growth but the impact is small.

Sadorsky [87] investigated the impact of financial development on energy consumption using data of 9 Central and Eastern European frontier economies. He reported that financial development increases energy demand once deposit money and bank assets to GDP, financial system deposits to GDP, liquid liabilities to GDP, and stock market capitalization are used as measures of financial development. Similarly, Shahbaz and Lean [88] examined energy demand for Tunisia and reported results showing that financial development increases energy demand resulting from economic growth. In the case of Malaysia, Tang and Tan [89] examined the relationship between financial development and energy consumption by incorporating relative prices and foreign direct investment (FDI) in energy demand function. They report bidirectional causality between financial development and energy consumption both in the short and the long runs. Islam et al. [90] reported that financial development, economic growth, and population are driving forces to increase energy demand in Malaysia. The feedback effect is also reported between financial development and energy consumption in the long run but financial development Granger causes energy demand in short run.

3. Motivation of the Study

This paper provides an investigation of the relationship between economic growth, financial development, and energy consumption using monthly data over the period of 1993–2010 in case of Lebanon. This is a country with a sectarian-based parliamentary republic located in the Middle East, with a population of approximately 4.2 million. The Lebanese economy is service oriented, with tourism and banking sectors being the main driving force, contributing over 70% of GDP and therefore considered the primary sectors for growth. The banking sector is one of the main pillars of Lebanese economy with a size equivalent to 350% of GDP as of 2009. Lebanese banks benefit from a strong net inflow from both expats and the Gulf States; domestic private sector credit growth has been 19% until October 2011 and the banking system's foreign assets have also grown rapidly, supporting the fact that the banking industry is not affected by the political unrest. About 18% of GDP is contributed by the industrial sector and about 5% by the agriculture. Net remittances from the Lebanese Diaspora living abroad, mainly in the Gulf region, also contribute 5% to the GDP. Tourism industry development in Lebanon dates back to the 1960s when Lebanon's capital, Beirut, was known as "the Paris of the Middle East." This sector contributes notably to the employment in the economy. Employment in that sector as a share of total employment stood at 31.2% in 2005 and is estimated at 38% in 2010. This is not surprising given that the percentage of tourism and travel in GDP stood at 31.2% in 2005 and is estimated at 37.6% in 2010 [91]. This is not unexpected in a country that is known for its diverse atmosphere, earliest history, ancient Roman ruins, preserved castles, and notable mosques and churches, as well as its stunning beaches in the Mediterranean Sea and rugged ski resorts.

The country's economy has faced much challenge owing to its continuous political unrest. The civil war (1975–1990) had a heavy unconstructive impact on the nation, causing the country to have a high budget deficit. Even more recent, the assassination of ex-Prime Minister Rafik Hariri in February 2005, the July 2006 war between Lebanon and Israel, the sit-ins, protests and clashes between the opposing government alliances in 2006 till 2008, and the constant instability and corruption within Lebanon contributed to the huge deficit and the increase in sectarianism. In 2010, growth in Lebanon was stimulated by rising nonresident deposits, an elevated number of tourist arrivals, and a vigorous real estate market. However, due to the regional political chaos, both tourist arrivals and expat's housing demand in the real estate sector have slowed down in 2011. Furthermore, the continued unrest in Syria seems to challenge Lebanon's economic prospects in 2012, especially when 25% of Lebanon's exports are to Syria and 11% of its imports are from Syria.

We think that examining the possible linkages between economic growth, financial development, and energy consumption in Lebanon is justified and needed for three reasons. First, the above features of Lebanese economy justify the need for this study as it will assist the policy makers in the country in assessing their priorities for resource allocations for the country's development. Second, the clear interface of tourism and energy use, in Lebanon, is another important factor for policy makers to consider given the outages of electricity and shortages of some fuels in the country. Third, the authors are not aware of any study of this issue for Lebanon with the exception of that of Dagher and Yacoubian [46]. Our study improves upon theirs, however, since we do not employ a bivariate framework as they did and since our sample is larger than theirs (1980–2009) and excludes the years of the civil war in Lebanon (1975–1990).

The findings should help better understand this relationship that underlies energy use, financial development, and economic growth nexus for Lebanon which will help identify an appropriate policy mix for these sectors in future economic planning for economic growth of that country.

4. The Data and Estimation Strategy

In this study our primary interest lies in the energy consumption-economic growth nexus, financial development-economic growth nexus, and energy consumption-financial development nexus.

The prime hypothesis of energy consumption-economic growth causality postulates that economic growth is impeded by energy conservation policies if causality runs from energy consumption to economic growth. Energy conservation policies do not have adverse impact on economic growth if causality is running from economic growth to energy consumption or no causality is found between both of the variables.

The second hypothesis deals with financial development and economic growth. Financial development boosts investment activities by directing financial resources to new and existing potential ventures which not only enhances domestic production but also raises the rate of economic growth. This implies that financial development drives economic growth. This unidirectional causality running from financial development to economic growth is called supply-side hypothesis. The rise in per capita income or economic growth will increase the demand of financial services for both customers (consumption purpose) and producers (investment purpose) which as a result raises financial development. This shows that economic growth leading to financial development is called demand-side hypothesis.

The third hypothesis deals with energy consumption and financial development nexus. Financial development results in an increase in funds availability to investors for new and existing investment and to consumers to purchase ticket items which directly can increase energy demand. In turn, a rise in demand for financial services associated with financial development can lead to increased demand for energy. On the other hand, an increase in energy consumption as a result of an increase in the level of income may also lead to further financial development as a larger percentage of the population is exposed to the need to use financial services. Finally financial development and energy use may be complementary in nature as a feedback effect may exist.

4.1. The Data. We have monthly frequency data over the period of $2000M_2$–$2010M_{12}$. We use M_2 as a measure of financial development (see Adu et al. [92] for more details). Economic growth is measured by the index of coincident indicator as a measure for real economic activity in Lebanon. The coincident indicator is a broad measure for real economic activity in Lebanon. It is based upon a linear combination of a set of indirect indicators. These consist of imports of petroleum products (18.2 per cent), electricity production (18.6 per cent), cement deliveries (16.5 per cent), number of foreign passengers (11 per cent), total international trade (11.8 per cent), value of checks clearance (12 per cent), and M3 money supply (12 per cent). Finally, we use energy use in millions of KWH as proxy energy consumption. We convert all series into natural logarithm to avoid sharpness and variations in the data. The log-linear specification provides efficient results as compared to simple linear specification. F_t denotes financial development, energy consumption is indicated by E_t, and Y_t is for economic growth in time period t. The data of M_2 and index of coincident indicator are obtained from The Central Bank of Lebanon while energy consumption data is obtained from the Central Administration of Statistics-Lebanon.

4.2. The ARDL Bounds Testing Approach. We employ the autoregressive distributed lag (ARDL) bounds testing approach to cointegration developed by Pesaran et al. [93] to explore the existence of long run relationship between economic growth, financial development, and energy consumption in the presence of structural break. This approach has multiple econometric advantages. The bounds testing approach is applicable irrespective of whether the variables are $I(0)$ or $I(1)$. The results of the ARDL approach are insensitive small sample data test. Moreover, a dynamic unrestricted error correction model (UECM) can be derived from the ARDL bounds testing through a simple linear transformation. The UECM integrates the short run dynamics with the long run equilibrium without losing any long run information. The UECM is expressed as follows:

$$\Delta \ln E_t = \alpha_1 + \alpha_T T + \alpha_E \ln E_{t-1} + \alpha_Y \ln Y_{t-1} + \alpha_F \ln F_{t-1}$$
$$+ \sum_{i=1}^{p} \alpha_i \Delta \ln E_{t-i} + \sum_{j=0}^{q} \alpha_j \Delta \ln Y_{t-j}$$
$$+ \sum_{k=0}^{r} \alpha_k \Delta \ln F_{t-k} \alpha_D D_1 + \mu_t,$$

$$\Delta \ln F_t = \alpha_1 + \alpha_T T + \alpha_E \ln E_{t-1} + \alpha_Y \ln Y_{t-1} + \alpha_F \ln F_{t-1}$$
$$+ \sum_{i=1}^{p} \vartheta_i \Delta \ln F_{t-i} + \sum_{j=0}^{q} \vartheta_j \Delta \ln E_{t-j}$$ (1)
$$+ \sum_{k=0}^{r} \vartheta_k \Delta \ln Y_{t-k} + \vartheta_D D_3 + \mu_t,$$

$$\Delta \ln Y_t = \alpha_1 + \alpha_T T + \alpha_E \ln E_{t-1} + \alpha_Y \ln Y_{t-1} + \alpha_F \ln F_{t-1}$$
$$+ \sum_{i=1}^{p} \beta_i \Delta \ln E_{t-i} + \sum_{j=0}^{q} \beta_j \Delta \ln Y_{t-j}$$
$$+ \sum_{k=0}^{r} \beta_k \Delta \ln F_{t-k} + \beta_D D_2 \mu_t,$$

where Δ is the first difference operator, $\ln Y_t$ is natural log of index of coincident indicator (economic growth), $\ln F_t$ is natural log of M_2 (financial development), and $\ln E_t$ is natural log of energy consumption. D is a dummy for structural break point and μ_t is error term assumed to be independently and identically distributed. The optimal lag structure of the first differenced regression is selected by the Akaike information criteria (AIC). Pesaran et al. [93] suggest F-test for joint significance of the coefficients of the lagged level of variables. For example, the null hypothesis of no long run relationship between the variables is $H_0 : \alpha_E = \alpha_F = \alpha_Y = 0$ against the alternative hypothesis of cointegration $H_a : \alpha_E \neq \alpha_F \neq \alpha_Y \neq 0^3$. Accordingly, Pesaran et al. [93] compute two sets of critical values (lower and upper critical bounds) for a given significance level. Lower critical bound is applied if the regressors are $I(0)$ and the upper critical bound is used for $I(1)$. If the F-statistic exceeds the upper critical value, we conclude in favor of a long run relationship. If the F-statistic falls below the lower critical bound, we cannot reject the null hypothesis of no cointegration. However, if the F-statistic lies between the lower and upper critical bounds, inference would be inconclusive. When the order of integration of all the series is known to be $I(1)$, then decision is made based on the upper critical bound. Similarly, if all the series are $I(0)$, then the decision is made based on the lower critical bound. To check the robustness of the ARDL model, we apply diagnostic tests. These tests are used to check for normality of error term, serial correlation, autoregressive conditional heteroskedasticity, white heteroskedasticity, and the functional form of the empirical model.

4.3. The VECM Granger Causality Test. After examining the long run relationship between the variables, we use the Granger causality test to determine the causality between the variables. If there is cointegration between the series, then the vector error correction method (VECM) can be developed as follows:

$$
\begin{bmatrix} \Delta \ln E_t \\ \Delta \ln Y_t \\ \Delta \ln F_t \end{bmatrix} = \begin{bmatrix} b_1 \\ b_2 \\ b_3 \\ b_4 \end{bmatrix} + \begin{bmatrix} B_{11,1} & B_{12,1} & B_{13,1} \\ B_{21,1} & B_{22,1} & B_{23,1} \\ B_{31,1} & B_{32,1} & B_{33,1} \end{bmatrix}
$$
$$
\times \begin{bmatrix} \Delta \ln E_{t-1} \\ \Delta \ln Y_{t-1} \\ \Delta \ln F_{t-1} \end{bmatrix} + \cdots + \begin{bmatrix} B_{11,m} & B_{12,m} & B_{13,m} \\ B_{21,m} & B_{22,m} & B_{23,m} \\ B_{31,m} & B_{32,m} & B_{33,m} \end{bmatrix}
$$
$$
\times \begin{bmatrix} \Delta \ln E_{t-1} \\ \Delta \ln Y_{t-1} \\ \Delta \ln F_{t-1} \end{bmatrix} + \begin{bmatrix} \zeta_1 \\ \zeta_3 \\ \zeta_3 \end{bmatrix} \times (\text{ECM}_{t-1}) + \begin{bmatrix} \mu_{1t} \\ \mu_{2t} \\ \mu_{3t} \end{bmatrix},
$$
$$(2)$$

where difference operator is Δ and ECM_{t-1} is the lagged error correction term, generated from the long run association. The long run causality is found by examining the significance of coefficient of lagged error correction term using t-test statistic. The existence of a significant relationship in first differences of the variables provides evidence on the direction of short run causality. The joint χ^2 statistic for the first differenced lagged independent variables is used to test the direction of short run causality between the variables. For example, $a_{12,i} \neq 0 \; \forall_i$ shows that economic growth Granger causes energy consumption and economic growth is Granger of cause of energy consumption if $a_{11,i} \neq 0 \; \forall_i$.

5. Results of the Study

The results of descriptive statistics and correlation matrix are reported in Table 1. The Jarque-Bera test statistics reveal that the series of economic growth, energy consumption, and financial development have normal distributions. Our empirical evidence finds that correlations between the variables are positive and strong. For instance, a positive correlation is found between energy consumption and economic growth. Financial development and economic growth are positively correlated and financial development and energy consumption also have positive correlation. The normal distribution of the series leads us to proceed for further analysis.

The next step is to test the unit root properties of the variables. The stationarity level of the variables is very important for policy implications. For example, if energy consumption series is stationary at level, it shows that innovations in energy use have transitory effects and series returns to its trend path; otherwise, innovations show permanent effect on energy consumption if energy consumption series is nonstationary. Similarly, the impact of financial policies adopted to improve financial sector efficiency has temporary effect on financial development if financial development series is stationary at level. Financial policies will have permanent impact on financial development if series is found to be integrated at $I(1)$. The shocks to economy by economic policies have permanent effects if economic growth series is nonstationary which implies that fiscal and/or monetary or any other stabilization policies would only have permanent effects on the real output levels. If economic growth series is stationary, then shocks to economy have transitory effect. Therefore, it is necessary to check the order of integration of the variables before applying the ARDL bounds testing to investigate the long run relationship among the series of interest. ADF and Ng-Perron unit root tests were therefore applied and results are disclosed in Table 2. The results reveal that energy consumption, economic growth, and financial development are nonstationary at level. All the variables are stationary at 1st difference with intercept and trend. This indicates that the series have unique order of integration, that is, $I(1)$.

However, the results of these traditional tests may be biased. These unit root tests do not have information about unknown structural break occurring in the series. The appropriate information about unknown structural breaks would help policy makers in designing a comprehensive energy, financial, and economic policy to enhance economic growth for the long run by considering these structural breaks. This issue is resolved by applying Clemente-Montanes-Reyes unit root test with single and two unknown structural breaks arising in the variables. Our empirical exercise indicated that all the series are nonstationary at level with single structural break in energy consumption, financial development, and economic growth in $2008M_1$, $2009M_1$, and $2006M_6$, respectively. (In this regard, we note that the 2006 Lebanon War started on July 12, 2006, and continued until a United Nations-brokered cease fire on August 14, 2006. Furthermore, Lebanon witnessed a series of protests and sit-ins that began on December 1, 2006. This was led by groups in Lebanon that opposed the US and Saudi-backed government of Prime Minister Fouad Siniora. This ended on May 21, 2008, following the Doha Agreement. On January 25, 2008, a bombing in the Lebanese capital, Beirut, killed a senior intelligence officer, who was involved in the investigation of assassination of former Prime Minister Rafiq Hariri who was killed in 2005. This was followed by a series of bombings and assassinations which have struck Lebanon, most of them occurring in and around the capital, Beirut, during the last few years. Finally, we note that 27 January 2009 marked a historical event in which Syria accepted Lebanon's first ambassador ever to Damascus.) We conclude that all the variables are stationary at first difference accommodating

TABLE 1: Descriptive statistic and correlation matrix.

Variable	Mean	Median	Maxi.	Mini.	Skewness	Kurtosis	Jarq.Bera	$\ln E_t$	$\ln Y_t$	$\ln M_t$
$\ln E_t$	6.7773	6.7719	7.0707	6.5132	0.2122	2.7366	1.3722	1.0000		
$\ln Y_t$	5.1574	5.1439	5.5861	4.7858	0.4778	2.3847	0.7105	0.7197	1.0000	
$\ln M_t$	10.1706	10.1084	10.9920	9.6700	0.9432	3.0181	1.9575	0.7017	0.9118	1.0000

TABLE 2: Unit root analysis.

Variable	ADF unit root at level		ADF unit root at 1st difference	
	T-statistic	Prob. value	T-statistic	Prob. value
$\ln E_t$	−2.9832 (3)	0.1394	−11.3974 (2)*	0.0000
$\ln F_t$	−2.2992 (4)	0.4321	8.8743 (4)*	0.0000
$\ln Y_t$	−1.9300 (4)	0.2547	−17.4153 (3)*	0.0000

Variable	Ng-Perron unit root test			
	MZa	MZt	MSB	MPT
$\ln E_t$	−8.0367 (5)	−1.9982	0.2486	11.3579
$\ln F_t$	−4.2960 (4)	−1.4656	0.3411	21.2112
$\ln Y_t$	−9.6893 (5)	−2.1405	0.2209	9.6774
$\Delta \ln E_t$	−46.2855 (2)*	−4.8075	0.1038	1.9851
$\Delta \ln F_t$	−25.3303 (4)	−3.5588	0.1405	3.5975
$\Delta \ln Y_t$	−24.4338 (5)*	−3.4612	0.1416	3.9365

Note: * represents significance at 1% levels. Lag order is shown in parentheses.

single and two unknown structural breaks confirmed by Clemente-Montanes-Reyes unit root test (see Table 3).

Given that, the computation of ARDL bounds testing is known to be sensitive to lag length selection. As such, inappropriate selection of lag length may produce biased results. Therefore, it is necessary to have exact information about lag order of the series to avoid the problem of biasedness of ARDL F-statistics [94]. We follow AIC criteria for selection of lag length where we found that lag order 4 is suitable for our data sample. The information about lag order is given in Table 4 following AIC criterion.

Table 5 reports the results of the ARDL bounds testing approach to cointegration. We followed the critical bounds produced by Pesaran et al. [93]. The critical bounds generated by Pesaran et al. [93] are suitable for large sample size (T = 500 to T = 40,000). Our findings reveal that calculated F-statistics seem to exceed upper critical bounds at 1% and 5%, respectively, once we treated energy consumption, financial development, and economic growth as dependent variables. This shows that there are three cointegrating vectors confirming the existence of long run relationship among the series in the presence of structural breaks. The ARDL models fulfill the assumptions of normality, ARCH, and functional forms of models. The findings note that error terms are normally distributed, there is no evidence of ARCH, and models are well articulated. (Although the ARDL bounds testing is widely used to analyze the relation between energy consumption and economic growth while ignoring the role of structural breaks in the series, we have overcome this issue by including dummy for structural break points. This confirms that our finds are more reliable and consistent than previous ones.)

The next step is to examine the long run relationship between the variables of which results are shown in Table 6. In energy consumption demand function, we find that financial development adds in energy consumption. A 1 per cent increase in financial development is linked with 0.1272 per cent increase in energy demand. The impact of economic growth on energy consumption is positive as expected and statistically significant at 5 per cent level of significance. We find that 0.2003 per cent energy consumption is increased due to 1 per cent increase in economic growth. Energy consumption has positive effect on financial development and it is statistically significant at 10 per cent level. A 1 per cent increase in energy demand raises financial development by 0.30 per cent. Economic growth has positive and statistically significant impact on financial development. A 1 per cent increase in economic growth is positively linked with financial development by 1.6 per cent. In economic growth empirical model, we find that energy consumption stimulates economic growth and it is statistically significant at 1 per cent level of significance. We note that a 1 per cent increase in energy consumption boosts economic growth by 0.2620 per cent. The positive relationship exists from financial development to economic growth at 1 per cent significance level. A 0.4239 increase in financial development leads to economic growth by 1 per cent. We find that energy consumption and financial development are complementary but energy consumption has stronger impact on financial development and the same inference is for economic growth to financial development. Energy consumption and economic growth are interdependent but economic growth depends on energy consumption. The assumptions of classical linear regression model (CLRM) are fulfilled by energy consumption, financial development, and economic growth models successfully. There is no evidence of nonnormality of error terms or serial correlation and no evidence of ARCH or white heteroskedasticity. The functional form of all the models is well specified.

In the short run (Table 7), we find that financial development and economic growth have positive impact on energy consumption at 5 per cent significance level. The impact of energy consumption and economic growth is also positive and it is statistically significant at 5 per cent level of significance. Finally, energy consumption stimulates economic growth at 1 per cent level of significance. Financial development also adds in economic growth at 5 per cent level. The negative and statistically significant estimates for each of the ECM_{t-1}, 0.6450, 0.0419, and −0.2693 (energy consumption, financial development, and economic growth), lend support to a long run relationship among the series. The short run deviations from the long run equilibrium are corrected by 66.50%, 4.19%, and 26.93% towards long run equilibrium path each month. The diagnostic tests show that error terms

TABLE 3: Clemente-Montanes-Reyes structural break unit root analysis.

Variable	Level data				First difference data			
	T_{B1}	T_{B2}	Test statistics	K	T_{B1}	T_{B2}	Test statistics	K
			Model: trend-break model					
$\ln E_t$	$2008M_4$	—	-2.053	3	$2006M_6$	—	-8.346^*	2
	$2004M_4$	$2008M_4$	-2.909	1	$2006M_6$	$2009M_8$	-7.510^*	1
$\ln F_t$	$2009M_1$	—	-2.372	3	$2005M_1$	—	-4.224^{**}	5
	$2003M_1$	$2008M_9$	-3.487	4	$2005M_1$	$2008M_1$	-5.643^{**}	6
$\ln Y_t$	$2006M_6$	—	-2.450	3	$2006M_6$	—	-6.465^*	5
	$2003M_9$	$2008M_3$	-5.270	4	$2006M_5$	$2006M_9$	-5.711^{**}	6

Note: T_{B1} and T_{B2} are the dates of the structural breaks; K is the lag length; $*$ and $**$ show significance at 1% and 5% levels, respectively.

TABLE 4: VAR lag order selection criteria.

Lag	$\log L$	LR	FPE	AIC	SC	HQ
			VAR lag order selection criteria			
0	245.5326	NA	$4.02e-06$	-3.911816	-3.843584	-3.884099
1	585.0314	657.0944	$1.94e-08$	-9.242442	-8.969512^*	-9.131571
2	600.3055	28.82361	$1.76e-08$	-9.343637	-8.866008	-9.149613
3	615.0278	27.07002^*	$1.60e-08$	-9.435932	-8.753605	-9.158755^*
4	624.1486	16.32916	$1.60e-08^*$	-9.437880^*	-8.550856	-9.077550
5	627.3032	5.495216	$1.76e-08$	-9.343600	-8.251878	-8.900117
6	633.6844	10.80681	$1.85e-08$	-9.301361	-8.004941	-8.774725
7	635.0608	2.264400	$2.10e-08$	-9.178400	-7.677282	-8.568610
8	642.5061	11.88850	$2.16e-08$	-9.153324	-7.447509	-8.460382

$*$ indicates lag order selected by the criterion; LR: sequential modified LR test statistic (each test at 5% level); FPE: final prediction error; AIC: Akaike information criterion; SC: Schwarz information criterion; HQ: Hannan-Quinn information criterion.

of short run models are normally distributed and are free of serial correlation, heteroskedasticity, and ARCH problems for all models. The Ramsey reset test suggests that functional form for the short run models is well specified.

The cumulative sum (CUSUM) and the cumulative sum of squares (CUSUMsq) tests suggest stability of the long and short run parameters (Figures 1 and 2). The graphs of the CUSUM and CUSUMsq test lie within the 5 per cent critical bounds which confirm stability of parameters [95] of the model which is also well specified.

5.1. The VECM Granger Causality. If cointegration is confirmed, there must be uni- or bidirectional causality among the series. We examine this relation within the VECM framework. Knowledge about causality can help craft appropriate energy and financial policies for sustainable economic growth. Table 8 reports results on the direction of long and short run causality. We find feedback relation between energy consumption and economic growth. This implies that economic growth depends upon energy consumption and rise in income per capita further increases energy demand. Therefore, adoption of energy conservation policies will have detrimental impact on economic growth. Our findings suggest the importance of encouraging energy exploring policies. In this regard, we praise the government of Lebanon in concluding its first offshore oil and gas rights auction which has drawn interest from about 100 companies. They have bought

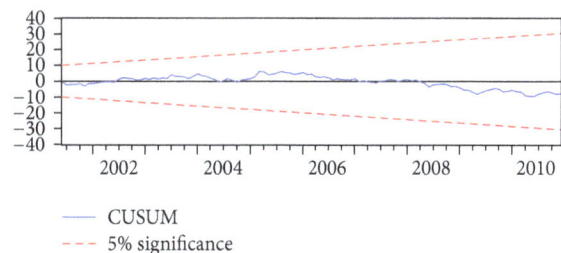

FIGURE 1: Plot of cumulative sum of recursive residuals. The straight lines represent critical bounds at 5% significance level.

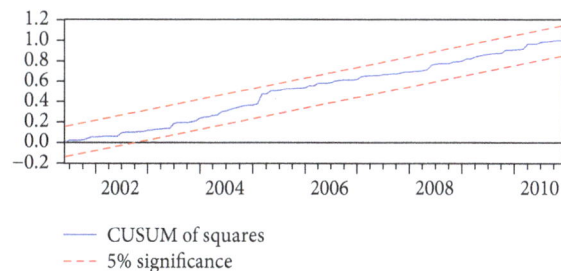

FIGURE 2: Plot of cumulative sum of squares of recursive residuals. The straight lines represent critical bounds at 5% significance level.

geophysical data in preparation for an upcoming bid round for Lebanese offshore energy production rights, which should

TABLE 5: The results of ARDL cointegration test.

| Estimated models | Bounds testing to cointegration | | | Diagnostic tests | | |
	Optimal lag length	F-statistics	Structural break	χ^2_{NORMAL}	χ^2_{ARCH}	χ^2_{RESET}
$F_E\,(E/F,Y)$	4, 4, 4	6.577**	$2008M_4$	0.8936	[1]: 0.3263	[3]: 1.7586
$F_F\,(F/E,Y)$	4, 4, 4	4.730**	$2009M_1$	0.1449	[1]: 0.2831	[2]: 0.4828
$F_Y\,(Y/E,F)$	4, 4, 4	8.719*	$2006M_6$	0.1229	[1]: 0.2005	[1]: 0.0237

| Significant level | Critical values ($T = 132$) | |
	Lower bounds $I\,(0)$	Upper bounds $I\,(1)$
1 per cent level	4.40	5.70
5 per cent level	3.47	4.57
10 per cent level	3.03	4.06

Note: * and ** denote the significance at 1 and 5 per cent levels, respectively. The optimal lag length is determined by AIC. [] is the order of diagnostic tests.

TABLE 6: Long run results.

| Variable | Dependent variable = $\ln E_t$ | | Dependent variable = $\ln M_t$ | | Dependent variable = $\ln Y_t$ | |
	Coefficient	T-statistic	Coefficient	T-statistic	Coefficient	T-statistic
Constant	4.4502*	22.1637	−0.0714	−0.0876	−0.9305**	−2.2596
$\ln E_t$	...	...	0.2968***	1.8336	0.2620*	3.3233
$\ln M_t$	0.1272*	2.6974	...	...	0.4239*	16.4255
$\ln Y_t$	0.2003**	2.2226	1.5958*	16.4255	...	...
R^2	0.5111		0.8358		0.8440	
Adj-R^2	0.5035		0.8332		0.8416	
F-statistic	67.4413*		32.8348*		34.9205*	
Diagnostic test	F-statistic	Probability	F-statistic	Probability	F-statistic	Probability
χ^2 NORMAL	0.2536	0.8808	0.5604	0.6066	4.0656	0.1309
χ^2 SERIAL	1.0040	0.8870	0.9786	0.8900	0.6545	0.5507
χ^2 ARCH	1.5812	0.1704	1.0796	0.4490	0.1962	0.9860
χ^2 WHITE	1.3436	0.2576	0.9372	0.4405	0.5576	0.5060
χ^2 REMSAY	0.1598	0.6899	0.8042	0.7659	2.6240	0.1077

Note: *, **, and *** denote the significance at 1%, 5%, and 10% levels, respectively. χ^2 NORMAL is for normality test, χ^2 SERIAL for LM serial correlation test, χ^2 ARCH for autoregressive conditional heteroskedasticity, χ^2 WHITE for white heteroskedasticity, and χ^2 REMSAY for Remsay Reset test.

TABLE 7: Short run results.

| Variable | Dependent variable = $\Delta \ln E_t$ | | Dependent variable = $\Delta \ln M_t$ | | Dependent variable = $\Delta \ln Y_t$ | |
	Coefficient	T-statistic	Coefficient	T-statistic	Coefficient	T-statistic
Constant	−0.0037	−0.5813	0.0073*	2.3969	0.0023	0.5188
$\Delta \ln E_t$	...	...	0.0639**	2.4847	0.1311*	2.6318
$\Delta \ln M_t$	0.4539**	2.5264	...	...	0.3246**	2.6012
$\Delta \ln Y_t$	0.2735**	2.3395	0.1442**	2.5082	...	...
ECM_{t-1}	−0.6450*	−7.9533	−0.0419*	−2.9695	−0.2693*	−4.5419
R^2	0.3756		0.0904		0.2002	
Adj-R^2	0.3608		0.0689		0.1813	
F-statistic	25.4650*		4.2105*		10.5974*	
Diagnostic test	F-statistic	Probability	F-statistic	Probability	F-statistic	Probability
χ^2 NORMAL	0.9044	0.6362	0.3808	0.6776	0.1352	0.9212
χ^2 SERIAL	0.2916	0.7416	0.2260	0.8500	0.8703	0.4212
χ^2 ARCH	0.0291	0.8646	1.6659	0.1931	0.5830	0.6065
χ^2 WHITE	0.7954	0.5752	0.2217	0.9691	0.4754	0.4012
χ^2 REMSAY	2.2072	0.1142	0.5671	0.5681	1.6040	0.1175

Note: * and ** denote the significance at 1% and 5% levels, respectively. χ^2 NORMAL is for normality test, χ^2 SERIAL for LM serial correlation test, χ^2 ARCH for autoregressive conditional heteroskedasticity, χ^2 WHITE for white heteroskedasticity, and χ^2 REMSAY for Resay Reset test.

TABLE 8: VECM Granger causality analysis.

Dependent variable	Direction of causality						
	Short run			Long run	Joint long-and-short run causality		
	$\Delta \ln E_{t-1}$	$\Delta \ln F_{t-1}$	$\Delta \ln Y_{t-1}$	ECT_{t-1}	$\Delta \ln E_{t-1}, ECT_{t-1}$	$\Delta \ln F_{t-1}, ECT_{t-1}$	$\Delta \ln Y_{t-1}, ECT_{t-1}$
$\Delta \ln E_t$	...	3.7677* [0.0068]	2.5274** [0.0444]	−0.8753* [−6.2656]	...	9.7481* [0.0000]	8.5766* [0.0000]
$\Delta \ln F_t$	2.2240*** [0.0707]	...	3.0709** [0.0191]	−0.0491** [−2.2649]	2.3876** [0.0422]	...	2.5597** [0.0301]
$\Delta \ln Y_t$	2.8445** [0.0256]	1.5706 [0.1868]	...	−0.2735* [−3.6323]	4.2855* [0.0013]	3.5162* [0.0054]	...

Note: *, **, and * * * show significance at 1, 5, and 10 per cent levels, respectively.

occur soon (Lebanon Explores Offshore Energy, By April Yee, 2/25/2013, http://www.moneyshow.com/investing/article/1/TheNational-30654/Lebanon-Explores-Offshore-Energy/). Furthermore, given that Lebanon is very close to some of the major producers of LNG in the world (Qatar, Nigeria, and Egypt), Lebanon should explore this option to meet its energy needs via this source in the short term given the close geographical location of these sources and the low cost of shipment as well as the good relations with these countries.

Financial development and energy consumption Granger cause each other, that is, supporting bidirectional causal relationship. Financial development results in lowering the cost of borrowing to consumers and producers for big durable items and for setting up new businesses. This raises energy demand which leads to further economic expansions. These further attract consumers and raise the demand for financial services and hence financial development. The bidirectional causality between financial development and economic growth also could show the importance of directing monetary policy in Lebanon to enhance capitalization, especially in the energy sector which is highly capital intensive. This is important given our earlier observation of the importance of encouraging energy exploring policies.

In the short run, bidirectional relationship between financial development and energy consumption is found. The feedback effect exists between energy consumption and economic growth and economic growth Granger causes financial development.

5.2. Variance Decomposition Method (VDM). We have used the generalized forecast error variance decomposition method (GFEVDM) using vector autoregressive (VAR) system to test the strength of causal relationship between energy consumption, financial development, and economic growth in case of Lebanon. This is due to the limitations associated with VECM Granger causality test which cannot capture the relative strength of causal relation between the variables beyond the selected time period. The GFEVDM indicates the magnitude of the predicted error variance for a series accounted for by innovations from each of the independent variables over different time-horizons beyond the selected time period. The main advantage of this approach is that it is insensitive with ordering of the variables because such ordering is uniquely determined by VAR system. Further,

the GFEVDM estimates the simultaneous shock effects. Engle and Granger [96] and Ibrahim [97] argued that, with VAR framework, variance decomposition approach produces better results as compared to other traditional approaches.

The results of variance decomposition approach are reported in Table 9. The results indicate that a 67.16 per cent portion of energy consumption is explained by its own innovative shocks while innovative shocks of financial development and economic growth contribute to energy consumption by 21.44 per cent and 11.39 per cent, respectively. The innovative shocks stemming in energy consumption contribute to financial development by 23.37 per cent. The contribution of economic growth to financial development is 15.68 per cent and the rest is explained by innovative shocks on financial development. Economic growth is 23.35 per cent and 39.78 per cent explained by innovative shocks in energy consumption and financial development.

Overall our results show that the feedback effect is found between financial development and energy consumption but it is stronger from energy consumption to financial development. The unidirectional causality is found running from energy consumption and financial development to economic growth, that is, energy-led growth and finance-led growth hypothesis.

The results of IRF are shown in Figure 3 which reveals that the response of energy consumption is positive but minimal after 10th and 6th time horizon due to one standard deviation shock stemming from financial development and economic growth. The response in financial development is positive and strong due to one standard deviation shock in energy consumption and economic growth. Energy consumption and financial development seem to contribute to economic growth. Overall our results are consistent with findings of variance decomposition approach.

6. Conclusion and Policy Implications

This paper explored the relationship between economic growth, financial development, and energy consumption in Lebanon. Using data for $2000M_2$–$2010M_{12}$, we have applied unit root test accommodating single unknown structural break stemming in the series. The ARDL bounds testing is applied to find out cointegration among the variables in the presence of structural breaks. The directions of causal

Response to generalized one S.D. innovations ±2 S.E.

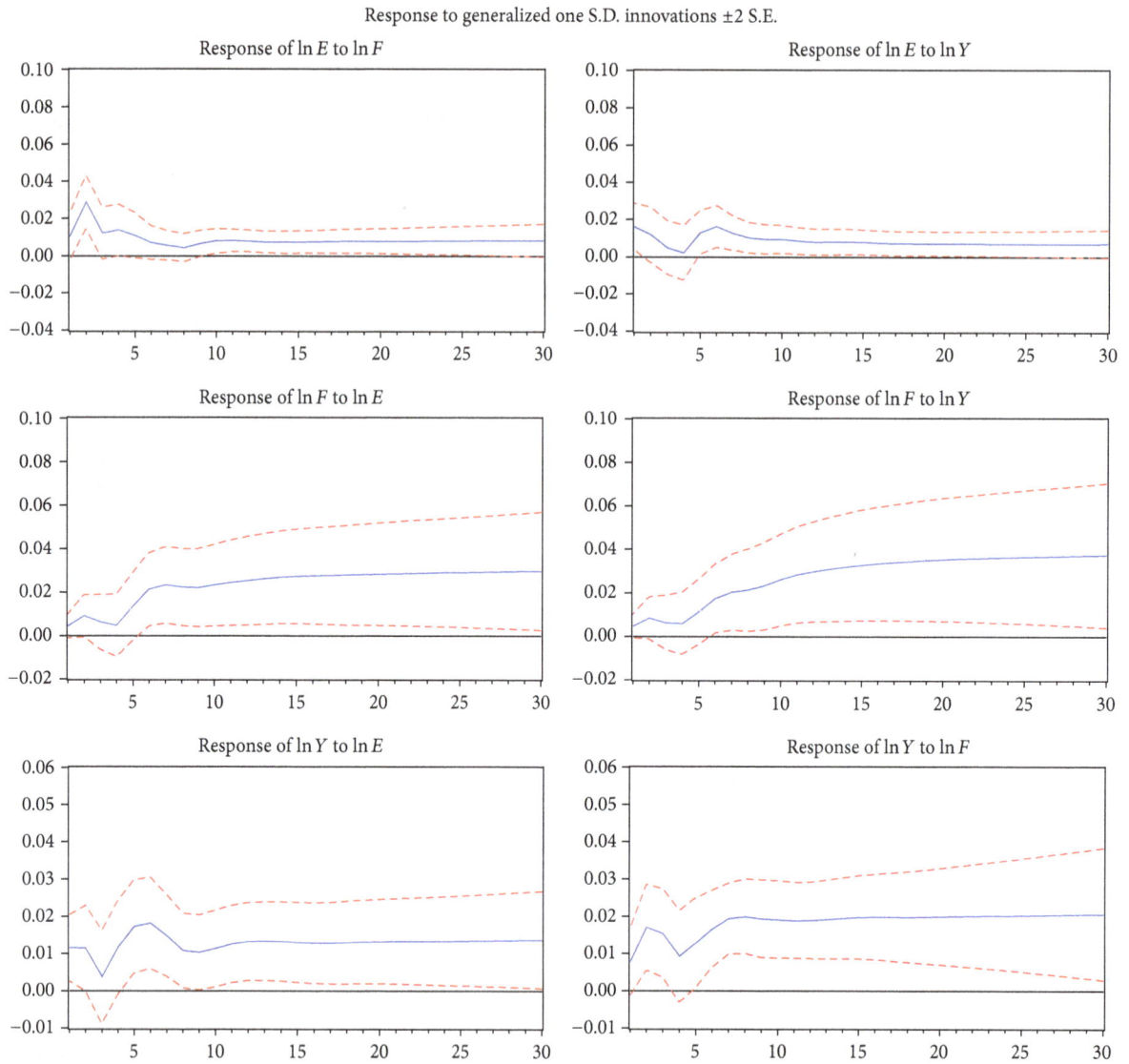

FIGURE 3: Impulse response function.

TABLE 9: Variance decomposition method (VDM).

Horizon	Variance decomposition of $\ln E_t$			Variance decomposition of $\ln F_t$			Variance decomposition of $\ln Y_t$		
	$\ln E_t$	$\ln F_t$	$\ln Y_t$	$\ln E_t$	$\ln F_t$	$\ln Y_t$	$\ln E_t$	$\ln F_t$	$\ln Y_t$
1	100.0000	0.0000	0.0000	2.0421	97.9578	0.0000	5.1366	1.4692	93.394
5	83.1278	14.6550	2.2170	4.0701	95.7476	0.1821	11.9018	11.2447	76.8533
10	77.8984	14.5835	7.5180	14.1938	81.9566	3.8495	17.8004	23.7391	58.4604
15	74.4234	16.3723	9.2042	18.5622	72.7660	8.6716	20.3784	30.3791	49.2424
20	71.7926	18.1021	10.1052	20.9817	66.9868	12.0313	21.6981	34.7860	43.5157
25	69.3898	19.8089	10.8012	22.4392	63.3360	14.2246	22.6609	37.7032	39.6358
26	68.9305	20.1451	10.9243	22.6571	62.7784	14.5644	22.8160	38.1738	39.0101
27	68.4776	20.4773	11.0450	22.8580	62.2635	14.8784	22.9631	38.6143	38.4224
28	68.0320	20.8044	11.1634	23.0436	61.7872	15.1691	23.1020	39.0285	37.8693
29	67.5941	21.1265	11.2793	23.2154	61.3459	15.4386	23.2326	39.4193	37.3480
30	67.1634	21.4439	11.3926	23.3747	60.9362	15.6890	23.3553	39.7885	36.8561

relationships between economic growth, financial development, and energy consumption were examined by applying the VECM Granger causality and robustness of causality results was tested by using innovative accounting approach.

Our results found cointegration between the series in the presence of structural breaks arising in the variables. We found that economic growth and financial development raise energy demand. Energy consumption boosts economic growth and financial development also contributes to economic growth by capitalization enhancing effect. Furthermore, energy consumption and economic growth increase the rate of financial development. The VECM Granger causality analysis revealed that bidirectional causality is found between energy consumption and economic growth. The feedback effect also exists between financial development and energy consumption. Economic growth and financial development are Granger causes of each other. The results by innovative accounting approach are different from the VECM Granger causality test. This may be due to difference in the methodological backgrounds of both techniques. The empirical analysis by innovation accounting approach shows the bidirectional causal relationship between financial development and energy consumption. Economic growth is Granger cause of energy consumption and financial development.

The study, therefore, recommends that, in the short run, policy makers should put more emphasis on developing strategies that would result in achieving higher mobilization of savings in order to boost Lebanese investors' confidence and to also attract more foreign investment in Lebanon. Furthermore, desired financial policy to encounter the rising demand for energy by enhancing the process of capitalization of the energy sector is also very desirable. Our results further caution of the use of policy tools geared towards restricting energy consumption in short run that is a part of national energy policy, as these may result in lower economic growth. Such conservation policies should be taken gradually and carefully as to not negatively impact the growth of the economy. Furthermore, given that Lebanon is very close to some of the major producers of LNG in the world (Qatar, Nigeria, and Egypt), it should explore this option to meet its energy needs via this source in the short term given the close geographical location of these sources and the low cost of shipment as well as the good relations with these countries. However, in the long run, the Lebanese government should shift its focus towards achieving higher economic growth, in order to boost its financial development and to sustain a steady flow of needed energy. In this regard, policy makers should put emphasis on the development of domestic energy resources to protect the country from any undesirable external energy shock given its extensive dependence on energy imports.

Conflict of Interests

The authors declare that there is no conflict of interests regarding the publication of this paper.

References

[1] J. Kraft and A. Kraft, "On the relationship between energy and GNP," *Journal of Energy and Development*, vol. 3, pp. 401–403, 1978.

[2] C.-C. Lee, C.-P. Chang, and P.-F. Chen, "Energy-income causality in OECD countries revisited: the key role of capital stock," *Energy Economics*, vol. 30, no. 5, pp. 2359–2373, 2008.

[3] J. B. Ang, "CO_2 emissions, energy consumption, and output in France," *Energy Policy*, vol. 35, no. 10, pp. 4772–4778, 2007.

[4] N. Apergis and J. E. Payne, "Energy consumption and economic growth: evidence from the Commonwealth of Independent States," *Energy Economics*, vol. 31, no. 5, pp. 641–647, 2009.

[5] N. Apergis and J. E. Payne, "CO_2 emissions, energy usage, and output in Central America," *Energy Policy*, vol. 37, no. 8, pp. 3282–3286, 2009.

[6] N. Apergis and J. E. Payne, "The causal dynamics between coal consumption and growth: evidence from emerging market economies," *Applied Energy*, vol. 87, no. 6, pp. 1972–1977, 2010.

[7] Y. Wolde-Rufael, "Energy consumption and economic growth: the experience of African countries revisited," *Energy Economics*, vol. 31, no. 2, pp. 217–224, 2009.

[8] M. Bartleet and R. Gounder, "Energy consumption and economic growth in New Zealand: results of trivariate and multivariate models," *Energy Policy*, vol. 38, no. 7, pp. 3508–3517, 2010.

[9] J. Q. J. Shan, "Does financial development lead economic growth? The case of China?" *Annals of Economics and Finance*, vol. 1, pp. 231–250, 2006.

[10] M. Shahbaz, H. H. Lean, and A. Farooq, "Natural gas consumption and economic growth in Pakistan," *Renewable and Sustainable Energy Reviews*, vol. 18, pp. 87–94, 2013.

[11] M. Shahbaz, S. M. A. Shamim, and N. Aamir, "Macroeconomic environment and financial sector's performance: econometric evidence from three traditional approaches," *The IUP Journal of Financial Economics*, vol. 1, pp. 103–123, 2010.

[12] C. B. L. Jumbe, "Cointegration and causality between electricity consumption and GDP: empirical evidence from Malawi," *Energy Economics*, vol. 26, no. 1, pp. 61–68, 2004.

[13] S. Abosedra and H. Baghestani, "New evidence on the causal relationship between U.S. energy consumption and gross national product," *Journal of Energy and Development*, vol. 14, pp. 285–292, 1991.

[14] B. S. Cheng and T. W. Lai, "An investigation of co-integration and causality between energy consumption and economic activity in Taiwan," *Energy Economics*, vol. 19, no. 4, pp. 435–444, 1997.

[15] U. Soytas and R. Sari, "Energy consumption and GDP: causality relationship in G-7 countries and emerging markets," *Energy Economics*, vol. 25, no. 1, pp. 33–37, 2003.

[16] W. Oh and K. Lee, "Causal relationship between energy consumption and GDP revisited: the case of Korea 1970–1999," *Energy Economics*, vol. 26, no. 1, pp. 51–59, 2004.

[17] K. Fatai, L. Oxley, and F. G. Scrimgeour, "Modelling the causal relationship between energy consumption and GDP in New Zealand, Australia, India, Indonesia, The Philippines and Thailand," *Mathematics and Computers in Simulation*, vol. 64, no. 3-4, pp. 431–445, 2004.

[18] C.-C. Lee, "Energy consumption and GDP in developing countries: a cointegrated panel analysis," *Energy Economics*, vol. 27, no. 3, pp. 415–427, 2005.

[19] C.-C. Lee, "The causality relationship between energy consumption and GDP in G-11 countries revisited," *Energy Policy*, vol. 34, no. 9, pp. 1086–1093, 2006.

[20] C.-C. Lee and C.-P. Chang, "Energy consumption and GDP revisited: a panel analysis of developed and developing countries," *Energy Economics*, vol. 29, no. 6, pp. 1206–1223, 2007.

[21] J. Chontanawat, L. C. Hunt, and R. Pierse, "Does energy consumption cause economic growth? Evidence from a systematic study of over 100 countries," *Journal of Policy Modeling*, vol. 30, no. 2, pp. 209–220, 2008.

[22] P. K. Narayan and R. Smyth, "Energy consumption and real GDP in G7 countries: new evidence from panel cointegration with structural breaks," *Energy Economics*, vol. 30, no. 5, pp. 2331–2341, 2008.

[23] N. Bowden and J. E. Payne, "The causal relationship between U.S. energy consumption and real output: a disaggregated analysis," *Journal of Policy Modeling*, vol. 31, no. 2, pp. 180–188, 2009.

[24] N. M. Odhiambo, "Energy consumption and economic growth nexus in Tanzania: an ARDL bounds testing approach," *Energy Policy*, vol. 37, no. 2, pp. 617–622, 2009.

[25] S. Z. Tsani, "Energy consumption and economic growth: a causality analysis for Greece," *Energy Economics*, vol. 32, no. 3, pp. 582–590, 2010.

[26] S. S. Wang, D. Q. Zhou, P. Zhou, and Q. W. Wang, "CO_2 emissions, energy consumption and economic growth in China: a panel data analysis," *Energy Policy*, vol. 39, no. 9, pp. 4870–4875, 2011.

[27] G. F. Yazdan and S. S. M. Hossein, "Causality between oil consumption and economic growth in Iran: an ARDL testing approach," *Asian Financial and Economic Review*, vol. 2, pp. 678–686, 2012.

[28] W. Lise and K. van Montfort, "Energy consumption and GDP in Turkey: is there a co-integration relationship?" *Energy Economics*, vol. 29, no. 6, pp. 1166–1178, 2007.

[29] G. Erdal, H. Erdal, and K. Esengün, "The causality between energy consumption and economic growth in Turkey," *Energy Policy*, vol. 36, no. 10, pp. 3838–3842, 2008.

[30] B.-N. Huang, M. J. Hwang, and C. W. Yang, "Causal relationship between energy consumption and GDP growth revisited: a dynamic panel data approach," *Ecological Economics*, vol. 67, no. 1, pp. 41–54, 2008.

[31] H. Mallick, "Examining the linkage between energy consumption and economic growth in India," *The Journal of Developing Areas*, vol. 43, pp. 249–280, 2009.

[32] S. Sa'ad, "Energy consumption and economic growth: causality relationship for Nigeria," *OPEC Energy Review*, vol. 34, pp. 15–24, 2010.

[33] P. T. Binh, "Energy consumption and economic growth in Vietnam: threshold cointegration and causality analysis," *International Journal of Energy Economics and Policy*, vol. 1, no. 1, pp. 1–17, 2011.

[34] A. Q. Qazi, K. Ahmed, and M. Mudassar, "Disaggregate energy consumption and industrial output in Pakistan: an empirical analysis," Discussion Paper, Wuhan University of Technology, 2012.

[35] I. O. Soile, "Energy-economy nexus in Indonesia: a bivariate cointegration," *Asian Journal of Empirical Research*, vol. 2, no. 6, pp. 205–218, 2012.

[36] Y. U. Glasure, "Energy and national income in Korea: further evidence on the role of omitted variables," *Energy Economics*, vol. 24, no. 4, pp. 355–365, 2002.

[37] A. Belke, F. Dobnik, and C. Dreger, "Energy consumption and economic growth: new insights into the cointegration relationship," *Energy Economics*, vol. 33, no. 5, pp. 782–789, 2011.

[38] J. C. Eggoh, C. Bangake, and C. Rault, "Energy consumption and economic growth revisited in African countries," *Energy Policy*, vol. 39, no. 11, pp. 7408–7421, 2011.

[39] A. C. Marques, J. A. Fuinhas, and J. P. Manso, "A quantile approach to identify factors promoting renewable energy in European countrie," *Environmental and Resource Economics*, vol. 49, no. 3, pp. 351–366, 2011.

[40] M. Kaplan, I. Ozturk, and H. Kalyoncu, "Energy consumption and economic growth in Turkey: co-integration and causality analysis," *Romanian Journal of Economic Forecasting*, vol. 2, pp. 31–41, 2011.

[41] M. Shahbaz, M. Zeshan, and T. Afza, "Is energy consumption effective to spur economic growth in Pakistan? New evidence from bounds test to level relationships and Granger causality tests," *Economic Modelling*, vol. 29, no. 6, pp. 2310–2319, 2012.

[42] J. A. Fuinhas and A. C. Marques, "Energy consumption and economic growth nexus in Portugal, Italy, Greece, Spain and Turkey: an ARDL bounds test approach (1965-2009)," *Energy Economics*, vol. 34, no. 2, pp. 511–517, 2012.

[43] M. Zeshan, "Finding the cointegration and causal linkages between the electricity production and economic growth in Pakistan," *Economic Modelling*, vol. 31, no. 1, pp. 344–350, 2013.

[44] U. Soytas, R. Sari, and B. T. Ewing, "Energy consumption, income, and carbon emissions in the United States," *Ecological Economics*, vol. 62, no. 3-4, pp. 482–489, 2007.

[45] C. Gross, "Explaining the (non-) causality between energy and economic growth in the U.S.—a multivariate sectoral analysis," *Energy Economics*, vol. 34, no. 2, pp. 489–499, 2012.

[46] L. Dagher and T. Yacoubian, "The causal relationship between energy consumption and economic growth in Lebanon," *Energy Policy*, vol. 50, pp. 795–801, 2012.

[47] S. Johansen, "Statistical analysis of cointegration vectors," *Journal of Economic Dynamics and Control*, vol. 12, no. 2-3, pp. 231–254, 1988.

[48] H. Y. Toda and T. Yamamoto, "Statistical inference in vector autoregressions with possibly integrated processes," *Journal of Econometrics*, vol. 66, no. 1-2, pp. 225–250, 1995.

[49] J. A. Schumpeter, *The Theory of Economic Development*, Harvard University Press, Cambridge, Mass, USA, 1932.

[50] R. Goldsmith, *Financial Structure and Development*, Yale University Press, New York, NY, USA, 1969.

[51] R. McKinnon, *Money and Capital in Economic Development*, The Brookings Institution, Washington, DC, USA, 1973.

[52] E. Shaw, *Financial Deepening in Economic Development*, Oxford University Press, New York, NY, USA, 1973.

[53] A. Demirguc-Kunt and R. Levine, "Stock markets, corporate finance, and economic growth: an overview," *World Bank Economic Review*, vol. 10, no. 2, pp. 223–239, 1996.

[54] R. Levine, "Financial development and economic growth: views and agenda," *Journal of Economic Literature*, vol. 35, no. 2, pp. 688–726, 1997.

[55] E. Güryay, O. Şafakli, and B. Tüzel, "Financial development and economic growth: evidence from Northern Cyprus," *International Research Journal of Finance and Economics*, vol. 8, pp. 57–61, 2007.

[56] Y. Wolde-Rufael, "Re-examining the financial development and economic growth nexus in Kenya," *Economic Modelling*, vol. 26, no. 6, pp. 1140–1146, 2009.

[57] M. Shahbaz, "A reassessment of finance-growth nexus for Pakistan: Under the investigation of FMOLS and DOLS techniques," *The IUP Journal of Applied Economics*, vol. 1, pp. 65–80, 2009.

[58] R. M. Townsend, "Optimal contracts and competitive markets with costly state verification," *Journal of Economic Theory*, vol. 21, no. 2, pp. 265–293, 1979.

[59] J. Greenwood and B. Jovanovic, "Financial development, growth, and the distribution of income," *Journal of Political Economy*, vol. 98, no. 5, pp. 1076–1107, 1990.

[60] S. Abu-Bader and A. S. Abu-Qarn, "Financial development and economic growth: the Egyptian experience," *Journal of Policy Modeling*, vol. 30, no. 5, pp. 887–898, 2008.

[61] M. H. Ibrahim, "The role of the financial sector in economic development: the Malaysian case," *International Review of Economics*, vol. 54, no. 4, pp. 463–483, 2007.

[62] A. Jalil and Y. Ma, "Financial development and economic growth: time series evidence from Pakistan and China," *Journal of Economic Cooperation Among Islamic Countries*, vol. 29, no. 2, pp. 29–68, 2008.

[63] P. Coccorese, "An investigation on the causal relationships between banking concentration and economic growth," *International Review of Financial Analysis*, vol. 17, no. 3, pp. 557–570, 2008.

[64] M. Masih, A. Al-Elg, and H. Madani, "Causality between financial development and economic growth: an application of vector error correction and variance decomposition methods to Saudi Arabia," *Applied Economics*, vol. 41, no. 13, pp. 1691–1699, 2009.

[65] S. Kar and K. Mandal, "Reexamining the finance—growth relationship for a developing economy: a time series analysis of post-reform India," IEG Working Paper No. 313, 2012.

[66] H. T. Patrick, "Financial development and economic growth in underdeveloped countries," *Economic Development and Cultural Change*, vol. 14, no. 1, pp. 174–189, 1966.

[67] Q. Liang and J.-Z. Teng, "Financial development and economic growth: evidence from China," *China Economic Review*, vol. 17, no. 4, pp. 395–411, 2006.

[68] M. K. Fung, "Financial development and economic growth: convergence or divergence?" *Journal of International Money and Finance*, vol. 28, no. 1, pp. 56–67, 2009.

[69] J. O. Chukwu and C. C. Agu, "Multivariate causality between financial development and economic growth in Nigeria," *African Review of Money Finance and Banking*, pp. 7–21, 2009.

[70] H. Amarathunga, "Finance-growth nexus: evidence from Sri Lanka," *Staff Studies*, vol. 40, pp. 1–36, 2010.

[71] N. M. Odhiambo, "Finance-investment-growth nexus in South Africa: an ARDL-bounds testing procedure," *Economic Change and Restructuring*, vol. 43, no. 3, pp. 205–219, 2010.

[72] N. M. Odhiambo, "Financial deepening, capital inflows and economic growth nexus in Tanzania: a multivariate model," *Journal of Social Science*, vol. 28, pp. 65–71, 2011.

[73] N. M. Odhiambo, "Financial deepening, capital inflows and economic growth nexus in Tanzania: a multivariate model," *Journal of Social Sciences*, vol. 28, no. 1, pp. 65–71, 2011.

[74] M. K. Hassan, B. Sanchez, and J.-S. Yu, "Financial development and economic growth: new evidence from panel data," *Quarterly Review of Economics and Finance*, vol. 51, no. 1, pp. 88–104, 2011.

[75] O. Ilhan, "Finance-growth nexus: empirical evidence from South Africa," *Economic Studies*, vol. 4, pp. 3–17, 2007.

[76] C. Zheng, X. Han, and L. He, "Financial development and economic growth: evidence from cointegration and granger causality tests," in *Proceedings of the 2nd IEEE International Conference on Advanced Management Science (ICAMS '10)*, pp. 237–241, Chengdu, China, July 2010.

[77] H.-A. N. Al-Malkawi, H. A. Marashdeh, and N. Abdullah, "Financial development and economic growth in the UAE: empirical assessment using ARDL approach to co-integration," *International Journal of Economics and Finance*, vol. 4, no. 5, pp. 105–115, 2012.

[78] K. Eslamloueyan and E. A. Sakhaei, "The short run and long run causality between financial development and economic growth in the Middle East," *Iranian Journal of Economic Research*, vol. 16, pp. 61–76, 2011.

[79] H.-A. N. Al-Malkawi, H. A. Marashdeh, and N. Abdullah, "Financial development and economic growth in the UAE: empirical assessment using ARDL approach to co-integration," *International Journal of Economics and Finance*, vol. 4, no. 5, pp. 105–115, 2012.

[80] A. Bakhouche, "Does the financial sector promote economic growth? A case of Algeria," *Savings and Development*, vol. 31, no. 1, pp. 23–44, 2007.

[81] E. R. Gantman and M. P. Dabós, "A fragile link? A new empirical analysis of the relationship between financial development and economic growth," *Oxford Development Studies*, vol. 40, no. 4, pp. 517–532, 2012.

[82] I. Love and L. Zicchino, "Financial development and dynamic investment behavior: evidence from panel VAR," *Quarterly Review of Economics and Finance*, vol. 46, no. 2, pp. 190–210, 2006.

[83] P. Sadorsky, "The impact of financial development on energy consumption in emerging economies," *Energy Policy*, vol. 38, no. 5, pp. 2528–2535, 2010.

[84] N. G. Mankiw and W. Scarth, *Macroeconomics*, Worth, New York, NY, USA, 2008.

[85] F. Karanfil, "How many times again will we examine the energy-income nexus using a limited range of traditional econometric tools?" *Energy Policy*, vol. 37, no. 4, pp. 1191–1194, 2009.

[86] Yandan and Z. Lijun, "Financial development and energy consumption: an empirical research based on Guangdong Province," in *Proceedings of the International Conference on Information Management, Innovation Management and Industrial Engineering (ICIII '09)*, pp. 102–105, December 2009.

[87] P. Sadorsky, "Financial development and energy consumption in Central and Eastern European frontier economies," *Energy Policy*, vol. 39, no. 2, pp. 999–1006, 2011.

[88] M. Shahbaz and H. H. Lean, "Does financial development increase energy consumption? The role of industrialization and urbanization in Tunisia," *Energy Policy*, vol. 40, no. 1, pp. 473–479, 2012.

[89] C. F. Tang and B. W. Tan, "The linkages among energy consumption, economic growth, relative price, foreign direct investment, and financial development in Malaysia," *Quality and Quantity*, vol. 48, no. 2, pp. 781–797, 2014.

[90] F. Islam, M. Shahbaz, A. U. Ahmed, and M. M. Alam, "Financial development and energy consumption nexus in Malaysia: a multivariate time series analysis," *Economic Modelling*, vol. 30, no. 1, pp. 435–441, 2013.

[91] R. Lanquar, "Tourism in the Mediterranean: scenarios up to, 2030," MEDPRO Report 1, CASE—Center for Social and Economic Research, 2011.

[92] G. Adu, G. Marbuah, and J. T. Mensah, "Financial development and economic growth in Ghana: does the measure of financial development matter?" *Review of Development Finance*, vol. 3, no. 4, pp. 192–203, 2013.

[93] M. H. Pesaran, Y. Shin, and R. J. Smith, "Bounds testing approaches to the analysis of level relationships," *Journal of Applied Econometrics*, vol. 16, no. 3, pp. 289–326, 2001.

[94] M. Shahbaz, "Income inequality-economic growth and non-linearity: a case of Pakistan," *International Journal of Social Economics*, vol. 37, no. 8, pp. 613–636, 2010.

[95] R. L. Brown, J. Durbin, and J. M. Ewans, "Techniques for testing the constance of regression relations overtime," *Journal of Royal Statistical Society*, vol. 37, pp. 149–172, 1975.

[96] R. F. Engle and C. W. Granger, "Co-integration and error correction: representation, estimation, and testing," *Econometrica*, vol. 55, no. 2, pp. 251–276, 1987.

[97] M. Ibrahim, "Sectoral effects of monetary policy: evidence from Malaysia," *Asian Economic Journal*, vol. 19, no. 1, pp. 83–102, 2005.

17

Gearbox Fault Diagnosis of Wind Turbine by KA and DRT

Mohammad Heidari

Department of Mechanical Engineering, Abadan Branch, Islamic Azad University, Abadan, Iran

Correspondence should be addressed to Mohammad Heidari; moh104337@yahoo.com

Academic Editor: Nand Kishor

The spectral kurtosis analysis (KA) is used to select the filter parameters (FPs) combined with the application of the demodulation resonance technique (DRT) for a gearbox fault diagnosis (FD) of wind turbine. Based on the proposed method, the FPs can be selected automatically according to the kurtosis maximization principle. By changing of the shaft speed under the variable loads conditions, the natural frequency (NF) of the gearbox will be shifted and will affect the accuracy of the detection of the faults. So, the effect of the external loads on the NF of the gearbox is examined based on the simulation of the gearbox. In addition, the fast kurtogram (FK) combined with the demodulated resonance technology is used to process the simulated faulty signal of a gearbox. The results show that the FD of the gearbox is modified by correcting the NF shifts due to the variation of the operating loads.

1. Introduction

For a fault signal with wide frequency band, the inherent vibration of the system can be induced. That is, the fault signal with wideband (WB) will resonate the structure and the sensor itself in its natural vibration (NV) mode; it must also include the fault source signal, the NV signal of the tested structure system, and the actual vibration signal of the sensor itself. Analyzing the signals needs to choose a high frequency (HF), NV as the work goal by using the band-pass filter (BPF) to separate NB and then through the envelope detection to separate the fault information, so the system faults can be diagnosed by spectrum analysis. In the process of selecting the BPF, the center frequency (CF) of the BPF should be equal to the corresponding NF. However, this artificial method has some limitations in the real applications. Zhang et al. [1] studied the characteristics of demodulated resonance technique and applied it on train and off train failure diagnosing on power car and passenger trains bogie. Barszcz and Sawalhi [2] presented a method for FD of rolling bearings based on combining EEMD adaptive denoising with adaptive demodulated resonance. In the analysis of a real model and nondamping model, the NF and the vibration mode will be affected by the variation of mass and stiffness of the system. For the FD of gear, the changes of loads and speeds will produce dynamic rotating prestress, and this will lead to the shift of the nature frequency in the gear system. In kurtogram research, simulation and experimental verification of FD for bearing were carried out by Wang [3]; they applied the FK algorithm to the DR successfully. McDonald et al. [4] solved the problem of the difficulty of choosing the parameters in traditional DR by the method of acoustic emission in which the SVD and FK algorithm were applied to bearing FD. Heidari et al. [5] applied the spectral kurtosis (SK) method based on LMD to gear fault diagnosis; during the process, LMD has been used to obtain the different time domain distribution of signal, followed by determining the maximum kurtosis of the different channels in the time domain, according to the kurtosis maximization principle to determine the FPs. However, the effect of operating conditions in the diagnostic process is rarely considered. The diagnosis accuracy will be improved by considering the influence of the operating conditions. In this study, the influence of varying load and rotational speed on the NF of wind turbine gearbox is studied by calculating the prestress mode of the gearbox. The automatic acquisition method of BPF in CF and bandwidth (BW) in the DRT is validated by using the SK analysis method. Signal spectrum kurtosis index can reflect the signal frequency and indicate the transient impact strength, so it can achieve the effect of self-adaptation; the envelope analysis is simplified; and the diagnosis result is more accurate.

TABLE 1: NF of the rotating system at different rotating speeds.

		Speed (RPM)							
		0	100	500	1500	2000	2500	5000	10000
	1	316.1	356.8	368.8	386.6	399.8	415.5	425.8	489.5
	2	324.5	369.7	389.7	388.5	407.5	439.6	458.8	509.5
	3	332.1	372.1	421.9	424.5	435.6	479.8	501.9	578.4
	4	355.2	383.2	463.1	463.6	486.5	512.6	555.7	599.7
NF of each order f/Hz	5	376.4	409.5	475.3	478.7	498.8	556.8	603.5	605.5
	6	426.5	434.8	527.2	519.6	588.3	637.1	678.6	751.4
	7	468.7	601.5	668.6	638.8	691.1	732.8	783.7	868.6
	8	529.4	612.3	706.8	660.4	703.3	789.1	808.2	932.6
	9	580.3	633.3	743.3	693.7	756.3	827.1	896.5	1031.3
	10	604.4	698.3	772.5	731.5	792.5	894.1	935.5	1072.3

2. The Influence of Prestress on the Inherent Mode of Gear

To determine the initial stress using the changing in load when the prestressed model analysis is carried out and the static structure analysis is required before the mode analysis of the prestress [7], the equation can be expressed as

$$[K]\{X\} = [F]. \tag{1}$$

In (1), $[K]$ is the stiffness matrix (SM) and $[F]$ is an external load matrix to determine the stress SM for structural analysis $\sigma_0 \to [S]$; the modal analysis equation with prestress can be expressed as follows [8]:

$$\left([K + S] - \omega_1^2 [M]\right) \varphi_1 = 0. \tag{2}$$

σ_0 and $[S]$ are the stress displacement based on static analysis and the prestress effect matrix, respectively. In the gearbox, the input speeds will be varied in the gear transmission system under variable load conditions. For the wind turbine, the impeller is driven by the wind energy to obtain the mechanical energy to drag the gearbox and then drive the generator shaft to rotate at high speed [9]; additional mass and rotational speed changes that were caused by wind impact will generate rotational prestress and change the NF [10]. So, for a gearbox, the transmission model of the gearbox is set up; the effects of rotational speed and other factors on the NF and vibration modes are studied under the premise of considering the inertia and the prestress of the system; the NFs and vibration modes of gear at different rotational speeds are calculated, that is, 100 RPM, 500 RPM, 800 RPM, 1200 RPM, 1500 RPM, 5000 RPM, and 10000 RPM. Table 1 shows the first 10 NFs of the gear at different speeds. Table 1 shows that the NF of each order is also varied alongside the changing of the operating speeds.

Table 1 demonstrates that the dynamic rotating prestress will lead to the variation of NF. Therefore, the influence of the variable loads on the NF of the system should be considered for the accuracy of FD results obtained. For the gearbox system, because of the random characteristics of the system, the input shaft speed and load of the gearbox are dynamic and will affect the characteristics of the NFs, so the changes of actual frequency during the analysis need to be taken into account.

3. RD and FK

3.1. RDT. RDT is a method developed from the vibration detection and analysis technology, which is based on the principle of DR for the impact FD [11]. In general, for the fault signal with WB, the intensity will be greatly improved at the NF of the structure, and the HF can be separated from the NV through a filter. Through analyzing the envelope of the signal after band-pass filtering, it can get a pulse whose frequency is consistent with the impact of failure. After the envelope processing, the signal is filtered to remove the residual HF interference signal, and the fault signal component with low frequency is reserved. Compared with the calculation of the characteristic frequency of the components, the specific reason for the failures can be judged using power spectrum analysis. The basic principle of DR is [12] as follows: (1) separating the HF vibration, because the wide frequency band signal is greatly enhanced at the resonant frequency of the acceleration sensor, with the design of a BPF whose CF is equal to the NF to separate the HF NB; (2) getting a pulse train from the filtered resonance signal by envelope demodulation whose frequency is equal to the fault signal; (3) filtering the envelope detection signal by low-pass filter (LPF), removing HF noise signal, and retaining the low frequency signal; (4) analyzing the power spectrum and fault characteristics, so as to extract the fault features. For the BPF design, the center frequency and BW need to be determined properly to obtain an accurate analysis result. However, the artificial method of determining the parameters of BPF has a great challenge and limitation. The FK algorithm can be adopted to realize the optimized parameters for the BPF design and hence could greatly improve the diagnostic efficiency and accuracy.

3.2. FK. The basic idea of SK is to calculate the kurtosis value for each frequency line so as to find the impact of the frequency band. The bigger the absolute value of kurtosis, the more serious the fault.

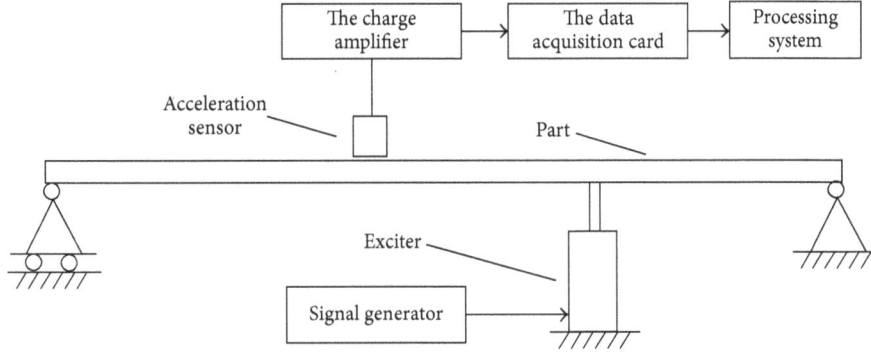

FIGURE 1: The diagram of test rig.

The calculation equation is as follows [13]:

$$\text{SK}(f) = \frac{\left\langle |H(t,f)|^4 \right\rangle}{\left\langle |H(t,f)|^2 \right\rangle^2} - 2 \quad (f \neq 0), \tag{3}$$

$$H(t,f) = \int_{-\infty}^{+\infty} \left[x(\tau) r^*(\tau - t) \right] e^{-j2\pi f\tau} dt. \tag{4}$$

In (4), $H(t,f)$ is the complex envelope of vibration signal in time-frequency domain, $r(t)$ is the time window, SK index is zero when $x(t)$ is a stationary signal, and $x(t)$ is a nonstationary signal when there is noise signal $b(t)$ interference. Equation (5) shows the calculation of SK:

$$K_{x+b}(f) = \frac{K_x(f)}{[1 + \rho(f)]^2}. \tag{5}$$

In (5), $\rho(f)$ is the reciprocal of signal-to-noise ratio (SNR) in the equation. Equation (5) shows that larger $\rho(f)$ will lead to smaller SK index that cannot reflect the features of shock. So, the kurtogram is introduced for the analysis. The basic idea is that because $K_{x+b}(f)$ is determined by f and the frequency resolution, this is the same as the principle of how to choose the CF and BW of the BPF in the DR. Therefore, the kurtogram algorithm is applied to the selection of the BPF parameters of the resonant demodulation. For the kurtogram algorithm, constructing a filter model whose BW is 1/4,

$$h_0(n) = h(n) e^{(jn\pi/4)} \quad \left(f \in \left[0, \frac{1}{4} \right] \right), \tag{6}$$

$$h_1(n) = h(n) e^{(j3n\pi/4)} \quad \left(f \in \left[\frac{1}{4}, \frac{1}{2} \right] \right). \tag{7}$$

In (6) and (7), $h_0(n)$, $h_1(n)$, and $h(n)$ are LPF, high-pass filter (HPF), and LPF whose cutoff frequency is 1/8, respectively. By filtering the signal $x(n)$ by the LPF and HPF, $c_k^i(n)$ is the short time Fourier transform coefficient of the first i filter in the k layer that is used as input again for filtering. $c_k^i(n)$ is the short time Fourier transform of $x(n)$ where CF is $f_i = (i + 2)^{-1} 2^{-(k+1)}$ and frequency resolution is $(\Delta f)_k = 2^{-(k+1)}$ and $i = (0, 1, \ldots, 2^{-(k+1)})$.

3.3. Applied FK to DR. Using $c_k^i(n)$ instead of $H(t,f)$ in (1),

$$\text{SK}(f) = \frac{\left\langle |c_k^i(n)|^4 \right\rangle}{\left\langle |c_k^i(n)|^2 \right\rangle} - 2 \quad \left(i = 0, 1, \ldots, 2^{k-1} \right). \tag{8}$$

The SK index becomes maximum through the FK calculation to obtain the CF and frequency resolution. SK index is used as the CF and frequency resolution of the BPF in DR, and then the fault characteristic frequency can be obtained by filtering and analysis [14].

4. Filtered CF Calibration

The CF of the filter which is automatically captured is close to the NF of the structure; modifying the filter CF according to the actual NF of the structure can make the results more accurate. Firstly, the quality and SM of the system are calculated when calculating the NF of the system, and SM is also divided into the time-varying SM and average SM. So, the average SM is used to find the NF of the system and the time-varying SM is used as the parameters of the system. Based on the above parameters, the NF can be calculated as in [15]. In order to measure the NF of the system more accurately, it is necessary to carry out the vibration test of the structure. Based on the measured amplitude-frequency curve, we determine the NF and damping ratio of the system. Figure 1 shows the test rig.

The exciter is applied to the excitation of different frequencies near the theoretical value of the NF, and the frequency is the NF of the system when the amplitude reaches the maximum value by referring to the theoretical value of NF in Figure 2 [16]. The excitation device gradually increases the excitation frequency on the measured part, and, with the frequency increasing, the amplitude of the measured part is gradually increased between 300 Hz and 550 Hz, and the amplitude of the sample is gradually reduced when the excitation frequency is greater than 550 Hz. Therefore, the amplitude of the test piece reaches its maximum value at 550 Hz, which inferred the notion that the NF of the measured part is about 550 Hz.

The NF of the system can be measured, and we extract the HF of NV with maximum value of kurtosis as the CF for the BPF and then extract the fault characteristic frequency

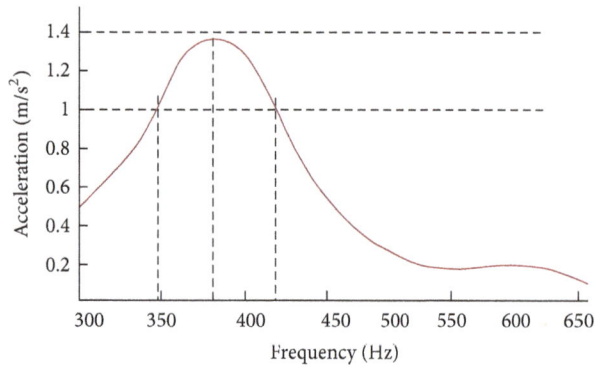

FIGURE 2: Vibration curve [6].

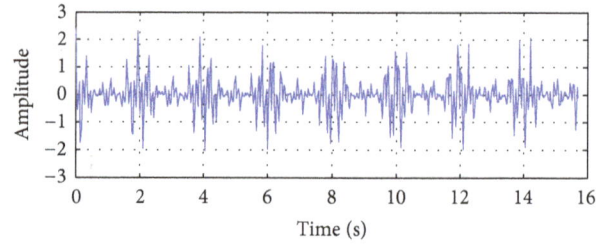

FIGURE 3: Impulse signal.

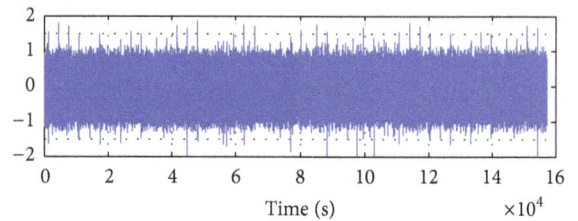

FIGURE 4: Vibration signal with noise.

to judge the feature of the faults. However, the NF of the system is measured under the condition of no stress state, and the data of resonance demodulation is measured under a certain working condition with stress; it is necessary to exclude the frequency increment when choosing the FPs by resonance demodulation to get more accurate results. With the increasing of the order, the compensation value of NF is increased, and the compensation value of each step also increased with the increasing of the prestress. The greater the prestress, the greater the difference that needs to be compensated.

5. Simulation Study

5.1. Establish Fault Model.
The model should be built to meet the following relationship with the typical case of 1.5 MW wind turbine gearbox: the modulus of the ring, the planet gear, and the sun gear must be the same in order to ensure that the gear train can have a correct assembly relationship [6]. In the simulation study, cut off one tooth of the planet gear to simulate the whole tooth broken fault.

5.2. Signal Simulation.
The model is imported into ADAMS and the material properties are 45 steel, the density is $7.801E-006 \, \text{kg/mm}^3$, the elastic modulus is $2.07E + 005 \, \text{N/mm}^2$, and Poisson's ratio is 0.29. Contact parameters are set as follows: IMPACT-function-based contact is used to determine the contact force and ADAMS/Solver using the IMPACT function in ADAMS library function is used to calculate the contact force; the principle of parameters is set as follows: (1) material stiffness is specified; the higher the stiffness value, the more difficult the integral solution; (2) exponent force is used to calculate the value of the contribution of the material stiffness to the instantaneous stress. Taking 1.5 or more than this value, the range becomes greater than or equal to 1. A value of 2 or even 3 is desirable for rubber, whereas for metals values are often taken from 1.3 to 1.5; (3) define the damping properties of the contact material whose range is greater than or equal to 0, usually taking from 0.1% to 1% of the value of the stiffness; (4) the penetration values are defined for full damping. The damping coefficient is zero when penetration is

zero; ADAMS/Solver uses the three STEP functions to solve the damping coefficient between the two points. Its range is greater than or equal to 0. Simulation is carried out under no-load conditions, where the input speed is set to 60 RPM, the simulation time is 1 s, the step size is 0.001, the average value of the output shaft speed is 2125 RPM, theoretical output average speed is 2130 RPM, and the error rate is 0.23%. This proved the accuracy of the simulation model. The impact signal in the time domain can be obtained by the optimization process as shown in Figure 3.

The fault simulation signal is processed through adding random noise with SNR of −9 db. This presents the notion that the background noise energy is much greater than the failure impact signal, and the fault signal is submerged by the noise signal, and the actual impact of the fault line is very difficult to observe. It is more similar to the actual measured signals in the real applications. The waveform of vibration signal with noise in time domain is shown in Figure 4.

5.3. Simulation and Verification.
The signal is calculated in Figure 4 by FK, and the calculated map is shown in Figure 5.

It can be seen from Figure 5 that the CF is 375.1 Hz where the spectrum kurtosis index is maximum, and the corresponding frequency resolution is 25.6 Hz. The analog signal is filtered using the 375.1 Hz and 25.6 Hz as the CF and BW of the resonant demodulation; the envelope of the filtered signal in the time domain is shown in Figure 6; and its spectrum is shown in Figure 7. Based on the spectrum of the resonance demodulation results, the fault frequency of the simulated signal can be obtained which is 27.6 Hz when the operating speed is 1500 RPM as shown in Figure 7.

Figure 8 shows the final demodulation spectrum at different rotation speeds of 2000 RPM, 2500 RPM, 5000 RPM, and 10000 RPM.

When the rotation speeds of the planetary gear are 2500 RPM, 5000 RPM, 2000 RPM, and 10000 RPM, respectively, the corresponding fault frequencies will be 36.5 Hz,

fb-kurt.2, K_{max} = 613.3 @ level 1, BW = 25.6 Hz,
f_c = 375.1 Hz

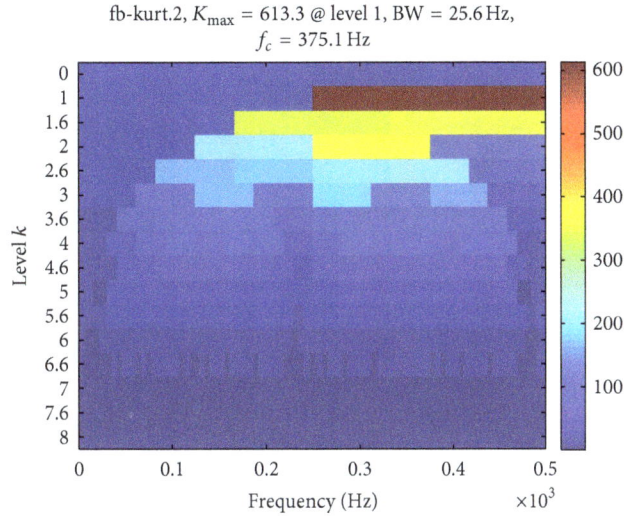

FIGURE 5: The calculation of the kurtosis.

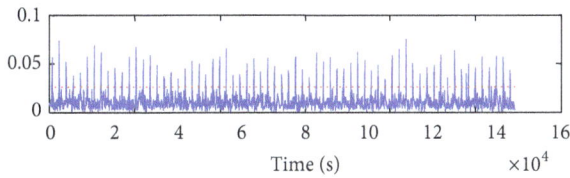

FIGURE 6: Time-frequency envelope.

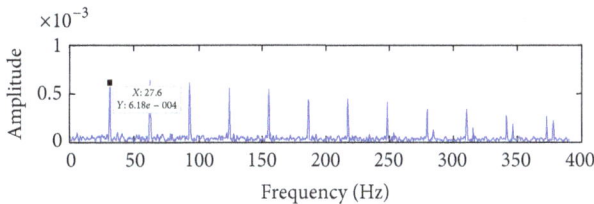

FIGURE 7: The spectrum of resonance demodulation.

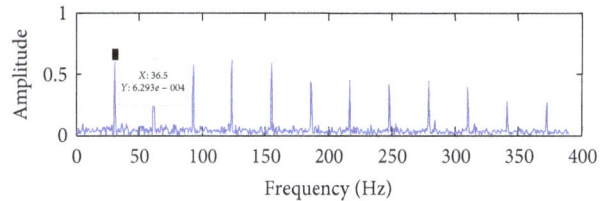

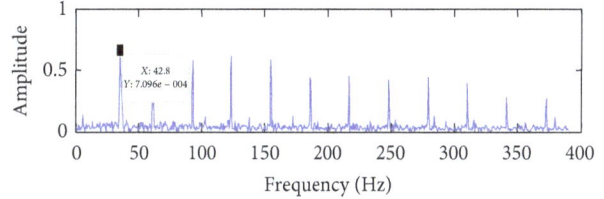

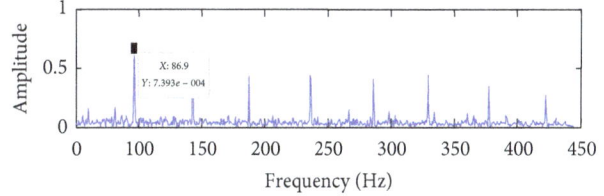

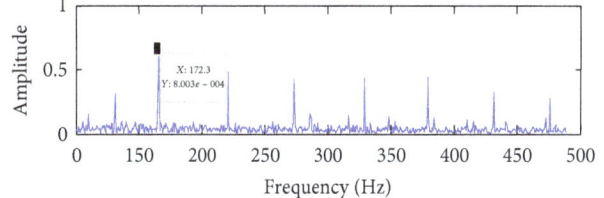

41.8 Hz, 86.9 Hz, and 172.3 Hz as shown in Figure 8. Calculate the fault frequency again after correction of the inherent frequency of the FPs in the process of SK calculation. The fault characteristic frequency at each speed is obtained in Figure 9. It can be seen from Figure 9 that when the rotation speeds are 2000 RPM, 1500 RPM, 2500 RPM, 5000 RPM, and 10000 RPM, the corrected corresponding fault frequencies will be 25.4 Hz, 34.8 Hz, 41.9 Hz, 84.6 Hz, and 168.2 Hz. And then the theoretical value of the fault characteristic frequency at different speeds is obtained according to the fault characteristic frequency calculation equation, compared to the unmodified fault characteristic frequency, the modified fault characteristic frequency, and the theoretical value. From the results, it can be seen that the modified fault characteristic frequency curve is closer to the theoretical value. The error of the correction value and the theoretical value is 1.86%. From the above, to a certain extent, we can know that the diagnosis is more accurate, for the error between the modified frequency value and the theoretical value is smaller.

FIGURE 8: FFT analysis of RD at 2000 RPM, 2500 RPM, 5000 RPM, and 10000 RPM.

6. Summary

In this paper, the effect of the rotating prestress on the NF of a wind turbine gearbox was studied, and the FK algorithm is applied to the resonance demodulation in the analysis of the impact of rotating machinery. It is shown that this method can obtain the central frequency and BW of the

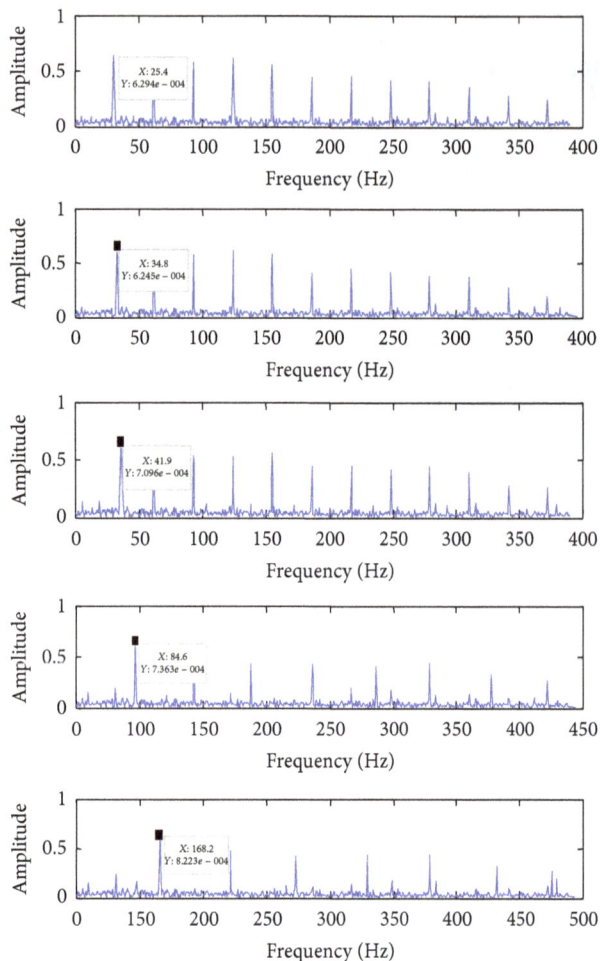

FIGURE 9: FFT analysis of RD after frequency correction at 1500 RPM, 2000 RPM, 2500 RPM, 5000 RPM, and 10000 RPM, respectively.

BPF automatically. Through the analysis, it is found that the prestress has a certain effect on the CF, which eliminated the effect of prestress on center frequency after it is automatically obtained; the analysis shows that the demodulation results are more accurate for the gearbox condition monitoring and FD.

Competing Interests

The author declares that there are no competing interests.

References

[1] X. Zhang, J. Kang, L. Xiao, J. Zhao, and H. Teng, "A new improved Kurtogram and its application to bearing fault diagnosis," *Shock and Vibration*, vol. 2015, Article ID 385412, 22 pages, 2015.

[2] T. Barszcz and N. Sawalhi, "Fault detection enhancement in rolling element bearings using the minimum entropy deconvolution," *Archives of Acoustics*, vol. 37, no. 2, pp. 131–141, 2012.

[3] H. Wang, J. Chen, G. Dong, and F. Cong, "Application of resonance demodulation in rolling bearing fault feature extraction based on fast computation of kurtogram," *Journal of Vibration and Shock*, vol. 32, no. 1, pp. 35–48, 2013.

[4] G. L. McDonald, Q. Zhao, and M. J. Zuo, "Maximum correlated Kurtosis deconvolution and application on gear tooth chip fault detection," *Mechanical Systems and Signal Processing*, vol. 33, pp. 237–255, 2012.

[5] M. Heidari, H. Homaei, H. Golestanian, and A. Heidari, "Fault diagnosis of gearboxes using wavelet support vector machine, least square support vector machine and wavelet packet transform," *Journal of Vibroengineering*, vol. 18, no. 2, pp. 860–875, 2016.

[6] D. Qin, Z. Xing, and J. Wang, "Optimization design of system parameters of the gear transmission of wind turbine based on dynamics and reliability," *Chinese Journal of Mechanical Engineering*, vol. 44, no. 7, pp. 24–31, 2008.

[7] L. Cai, S. Ma, Y. Zhao, Z. Liu, and W. Yang, "Finite element modeling and modal analysis of heavy-duty mechanical spindle under multiple constraints," *Journal of Mechanical Engineering*, vol. 48, no. 3, pp. 165–173, 2012.

[8] X. Qiu, Q. Han, and F. Chu, "Review on dynamic analysis of wind turbine geared transmission systems," *Journal of Mechanical Engineering*, vol. 50, no. 11, pp. 23–36, 2014.

[9] X. Chen, J. Li, H. Cheng, B. Li, and Z. He, "Research and application of condition monitoring and fault diagnosis technology in wind turbines," *Journal of Mechanical Engineering*, vol. 47, no. 9, pp. 45–52, 2011.

[10] S. Zhang, J. Wu, S. Qin, and X. Yuan, "Dynamic characteristics analysis of mixed-flow pump runner based on ANSYS," *Water Resources and Power*, vol. 28, no. 10, pp. 107–108, 2010.

[11] H. Endo and R. B. Randall, "Enhancement of autoregressive model based gear tooth fault detection technique by the use of minimum entropy deconvolution filter," *Mechanical Systems and Signal Processing*, vol. 21, no. 2, pp. 906–919, 2007.

[12] J. Antoni and R. B. Randall, "The spectral kurtosis: application to the vibratory surveillance and diagnostics of rotating machines," *Mechanical Systems and Signal Processing*, vol. 20, no. 2, pp. 308–331, 2006.

[13] F. Combet and L. Gelman, "Optimal filtering of gear signals for early damage detection based on the spectral kurtosis," *Mechanical Systems and Signal Processing*, vol. 23, no. 3, pp. 652–668, 2009.

[14] J. Antoni, "Fast computation of the kurtogram for the detection of transient faults," *Mechanical Systems and Signal Processing*, vol. 21, no. 1, pp. 108–124, 2007.

[15] H. Hu, J. Wang, S. Qian, Y. Li, N. Shen, and G. Yan, "Dynamic modeling and its sliding controller of MR shock absorber under impact load," *Journal of Mechanical Engineering*, vol. 47, no. 13, pp. 84–91, 2011.

[16] H. Endo, R. B. Randall, and C. Gosselin, "Differential diagnosis of spall vs. cracks in the gear tooth fillet region: experimental validation," *Mechanical Systems and Signal Processing*, vol. 23, no. 3, pp. 636–651, 2009.

Permissions

List of Contributors

Bhavna Jain, Sameer Singh, Shailendra Jain, and R. K. Nema
MANIT, Bhopal 462003, India

Vincent Anayochukwu Ani
Department of Electronic Engineering, University of Nigeria, Nsukka 410001, Nigeria

Egill Thorbergsson and Tomas Grönstedt
Division of Fluid Dynamics, Department of Applied Mechanics, Chalmers University of Technology, 412 96 Gothenburg, Sweden

Omar Farrok
Department of Electrical and Electronic Engineering, Ahsanullah University of Science & Technology, Dhaka 1208, Bangladesh
Department of Electrical and Electronic Engineering, Rajshahi University of Engineering & Technology, Rajshahi 6204, Bangladesh

Md. Rabiul Islam and Md. Rafiqul Islam Sheikh
Department of Electrical and Electronic Engineering, Rajshahi University of Engineering & Technology, Rajshahi 6204, Bangladesh

Ali A. Rabah, Hassan B. Nimer, Kamal R. Doud and Quosay A. Ahmed
Energy Research Centre, Faculty of Engineering, University of Khartoum, P.O. Box 321, Khartoum, Sudan

Vincent Anayochukwu Ani
Department of Electronic Engineering, University of Nigeria (UNN), Nsukka 410001, Nigeria

Bahijjahtu Abubakar
Renewable Energy Programme, Federal Ministry of Environment, Abuja 900284, Nigeria

Amevi Acakpovi
Department of Electrical/Electronics Engineering, Accra Polytechnic, P.O. Box GP561, Accra, Ghana

O. Tsakiridis and E. Zervas
Department of Electronics, TEI of Athens, Egaleo, 12210 Athens, Greece

J. Stonham and D. Sklavounos
Department of Engineering and Design, Brunel University, Kingston Lane, Middlesex UB8 3PH, UK

Pablo del Río
Consejo Superior de Investigaciones Cient´ificas, C/ Albasanz 26-28, 28037 Madrid, Spain

Luis Janeiro
Ecofys, Kanaalweg 15-G, 3526 KL Utrecht, Netherlands

K. T. M. U. Hemapala and Lilantha Neelawala
Department of Electrical Engineering, University of Moratuwa, 10400 Moratuwa, Sri Lanka

Wail Aladayleh and Ali Alahmer
Department ofMechanical Engineering, Tafila TechnicalUniversity, P.O. Box 179, Tafila 66110, Jordan

Vaishali Sohoni, S. C. Gupta and R. K. Nema
Department of Electrical Engineering, Maulana Azad National Institute of Technology, Bhopal 462051, India

Albert Ayang
Higher Institute of the Sahel, Department of Renewable Energy, University of Maroua, P.O. Box 46, Maroua, Cameroon

Paul-Salomon Ngohe-Ekam
National Advanced School of Engineering, Energy and Automatic Laboratory, University of Yaounde I, P.O. Box 8390, Yaounde, Cameroon

Bossou Videme
Higher Institute of the Sahel, Department of Information and Telecommunication, University of Maroua, P.O. Box 46, Maroua, Cameroon

Jean Temga
Ecole Polytechnique Montreal, Polygrames Laboratory, P.O. Box 2500, Chemin Montreal, M6106, Canada H3T1J4

Okeolu Samuel Omogoye, Ayoade Benson Ogundare and Ibrahim Olawale Akanji
Electrical and Electronics Engineering Department, Lagos State Polytechnic, Ikorodu Campus, PMB 21606, Lagos, Nigeria

Baraka Kichonge
Mechanical Engineering Department, Arusha Technical College (ATC), P.O. Box 296, Arusha, Tanzania

Iddi S. N. Mkilaha and Geoffrey R. John
College of Engineering and Technology (CoET), University of Dar es Salaam (UDSM), P.O. Box 35131, Dar es Salaam, Tanzania

Sameer Hameer
Nelson Mandela African Institution of Science and Technology (NM-AIST), P.O. Box 447, Arusha, Tanzania

Salah Abosedra
College of Business Administration, American University in the Emirates, Academic City, Dubai, UAE

Muhammad Shahbaz
Department of Management Sciences, COMSATS Institute of Information Technology, 1.5 km Defence Road, off Raiwind Road, Defence Road, Lahore 54000, Pakistan

Rashid Sbia
Ministry of Finance, P.O. Box 333, Manama, Bahrain Department of Applied Economics, Free University of Brussels, Avenue F. Roosevelt 50, CP 140, 1050 Brussels, Belgium

Mohammad Heidari
Department of Mechanical Engineering, Abadan Branch, Islamic Azad University, Abadan, Iran

Index